Lecture Notes in Computer Science 5363

Commenced Publication in 1973
Founding and Former Series Editors:
Gerhard Goos, Juris Hartmanis, and Jan van Leeuwen

Giovanni Sommaruga (Ed.)

Formal Theories
of Information

From Shannon to Semantic Information Theory
and General Concepts of Information

 Springer

Volume Editor

Giovanni Sommaruga
University of Fribourg
CH-1700 Fribourg, Switzerland
E-mail: giovanni.sommaruga@unifr.ch

Library of Congress Control Number: Applied for

CR Subject Classification (1998): E.4, H.1.1, F.4

LNCS Sublibrary: SL 1 – Theoretical Computer Science and General Issues

ISSN   0302-9743
ISBN-10   3-642-00658-2 Springer Berlin Heidelberg New York
ISBN-13   978-3-642-00658-6 Springer Berlin Heidelberg New York

springer.com

© Springer-Verlag Berlin Heidelberg 2009
Printed in Germany

Typesetting: Camera-ready by author, data conversion by Scientific Publishing Services, Chennai, India
Printed on acid-free paper   SPIN: 12600021   06/3180   5 4 3 2 1 0

Preface

It is commonly assumed that computers process information. But what is information? In a technical, important, but nevertheless rather narrow sense, Shannon's information theory gives a first answer to this question. This theory focuses on measuring the information content of a message. Essentially this measure is the reduction of the uncertainty obtained by receiving a message. The uncertainty of a situation of ignorance in turn is measured by entropy. This theory has had an immense impact on the technology of information storage, data compression, information transmission and coding and still is a very active domain of research.

Shannon's theory has also attracted much interest in a more philosophic look at information, although it was readily remarked that it is only a "syntactic" theory of information and neglects "semantic" issues. Several attempts have been made in philosophy to give information theory a semantic flavor, but still mostly based on or at least linked to Shannon's theory. Approaches to semantic information theory also very often make use of formal logic. Thereby, information is linked to reasoning, deduction and inference, as well as to decision making.

Further, entropy and related measure were soon found to have important connotations with regard to statistical inference. Surely, statistical data and observation represent information, information about unknown, hidden parameters. Thus a whole branch of statistics developed around concepts of Shannon's information theory or derived from them. Also some proper measurements appropriate for statistics, like Fisher's information, were proposed.

Algorithmic information theory introduced by Kolmogorov, Solomonoff and Chaitin provides a new look at the concept of information. It is again basically a theory of measuring information content. Here the information content of an information object, for instance, a binary string, is measured by the length of the shortest program which computes this object. It is based on Turing machines. A main result of this approach to information is the clarification of the concept of randomness and probability. Therefore it is not too surprising that algorithmic information theory reproduces Shannon's results although in a rather different context.

Not too long ago it was noted that information is related to questions. Information represents answers to such questions. Or it was remarked that pieces of information shed light on a given context, and that this information might possibly also be transported through channels to other contexts. The problems of questions and of information related to questions were considered by Groenendijk and Stockhoff. Further, the flow of information between different contexts was studied by Barwise and Seligman. From quite a different point of view, similar issues were captured by the Fribourg school which introduces the concept of information algebras. Pieces of information come from different sources, concern different questions, can be combined or aggregated and focused on the questions of interest. These algebraic structures also provide a rigorous foundation for a theory of uncertain information based on probability theory. Furthermore they offer sufficient conditions for efficient generic methods of inference covering diverse domains such as relational databases, probability networks, logic systems, constraint programming, discrete transforms and many more.

Under the title "Information and Knowledge," research groups of the Computer Science departments of the universities of Berne, Fribourg and Neuchtel collaborated over several years on issues of logic, probability, inference and deduction. Given the different approaches to the concept of information and its basic nature, one of the traditional Muenchenwiler seminars in May 2006 was devoted to an exchange of views between experts from the different schools mentioned above. The goal was to examine whether there is some common ground between these different formal theories. The contributions of the invited participants (with the exception of Robert van Roij, who was afterwards invited to contribute) are collected in this volume. The volume editor, Giovanni Sommaruga, discusses the question of whether there are one or several concepts of information as a first attempt to summarize the results of the seminar. It is up to the reader to continue in the direction of a possible unification of the different theories.

As the organizer of the May 2006 Muenchenwiler seminar, I would like to thank the authors for their participation in the seminar, their contributions to this volume and the patience they had to exercise during the editing process. My sincere thank goes to the editor of this volume, Giovanni Sommaruga, for all the work this implied and especially for his effort to compare the different approaches to information in search of a common thread. Thanks to Cris Calude for establishing the contacts with Springer for the publication of the volume. I am grateful to Cesar Schneuwly for the final typesetting preparations. Finally I thank the Swiss National Foundation for supporting several research projects on the subject of "logic and probability" and "information and knowledge," as well as the Swiss Confederation which supported the collaboration project between the universities of Berne, Fribourg and Neuchtel under the title of "deduction and inference." The Muenchenwiler seminar of May 2006, as well as many others, and the present volume are fruits of this encouragement.

Jürg Kohlas
Department of Computer Science
University of Fribourg (Switzerland)

Table of Contents

Introduction

Giovanni Sommaruga

Part I

This book's topic are formal theories of information. It may be useful to start explaining this topic a little further, and then to say what the structure of the book is like and what motivates it, and finally how this book compares with other works grappling with the same or a similar topic.

What is meant by the term 'formal' in 'formal theories of information'? All of the formal theories presented or discussed in the sequel either have a strongly mathematical or logical flavor or are downright mathematical. And how should the term 'information' be understood in the expression 'formal theories of information'? A first clarification of this term is provided by L. Floridi's introductory philosophical considerations in this book; a second attempt at a clarification is made by G. Sommaruga's concluding remarks.

What is the structure of this book? This book's structure could be represented by some sort of a circular model: The innermost circle will be called the syntactical one: it constitutes the basic skeleton or the set of essential components of any formal theory of information. The second, larger circle is called the semantical one: it adds the crucial feature of meaning to the information-theoretical consideration of mere signs (or well-structured data) in the smallest, innermost circle. The third, even larger, outermost circle might be called the pragmatic one: it adds the crucial feature of real-life usage of meaningful signs by humans to the information-theoretical consideration of mere meaningful signs in the intermediate circle.

This structure is motivated by a doubly unificatory purpose: on the one hand by the question of 'unification' of different approaches inside a given circle; on the other hand by the question of 'unification' underlying the various circles: is it possible to think of one unique concept of information which is gradually built up, developed over several stages represented by the different circles?[1]

K. Kornwachs and K. Jacoby's *Information. New Questions to a Multidisciplinary Concept*(1996) seems to pursue a similar unificatory purpose. The two editors reason in the introduction to their book as follows: There appear to be

---

[1] An alternative structure of this volume could have been the result of interchanging its second ('the syntactical approach') and its third part ('the semantical approach') for the following reasons: as argued for in sect. 3.3 of my contribution to this volume, the center concept of information is the semantical one which can be phrased in terms of questions and answers. A very sensible way of presenting the following articles would have been to start with the contributions to this center concept and to carry on with two extensions of it: the technical extensions of this center concept (i.e. the syntactical approach) and a pragmatical extension of it (i.e. beyond the semantical approach). I owe this interesting suggestion to Jürg Kohlas.

G. Sommaruga (Ed.): Formal Theories of Information, LNCS 5363, pp. 1–12, 2009.

only the following three kinds of concept of information: a) Shannon's (syntactical) concept modified in many different but not essential ways. It is so limited as to be almost uninteresting (according to a comment by E.U. v. Weizsäcker). b) a very vague concept in everyday language. It is so broad that it is just about meaningless. c) an economical concept of information as a commodity which, however, until now defied any attempt to define it. Thus, Kornwachs and Jacoby reach the following conclusion: The search for a unified concept of information is a hopeless endeavor; information is a multidisciplinary concept, i.e. every scientific discipline has its own concept of information. (1996:1ff) They carry on with the following observation: Scarce applications of Shannon's, i.e. statistical information theory could be made in cognitive science (psychology), biology, system theory, philosophy of science, linguistics and the social sciences. Therefore, all these sciences have started developing their own concept of information. That is why there is no unified concept of information available. Kornwachs and Jacoby continue by making a claim which seems to contradict their earlier conclusion. Claim: A unified concept of information can be reached by a multidisciplinary approach only. What does that claim mean? Does it mean that a unified concept of information has to account for the different concepts of information used in the different scientific disciplines? Is this what is meant by a multidisciplinary approach being a necessary condition for a unified concept? And what does the expression 'account for' imply in this context? Their explanations following their claim are by no means illuminating. Kornwachs and Jacoby's book amounts eventually to presenting various aspects of the concept of information and discussing various uses of the term 'information' in physics, biology, system theory, philosophy of science, philosophy and linguistics, all of this in agreement with their original conclusion, namely that information is (and cannot be but) a multidisciplinary concept.

Another weighty attempt at providing a unified theory of information is provided by W. Hofkirchner's *The Quest for a Unified Theory of Information. Proceedings of the Second International Conference on the Foundations of Information Science*(1999). In his introduction to this volume Hofkirchner raises several questions. The first of these questions is: Which are 'the philosophical and/or formal scientific suppositions [that] seem best suited to serve as a basis for a unified theory of information (UTI)'? (1999:xxi) Hofkirchner answers this question as follows: a UTI ought to be conceived of as a general theory of information-generating systems. (1999:xxii) This answer appears unsatisfactory for at least two reasons: First, in order to identify and construct theories about information-generating systems one has to know what information is, i.e. one has to know the concept of information. Hence Hofkirchner's answer is somewhat viciously circular. Second, these information-generating systems are (according to Hofkirchner) to be considered as particular kinds of systems, as physical, chemical, biotical etc. systems, depending on the material context. This means that UTI has to be conceived as a 'material' theory of information. And this conception implies that the underlying concept of information will at best be an analogous one and at worst equivocal. This consequence is hardly in the spirit of a UTI. Hofkirchner seems to point at a way out of these difficulties: A concept of information should be flexible enough to perform two functions: 'It must relate to the most various manifestations of information, thus enabling all scientific disciplines to use a common concept; at the same time, it must be precise enough to fit the unique requirement of each individual branch of science.' (1999:xxii) No theory can fulfil these two requirements other than a formal (mathematical) theory of information. The term 'formal' should not be understood in a purely formalistic sense, but at most in a sense that S. Shapiro calls deductivist. (cf. (Shapiro 2000:ch.6.2)) It is a logical mathematical theory of information, expressing or incorporating a formal concept of information and applicable to a wide range of scientific disciplines. It thus comes as no surprise when Hofkirchner writes: '[ ] the conference was unable to answer unambiguously the question of whether a UTI is possible at all, and, if so, if a theory of evolutionary systems represents suitable foundations for this; in which way different properties of information-generating self-organizing systems can be subsumed;...' (1999:xxiii).[2][3]

---

[2] D.F. Flückiger distinguishes in his ph.d. thesis *Beiträge zur Entwicklung eines vereinheitlichten Informations-Begriffs*(1995) two types of information theory: the so-called structural-attributive ones whose prototype is D. MacKay's descriptive information theory, and the so-called functional-cybernetic ones whose prototype is Shannon's statistical information theory. (1995:2,69; cf. also his (1999)) He makes an attempt at combining two essential perspectives on information, namely the perspective of information transmission (Shannon) and the perspective of information accumulation (Nauta). Flückiger's goal is to find a (consistent) concept of information underlying both these perspectives (1995:63). On the way to finding such a concept, he makes extensive use of modern brain biology. Flückiger's approach has a similar objective as the Barwise-Seligman theory of information and information flow, but unlike the latter one, it suffers from the same flaw as Hofkirchner's approach, namely from not being a really formal theory.

[3] The objectives of P. Keller's thesis *Information Flow. Logics for the (r)age of information*(2002) are somewhat similar to those of this book: (i) 'to give a conceptual analysis of the notions of information, data and knowledge and their interrelations' – where in this book the concept of knowledge plays no role whatsoever –, and (ii) 'to apply this analysis to the theory of information flow' (2002:I) – where in this book, the analysis is partially applied, partially extracted from the theory of information flow and other formal theories of information –. Keller carries out task (i) by comparing different theories of information with each other, such as Dretske's philosophical theory of information, situation-theoretic information theory and epistemic modal logic of information. He mentions three possible reasons for the apparent fact that the different theories of information considered by him are incommensurable. (2002:VII/VIII) One may be tempted to add a fourth reason, namely that Keller's choice of theories to be compared with each other wasn't particularly fortunate, or say, too heterogeneous. His conclusion at the end of his thesis is disappointed and delusive: 'the concept [of information, G.S.] is elusive and there is not much to be hoped from a 'philosophy' of information' (2002:240), and by no means shared by the editor of this book. It is one among other objectives of this book that the reader may come, after reading this book, to the opposite conclusion.

Part II: The Individual Contributions

In his contribution **Philosophical Conceptions of Semantic Information**, L. Floridi sets out to explore and clarify the wide and messy conceptual field surrounding the concept of information. Even though he starts by declaring that there is a considerable number of concepts of information depending on the level of abstraction and the requirements of one's perspective, he essentially zooms in on three fundamental concepts: information as (well-structured) data, information as meaningful well-structured data (meaningful content), and information as truthful meaningful well-structured data. He then provides a philosophical discussion of the nature of (well-structured) data. After a brief philosophical presentation of statistical information theory (called MTC by Floridi), he examines the concept of information as semantic content and especially the one he calls factual semantic content (factual information) and he presents a sketch of the debate on whether factual information ought to be truthful or not in order to correctly be called information. At the end, he considers the relationship between MTC and a semantic theory of information, thereby continuing the previous sketch on the level of theories: for the weakly semantic theories of information, information as semantic content is alethically neutral, whereas for the strongly semantic theories, information as semantic content has to be truthful.

The canonical measure of probabilistic uncertainty is *Shannon's entropy* (1948), whose properties and applications constitute *Information Theory*. In Information Theory, the entropy of a message limits its minimum coding length, in the same way that, more generally, the complexity of the message determines its compressibility in the Kolmogorov-Chaitin-Solomonov *Algorithmic Information Theory*. In his contribution **Information Theory, relative Entropy and Statistics**, F. Bavaud summarizes and revisits the classical Shannonian framework from a statistical inferential perspective: besides coding and compressibility interpretations, the relative entropy $K(f\|g)$ (or Kullback-Leibler divergence) possesses a direct probabilistic meaning, and measures the badness-of-fit between an empirical distribution f and a model distribution g - a theme first explored by authors such as Kullback, Sanov, Jaynes, Billingsley, Csiszár, and Cover among others. Through about twenty examples, Bavaud illustrates a few formal properties of the functional $K(f\|g)$, rich enough to capture the various aspects of the confrontation between models (= what we believe) and data (=what we observe), that is the art of classical statistical inference, including Popper's falsificationism as a special case. In particular, the asymmetry of $K(f\|g)$ nicely matches the epistemological asymmetry between data and models, as illustrated by Fisher's single hypothesis testing, the Neyman-Pearson testing between two hypotheses, and Bayesian model selection. Also, the exact additive decomposition of the relative entropy holds in two dual contexts, namely for convex families of empirical distributions, or for exponential families of model distributions. Moreover, the principles of Maximum Likelihood and Maximum Entropy clearly emerge as dual to each other, which clarifies the (often misunderstood) epistemological meaning of the former, namely as a method of reconstructing under incomplete observation the most likely data under some prior model - which is highlighted in the so-called EM algorithm consisting of an alternating use of both principles. In the last section, Bavaud demonstrates how the relative entropy formalism extends beyond independence models, and can be used to test independence or to test the order of a Markov chain. His conclusion, in the spirit of convex and exponential models, illustrates the heating and cooling of texts by a few textual simulations, and the mixing (in an additive or multiplicative way) of English and French texts.

C.S. Calude's contribution **Information: The Algorithmic Paradigm** has almost the form of a dialogue: questions are raised, answers are given which in turn may raise new questions etc. Moreover, a central theme of Calude's with variations is incompleteness. After introducing bits, i.e. binary digits, and bits-strings, Calude raises the question: How efficiently can all the non-negative numbers be coded? In order to answer this question, he introduces a special type of Turing machine, namely the self-delimiting universal Turing machine U, and he also explains the following coding problem: If one considers all Turing machines of length at most n, i.e. $2^{n+1} - 1$ Turing machines, some Turing machines halt on a certain input x, others don't. If all the Turing machines of length n are ordered lexicographically and for each Turing machine, one asks whether it stops or not, one gets a bit-string of length $2^{n+1} - 1$ encoding the whole information. Can the same amount of information be encoded with fewer bits? The answer is yes, and expressed by the Omega number Ω_U whose binary expansion is $0.\omega_1\omega_2\ldots\omega_m\ldots$. The halting information for all Turing machines p s.t. $n \geq |p|$ can then be compressed into a string of length $n : \omega_1\omega_2\ldots\omega_n$. It is now possible to answer the original question: The most efficient coding of all the non-negative numbers is provided by the domain of a self-delimiting universal Turing machine. Calude continues showing that many problems in mathematics can be rephrased in terms of the halting/non-halting status of appropriately constructed self-delimiting Turing machines. The next question to be discussed is whether computers can produce new information? The amount of information $H_U(x)$ contained in a bit-string x is the smallest length of a Turing machine by means of which a self-delimiting universal Turing machine U produces x. To produce new information then means to start with an input x and produce an output y s.t. $H_U(x) < H_U(y)$. The question just asked becomes: Is there any computable process which can produce infinitely many outputs each having more information than its corresponding input? Calude demonstrates that the answer is essentially negative: a computer cannot create much new information. The ensuing question is: But how much can one expect to be created? If a Gödelian theory is roughly speaking a formal theory for which Gödel's incompleteness theorems hold, such a theory can be used to prove theorems having a bit more information than the theory itself. The next point raised concerns a link between algorithmic and statistical information theory: Calude presents an algorithmic version of Shannon's noiseless coding theorem. Next, he treats the relationship between algorithmic randomness and incompleteness: An infinite sequence $x_1x_2\ldots x_n\ldots$ is algorithmically random if there exists a positive constant c s.t. $H_U(x_1x_2\ldots x_n) \geq n - c$. It has then been proved that (the sequence of bits of) Ω_U is algorithmically random. Questions: i) Are there other natural algorithmically random sequences? And ii) Are there any computably enumerable and algorithmically random numbers other than Ω_U? The answer to the first question is positive: If ζ_U is the so-called zeta number of a self-delimiting universal Turing machine U, ζ_U can be shown to be algorithmically random. The answer to the second question, however, is negative: for one can prove that a real number $\alpha \in (0,1)$ is computably enumerable and algorithmically random iff there exists a self-delimiting universal Turing machine U s.t. $\alpha = \Omega_U$. The link to incompleteness is establised by the following result: A Gödelian theory cannot determine more than finitely many digits of Ω_U. Calude also comes back to incompleteness in his last point: If one expresses the property of Ω_U being algorithmically random as the uncertainty relation $\Delta_s \cdot \Delta(\omega_1 \ldots \omega_s) \geq 1$, one can derive from it Gödel's incompleteness theorem, that is, uncertainty implies incompleteness.

J. Kohlas starts his article **Information Algebra** by explaining intuitively the basic components and ideas of his algebraic theory of information. In the second section, he gives an axiomatic presentation of the algebra of information which he motivates by showing that the relational algebra associated with relational databases is its prototype. In the following subsections, he reinterprets the projection operation of an information algebra in two ways: by interpreting it as variable elimination, he points to the connection between information algebra and logic; by interpreting it as a transport operation, he prepares the ground for the definition of an interesting equivalent version of the information algebra, i.e. the so-called domain-free one. The third section is dedicated to a variety of examples, non-logical and logical, of information algebras: fuzzy set theory (or parts thereof) can be conceived of as an information algebra, and as for the logical examples, propositional logic, first order logic, and so-called contexts can equally be conceived of as information algebras. The contexts are designed as a more general logical framework for obtaining information algebras. The fourth and last section links information algebra to statistical information theory. The first subsection explains how information algebra gives rise to a natural partial order of information content. In the second subsection, attention is drawn to the fact that in a relational (information) algebra, the information content of a relation depends on the question of one's interest. Since this fact is related to the Boolean character of a relational algebra, the so-called Boolean information algebras are introduced.The next subsection shows how this partial order of information content can be used to define particular information algebras based on basic, finest information pieces, called atoms; these algebras are subsequently called atomic information algebras. The fourth and last subsection deals with the measurement of information content in the case of atomic information algebras, using Hartley's measure. This quantitative information measure measures the reduction of uncertainty by an information element of an atomic information algebra and it also respects the qualitative, partial order of information content. Despite these connections between information algebra and statistical information theory, Kohlas keeps emphasizing that atomic information algebra is not a statistical information theory as entropy doesn't develop its full power in it.

In information algebra, information is represented by an algebraic structure in which pieces of information refer to precise questions and can be combined and focussed on other questions. Uncertain information arises when a piece of information is known to be true under certain assumptions which themselves are not necessarily known to be true. Varying these assumptions leads to different information by means of assumption-based reasoning. If the likelihood of different assumptions is varied and described by a probability measure, it is possible to measure the degree of support of a piece of information in terms of the probability of those assumptions supporting this piece of information. This is the basic tenet of J. Kohlas and Ch. Eichenbergers approach **Uncertain Information**. Section two starts off with a presentation of functional models describing the process whereby a data (an answer) is generated from a parameter (question) and some random element (an assumption). The basic idea of assumption-based reasoning is to suppose that a random element generated some data and then to determine the consequences of this supposition on the parameter (and to determine these consequences in terms of the probabilities of the resp. random element(s)). The last technical term introduced in this section is the one of a hint: A hint is essentially a mapping from a probability space into a certain set, and, more precisely, a mapping of an assumption to the smallest set of possible answers to a given question, containing for sure the right answer. Intuitively, a hint represents a piece of information concerning the right answer to some question, if this answer depends on certain assumptions. Section three introduces a generalisation of the hints, namely random variables with values in an information algebra: A (simple) random variable is a mapping from a (finite) probability space whose elements represent uncertain assumptions into an information algebra. Since it is shown that (simple) random variables form themselves an information algebra, they are on the hand information, and, due to their relation to a probability space, they are on the other hand uncertain information. Section four associates random variables with probability distributions: These probability distributions arise from the probabilities of the assumptions supporting the answers to some question. A degree of support of an answer to some question, as well as a degree of possibility (or plausibility) of some answer are defined by means of the random variables. The support and the possibility function actually represent distribution functions of the random variables and can, in the case of simple random variables, be defined in terms of basic probability assignments. In the basically last section five, the fact is exploited that uncertain, i.e. assumption-based, information is also information, i.e. constitutes an information algebra of random variables. This fact allows for the definition of an order between the elements of this information algebra. This order is induced by the algebra and reflects a comparison of random variables w.r.t. information content taking into account that this information content is also related to assumptions. If the information algebra of random variables is Boolean, it can be generalised in such a way as to admit also of varying probability spaces of assumptions. In this latter case, a measure of information content can be introduced in a way analogous to the one presented in the article Information Algebra provided the respective Boolean information algebra is atomic. This measure of information content can be regarded as the reduction of uncertainty by the random variable w.r.t. full ignorance (where uncertainty is measured by Shannon's concept of entropy adapted to the information algebra of random variables). Thereby, a link to Shannon's theory of information is established.

The general theme of R. van Rooij's article **Comparing Questions and Answers: A bit of Logic, a bit of Language, and some bits of Information** is the informative value of questions and answers and its measurement. Van Rooij's contribution is set up in a dialectical way: He begins with a first definition of this informational value and then points out its limitations. He goes on giving a second definition which takes into account the crticicism of the first, but then he points out the limitations of this second attempt. And so he carries on presenting a third definition etc. Van Rooij starts (in the second section) by explaining the meaning of questions as well as the entailment relation between questions within the framework of Groenendijk and Stokhof's partition semantics. In the third section, he first discusses Groenendijk and Stokhof's semantic comparison of (relevant) answers and questions and then observes that the state of information of a questioner influences the relevance of questions and answers. This observation leads to a pragmatic comparison of questions relevant w.r.t. an information state K as well as a comparison of relevant answers to a question w.r.t. K. Van Rooij ends this section by pointing out that the qualitative notion of relevance in the pragmatic comparions is too rough, and that the partial ordering relations between questions and answers should be extended to total orderings by measuring the informativity and relevance of answers and questions in a quantitative way. The next fourth section sets out to explain how this could be achieved. Van Rooij follows the lead of Carnap and Bar-Hillel by defining the informational value of an answer (a proposition) A, $\inf(A)$, as the negative logarithm (base 2) of its probability. As the inf-function is monotone increasing w.r.t. the entailment relation between propositions, the total ordering relation induced by the inf-function is exactly an extension of the partial ordering induced by the entailment relation. He then defines the informational value (or entropy) of a question in a formally analogous way to Shannon's definition of the entropy of a coding system as the average informational value of its answers. This definition allows likewise to extend the partial ordering on questions mentioned earlier to a total ordering. Let B be the question (partition) an answer to which provides total information about the world; B has a certain entropy. Van Rooij now defines the informational value of an answer q to question B as the reduction of entropy of B upon learning q, and the informational value of question Q w.r.t. question B as the average reduction of entropy of B upon learning an answer to Q. As soon as question B is replaced by a mutually exclusive and exhaustive set of hypotheses H (a partition), the use of Shannon's conditional entropy becomes unavoidable. The informational value of question Q w.r.t. question H serves to define the informational usefulness of Q w.r.t. H, which in turn is important when an agent is faced with the decision problem which of the hypotheses to choose. Something analogous can be done for answers. But then van Rooij points out the limitations of this approach: the measure of usefulness of questions and answers w.r.t. a decision problem is reasonable only for those kinds of decision problem where the decisions depend alone on the probabilities involved. As soon as desirabilities or utilities influence the actions to be chosen, the approach followed up to now is not appropriate any longer. To take into consideration not only probabilities, but also desirabilities is the approach presented in the fifth and last section. Suppose an agent has to deal with a decision problem. W.r.t. assertions (answers), van Rooij distinguishes between the highest expected utility according to the original decision problem and the utility value of making an informed decision conditional on learning a certain assertion, and defines the utility value of the assertion by the difference between these two values. The expected utility value of a question can then be defined in terms of the utility values of the possible answers. He carries on quoting a theorem according to which measuring the expected utility value of a question w.r.t. a decision problem corresponds to the qualitative 'measurement' of the resp. question. It might now be expected that something similar holds w.r.t. assertions. This, however, is not the case: the utility value of an assertion or answer resp. does not only not behave monotone increasing w.r.t. the entailment relation between propositions, it also doesn't behave monotone increasing w.r.t. the informational value of an assertion w.r.t. a set of hypotheses. The last subsection's starting point is the observation that not only is there in general no connection between the utility value of an assertion and the informational value of an assertion w.r.t. a set of hypotheses, there is in general no connection between the expected utility value of a question and the expected informational value of a question (w.r.t. the most fine-grained partition) either.

J. Seligman starts in his article **Channels: from Logic to Probability** from the assumption that information arises in conditions of uncertainty: uncertainty is reduced by gaining information. Essential for any mathematical model of information is the representation of a state of uncertainty and the change of state induced by the acquisiton of a piece of information. Probability theory provides one such model: Shannon showed how this theory can be used to give a precise measure of uncertainty and to model the movement of information in a system of communication channels. Dretske tried to extend Shannon's model to an information-based semantics and epistemology, which was developed by Barwise, Perry and others as 'situation semantics' and 'situation theory'. Formal logic provides another such model: Barwise and Seligman worked out an account of information flow using a more abstract model of channels and based on formal logic. This account is called the Barwise-Seligman theory. Seligmans aim is to adapt the Barwise-Seligman theory in order to give a similarly abstract account of Dretske's conception of information. The Barwise-Seligman theory of information and information flow makes use of various structures called 'classifications', maps between classifications called 'infomorphisms' and combinations of infomorphisms called 'channels'. In sect. 1, Seligman presents all these basic structures as well as different types of channel related to different fields. If two classifications combined by two infomorphisms satisfying certain conditions are probability spaces, the resulting binary channel is called a 'Shannon channel'; if the two classifications similarly combined by two infomorphisms are formal language classifications, the resulting channel is called a 'Tarski channel'; and if the tokens of classifications are actual concrete events rather than possible configurations, the channel constructed from them is called a 'concrete event channel'. After a brief review in sect. 2 of Shannon's definition of information flow in Shannon channels and Dretske's information-based account of knowledge and belief, Seligman points out a structural similarity between information flow in Shannon channels and information flow in Tarski channels, but he also demonstrates i) that this similarity cannot be formulated within the Barwise-Seligman theory in terms of strong information flow: the infinite Shannon channels elude this attempt (the Strength Problem); and ii) that the model underlying this similarity cannot simply be adapted to information flow in concrete event channels (required for Dretske's epistemological project)(the Modality and the Context Problem). To solve the Modality and the Context Problem, Seligman needs on the one hand a suitable 'linking relation' between sets of types in the core of a concrete event channel to model the regularities on which information flow depends, and on the other hand a suitable set of 'normal tokens' to characterise the contextual connections between particular events: this linking relation and this set of normal tokens are used to define the concept of link on that classification, and ultimately to define information flow relative to a link. At the level of types and tokens this means: There is information flow relative to a link if both components are given: a receiver event type 'indicates' a source event type, and a particular receiver event 'signals' a particular source event within a (core of a) channel C. Sect. 3 serves to determine the value of the link, introduced in the previous sect. The ultimate philosophical goal is to actually find a definition of information flow relative to a link determined by any theory whatsoever, while avoiding the 3 just mentioned problems. Now, a set of pairs of subsets of the set of types of a formal language classification A satisfying certain conditions is called a theory or Tarski theory of A. If the classification is of a probability space P, the resp. theory is called a Dretske theory of P. Seligman then axiomatically characterises the so-called 'Gentzen theories' and shows that all Tarski and all Dretske theories are Gentzen theories, and he succeeds in characterising the Tarski theories. He subsequently raises the question whether the relationship between Gentzen, Tarski and Dretske theories is duplicated w.r.t. the links they determine and he answers it in a negative way: the reason being that a link can be determined by more than one theory. In sect.s 4 and 5, Seligman sets out to find a characterisation of the Dretske theories, both axiomatically and situation semantically. In sect. 4, he characterises the Dretske theories as the theories of extensional Barwise structures satisfying the principle of No Countable Mystery. In sect. 5, he discovers a few properties characterising the class of Dretske theories of a probability space P. In the last sect. 6, Seligman calls his analysis of information flow developed throughout the sections 2-5 the signalling/indicating model of information flow, and he compares it to a model presented in his joint book with J. Barwise (1997), which he calls the logic-movement model of information flow. He notes that in this book, information flow is not modelled as a relation between individual types (or tokens) in the source and receiver, but as a movement of local logics around a network of classifications, whereby a local (Tarski) logic is roughly speaking a Tarski theory on a classification A restricted to a subset (of normal tokens) of the set of tokens of A. Local logics on a classification represent information about the regularities within it. Seligman finally observes that movement of local logics yields a more coherent model of information flow in concrete event channels than the signalling/indicating model, which can be seen as a special case.

In his contribution **Modeling Real Reasoning** K. Devlin sets out to develop a mathematical model for real-life logical reasoning analogous to classical formal logic as a mathematical model for formal reasoning in pure mathematics. He starts off by presenting a couple of reasons why such a model cannot consist of an application of classical formal logic or simple modifications thereof. Next, Devlin treats the topic of information which is related to real-life logical reasoning (and also to other forms of reasoning) in the following way: reasoning is a specific and very important form of purposeful information gathering and information processing. In virtue of the following general observations concerning information, namely that information can arise by virtue of regularities in the world, and that anything can be used to represent information, two tasks have to be tackled with: first, provide a precise, representation-free definition of information, and second, study the nature of the regularities whereby things in the world represent information. These two tasks have been the main focus of attention and the main subject of situation theory (or situation-theoretical information theory). Next, Devlin provides a concise and elegant survey of parts of situation-theory. In the following section, he uses situation theory to model real-life logical reasoning. The basic evidential reasoning element in his model is reminiscent of a proposition in the situation-theoretical sense, and the evidential reasoning process of a proof in the formal logical sense. An evidential reasoning process is constituted by a certain number of evidential reasoning elements some of which are the result of basic reasoning steps. Devlin presents and explains several of these basic reasoning steps. By making explicit in the model the features of the context situation that provide direct support for the items of information considered in the reasoning, and by accounting for various aspects of the reasoning process, Devlin's model clearly goes beyond situation theory and makes it possible to obtain a finer-grained analysis of a specific reasoning process than could be obtained by situation theory. Next, Devlin applies his situation-theoretic model of real-life logical reasoning to three special cases, namely to mathematical reasoning, to reasoning from a common source, and to Bayesian reasoning. In the last section, Devlin motivates on the one hand his model against the background of situation theory, and on the other he briefly discusses ways other than understanding and analysing real reasoning processes that his model could be used for.

In his article **One or Many Concepts of Information?** G. Sommaruga carries on the conceptual work already carried out by L. Floridi, but at the same time trying to take stock of the articles on the various formal theories of information. In his first section, he introduces the distinction between ordinary language concepts, informal theoretical concepts and formal theoretical concepts and he applies this distinction to the concept of information and to the title question in particular. The second section consists of applying the conceptual apparatus developed in the first part to the formal theories of information. This application leads up to an information-theoretical analogue **(T)** to Church's Thesis **(CT)**. The remainder of section two is devoted to a philosophical reflection on **(T)** and to an attempt to provide evidence for **(T)**. The third and last section draws a few conclusions from the previous conceptual analyses and considerations: The most appropriate point of view w.r.t. the title question may very well be a centralized (but not a reductionist) one, and adopting such a point of view may also provide some directions for future work on formal theories of information. As I vaguely recall having read in one of Donald Davidson's articles: It's good to know that we won't run out of work.

References

[Barwise/Seligman (1997)]	Barwise, J., Seligman, J.: Information Flow. The Logic of Distributed Systems. Cambridge University Press, Cambridge (1997)
[Flückiger (1995)]	Flückiger, F.: Beiträge zur Entwicklung eines vereinheitlichten Informationsbegriffs. Berne, ph.d. thesis, Berne University (1995), http://splendor.unibe.ch/Federico.Flueckiger
[Flückiger (1999)]	Flückiger, F.: Towards a Unified Concept of Information: Presentation of a New Approach in Hofkirchner, pp. 101–111 (1999)
[Hofkirchner (1999)]	Hofkirchner, W.: The Quest for a Unified Theory of Information. In: Proceedings of the Second International Conference on the Foundations of Information Science. Gordon and Breach Publ., Amsterdam (1999)
[Keller (2002)]	Keller, P.: Information Flow. Logics for the (r)age of information. Berne University, math. diploma thesis (2002), http://www.unige.ch/lettres/philo/enseignants/philipp/research/info.pdf
[Kornwachs/Jacoby (1996)]	Kornwachs, K., Jacoby, K.: Information:New Questions to a Multidisciplinary Concept. Akademie Verlag, Berlin (1996)
[Shapiro (2000)]	Shapiro, S.: Thinking About Mathematics. The Philosophy of Mathematics. Oxford University Press, Oxford (2000)

Philosophical Conceptions of Information

Luciano Floridi[1,2]

[1] University of Hertfordshire
[2] University of Oxford

1 Introduction

> I love information upon all subjects that come in my way, and espe-
> cially upon those that are most important.

Thus boldly declares Euphranor, one of the defenders of Christian faith in
Berkley's *Alciphron* (Berkeley, (1732), Dialogue 1, Section 5, Paragraph 6/10).
Evidently, information has been an object of philosophical desire for some time,
well before the computer revolution, Internet or the dot.com pandemonium (see
for example Dunn (2001) and Adams (2003)). Yet what does Euphranor love,
exactly? *What is information?* The question has received many answers in different fields. Unsurprisingly, several surveys do not even converge on a single,
unified definition of information (see for example Braman 1989, Losee (1997),
Machlup and Mansfield (1983), Debons and Cameron (1975), Larson and Debons
(1983)).

Information is notoriously a polymorphic phenomenon and a polysemantic
concept so, as an explicandum, it can be associated with several explanations,
depending on the level of abstraction adopted and the cluster of requirements
and desiderata orientating a theory. The reader may wish to keep this in mind
while reading this article, where some schematic simplifications and interpreta-
tive decisions will be inevitable. Claude E. Shannon, for one, was very cautious:

> The word 'information' has been given different meanings by various
> writers in the general field of information theory. It is likely that at least
> a number of these will prove sufficiently useful in certain applications
> to deserve further study and permanent recognition. *It is hardly to be
> expected that a single concept of information would satisfactorily account
> for the numerous possible applications of this general field.* (italics added)
> (Shannon (1993), p. 180).

Thus, following Shannon, Weaver (1949) supported a tripartite analysis of
information in terms of (1) technical problems concerning the quantification of
information and dealt with by Shannon's theory; (2) semantic problems relating
to meaning and truth; and (3) what he called "influential" problems concerning the impact and effectiveness of information on human behaviour, which he
thought had to play an equally important role. And these are only two early
examples of the problems raised by any analysis of information.

G. Sommaruga (Ed.): Formal Theories of Information, LNCS 5363, pp. 13–53, 2009.

Indeed, the plethora of different analyses can be confusing. Complaints about misunderstandings and misuses of the very idea of information are frequently expressed, even if to no apparent avail. Sayre (1976), for example, criticised the "laxity in use of the term 'information'" in Armstrong (1968) (see now Armstrong (1993)) and in Dennett (1969) (see now Dennett (1986), despite appreciating several other aspects of their work. More recently, Harms (1998) pointed out similar confusions in Chalmers (1996), who

> seems to think that the information theoretic notion of information [see section 3, my addition] is a matter of what possible states there are, and how they are related or structured ... rather than of how probabilities are distributed among them (p. 480).

In order to try to avoid similar pitfalls, this article has been organised into three main parts.

Section two attempts to draw a map of the main senses in which one may speak of *semantic information*, and does so by relying on the analysis of the concept of *data* (Fig. 1). Sometimes the several concepts of information organised in the map can be variously coupled together. This should not be taken as necessarily a sign of confusion, for in some philosophers it may be the result of an intentional bridging. The map is not exhaustive and it is there mainly in order to avoid some obvious pitfalls and to narrow the scope of this article, which otherwise could easily turn into a short version of the Encyclopedia Britannica. Its schematism is only a starting point for further research.

After this initial orientation, section three provides a brief introduction to information theory, that is, to the mathematical theory of communication (MTC). MTC deserves a space of its own because it is the quantitative approach to the analysis of information that has been most influential among several philosophers. It provides the necessary background to understand several contemporary theories of semantic information, especially Bar-Hillel and Carnap (1953), Dretske (1981) and Floridi (2004b)).

Section four focuses entirely on the philosophical understanding of semantic information, what Euphranor really loves.

The reader must also be warned that an initial account of semantic information as *meaningful data* will be used as yardstick to outline other approaches. Unfortunately, even such a minimalist account is open to disagreement. In favour of this approach one may say that at least it is less controversial than others. Of course, a conceptual analysis must start somewhere. This often means adopting some working definition of the object under scrutiny. But it is not this commonplace that one needs to emphasize here. The difficulty is rather more daunting. Philosophical work on the concept of (semantic) information is still at that lamentable stage when disagreement affects even the way in which the problems themselves are provisionally phrased and framed. Nothing comparable to the well-polished nature of the Gettier problem is yet available, for example. So the "you are here" signal provided in this article might be placed elsewhere by other philosophers. The whole purpose is to put the concept of semantic information firmly on the philosophical map. Further adjustments will then become possible.

2   An Informational Map

Information is a conceptual labyrinth, and in this section we shall begin to have a look at a general map of one of its regions, with the purpose of placing ourselves squarely in the semantic area. Fig. 1 summarises the main distinctions that are going to be introduced.

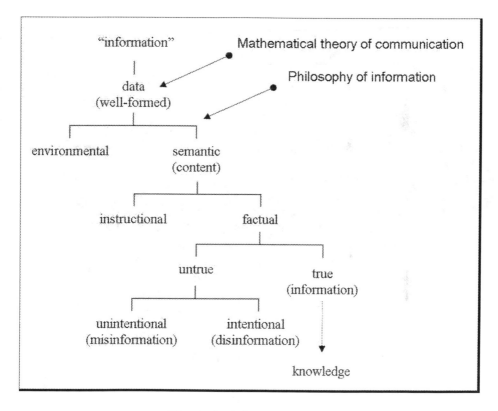

Fig. 1. An informational map

Clearly, percolating through the various points in the map will not make for a linear journey. Using a few basic examples, to illustrate the less oblivious steps, will also help to keep our orientation. So let me introduce immediately the one to which we shall return more often.

2.1   An Everyday Example of Information

Monday morning. You turn on the ignition key of your car, but nothing happens: the engine does not even cough. The silence of the engine worries you. Unsurprisingly, you also notice that the red light of the low battery indicator is flashing. After a few more attempts, you give up and ring the garage. You explain that your husband forgot to switch off the lights of the car last night it is a lie, you did, but you are too ashamed to confess it and now the battery is flat. The mechanic tells you that the instruction manual of your car explains how to use jump leads to start the engine. Luckily, your neighbour has everything you need. You read the manual, look at the illustrations, follow the instructions, solve the problem and finally drive to the office.

This everyday episode will be our "fruit fly". Although it is simple and intuitive, it provides enough details to illustrate the many ways in which we understand one of our most important resources: *information.*

2.2   The Data-Based Definition of Information

It is common to think of information as consisting of *data*. It certainly helps, if only to a limited extent. For, unfortunately, the nature of data is not well-understood philosophically either, despite the fact that some important past debates - such as the one on the given and the one on sense data - have provided at least some initial insights. There still remains the advantage, however, that the concept of data is less rich, obscure and slippery than that of information, and hence easier to handle. So a data-based definition of information seems to be a good starting point.

Over the last three decades, several analyses in Information Science, in Information Systems Theory, Methodology, Analysis and Design, in Information (Systems) Management, in Database Design and in Decision Theory have adopted a *General Definition of Information* (GDI) in terms of *data + meaning* (see Floridi 2005b) for an extended bibliography). GDI has become an operational standard, especially in fields that treat data and information as reified entities (consider, for example, the now common expressions "data mining" and "information management"). Recently, GDI has begun to influence the philosophy of computing and information (Floridi (1999) and Mingers (1997)).

A clear way of formulating GDI is as a tripartite defintion (Fig. 2):

GDI) σ is an instance of information, understood as semantic content, if and only if:
    GDI.1) σ consists of n *data* (d), for n ≥ 1;
    GDI.2) the data are *well-formed* (wfd);
    GDI.3) the wfd are *meaningful* (mwfd = δ).

Fig. 2. The General Definition of Information (GDI)

GDI requires a definition of data. This will be provided in the next section. Before, a brief comment on each clause is in order.

According to (GDI.1), data are the stuff of which information is made. We shall see that things can soon get more complicated.

In (GDI.2), "well-formed" means that the data are clustered together correctly, according to the rules (*syntax*) that govern the chosen system, code or language being analysed. Syntax here must be understood broadly (not just linguistically), as what determines the form, construction, composition or

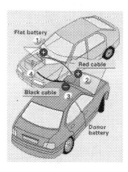

Fig. 3. How to jump start your car ©Copyright Bosh 2005

structuring of something (engineers, film directors, painters, chess players and gardeners speak of syntax in this broad sense). For example, the manual of your car may show (see Fig. 3) a two dimensional picture of the two cars placed one near the other, not one on top of the other.

This pictorial syntax (including the linear perspective that represents space by converging parallel lines) makes the illustrations potentially meaningful to the user. Using the same example, the actual battery needs to be connected to the engine in a correct way to function: this is still syntax, in terms of correct physical architecture of the system (thus a disconnected battery is a syntactic problem). And of course the conversation you carry on with your neighbour follows the grammatical rules of English: this is syntax in the ordinary linguistic sense.

Regarding (GDI.3), this is where semantics finally occurs. "Meaningful" means that the data must comply with the meanings (*semantics*) of the chosen system, code or language in question. However, let us not forget that semantic information is not necessarily linguistic. For example, in the case of the manual of the car, the illustrations are such as to be visually meaningful to the reader.

2.3   A Definition of Data

According to GDI, information cannot be dataless but, in the simplest case, it can consist of a single datum (d). Now a datum is reducible to just a lack of uniformity (*diaphora* is the Greek word for "difference), so a general definition of a datum is (Fig. 4):

Depending on philosophical inclinations, the diaphoric definition of data can be applied at three levels:

1. data as diaphora *de re*, that is, as lacks of uniformity in the real world out there. There is no specific name for such "data in the wild". A possible suggestion is to refer to them as *dedomena* ("data" in Greek; note that our word "data comes from the Latin translation of a work by Euclid entitled *Dedomena*). Dedomena are not to be confused with environmental data (see section 2.7.1). They are pure data or proto-epistemic data, that is, data before they are epistemically interpreted. As "fractures in the fabric of being" they can

> Dd) datum = x *being distinct* (≠) from y
> where the x and the y are two uninterpreted variables and the domain is left open to
> further interpretation.

Fig. 4. The diaphoric definition of data

only be posited as an external anchor of our information, for dedomena are never accessed or elaborated independently of a *level of abstraction* (more on this in section 4.2.2). They can be reconstructed as ontological requirements, like Kant's *noumena* or Locke's *substance*: they are not epistemically experienced but their presence is empirically inferred from (and required by) experience. Of course, no example can be provided, but dedomena are whatever lack of uniformity in the world is the source of (what looks to information systems like us as) as data, e.g. a red light against a dark background. Note that the point here is not to argue for the existence of such pure data in the wild, but to provide a distinction that (in section 2.6) will help to clarify why some philosophers have been able to accept the thesis that there can be no information without data representation while rejecting the thesis that information requires physical implementation;

2. data as diaphora *de signo*, that is, lacks of uniformity between (the perception of) at least two physical states, such as a higher or lower charge in a battery, a variable electrical signal in a telephone conversation, or the dot and the line in the Morse alphabet; and
3. data as diaphora *de dicto*, that is, lacks of uniformity between two symbols, for example the letters A and B in the Latin alphabet.

Depending on one's position with respect to the thesis of ontological neutrality (section 2.6) and the nature of environmental information (section 2.7.1) dedomena in (1) may be either identical with, or what makes possible signals in (2), and signals in (2) are what make possible the coding of symbols in (3).

The dependence of information on the occurrence of syntactically well-formed data, and of data on the occurrence of differences variously implementable physically, explain why information can so easily be decoupled from its support. The actual *format*, *medium* and *language* in which semantic information is encoded is often irrelevant and hence disregardable. In particular, the same semantic information may be analog or digital, printed on paper or viewed on a screen, in English or in some other language, expressed in words or pictures. Interpretations of this support-independence can vary quite radically. For Dd (see Fig. 4 above) leaves underdetermined

- the classification of the relata (*taxonomic neutrality*);
- the logical type to which the relata belong (*typological neutrality*);
- the kind of support required for the implementation of their inequality (*ontological neutrality*); and
- the dependence of their semantics on a producer (*genetic neutrality*).

We shall now look at each form of neutrality in turn.

2.4   Taxonomic Neutrality

A datum is usually classified as the entity exhibiting the anomaly, often because the latter is perceptually more conspicuous or less redundant than the background conditions. However, the relation of inequality is binary and symmetric. A white sheet of paper is not just the necessary background condition for the occurrence of a black dot as a datum, it is a constitutive part of the [black-dot-on-white-sheet] datum itself, together with the fundamental relation of inequality that couples it with the dot. Nothing seems to be a datum *per se*. Rather, being a datum is an external property. So GDI endorses the following thesis:

TaN) a datum is a relational entity.
The slogan is "data are relata", but GDI is neutral with respect to the identification of data with *specific* relata. In our example, GDI refrains from identifying either the red light or the white background as the datum. To understand why there cannot be "dataless information", we shall now look at the typological neutrality of GDI.

2.5   Typological Neutrality

According to GDI, information can consist of different types of data as relata (δ). Five classifications are quite common, although the terminology is not yet standard or fixed (but see Floridi (1999)). They are not mutually exclusive, and one should not understand them as rigid: depending on circumstances, on the sort of analysis conducted and on the level of abstraction adopted, the same data may fit different classifications.

$\delta 1$ *Primary data.* These are the principal data stored e.g. in a database, for example a simple array of numbers. They are the data an information-management system such as the one used in the car to indicate that the battery needs to be charged is generally designed to convey (in the form of information) to the user in the first place. Normally, when speaking of data, and of the corresponding information they constitute, one implicitly assumes that *primary* data/information is what is in question. So, by default, the red light of the low battery indicator flashing is assumed to be an instance of primary data conveying primary information.

$\delta 2$ *Secondary data.* These are the converse of primary data, constituted by their absence (one could call them anti-data). Recall how you first suspected that the battery was flat: the engine failed to make any of the usual noise. Likewise, in Silver Blaze, Sherlock Holmes solves the case by noting something that has escaped everybody else: the unusual silence of the dog. Clearly, silence may be very informative. This is a peculiarity of information: its absence may also be informative. When it is, the point is stressed by speaking of *secondary* information.

$\delta 3$ *Metadata.* These are indications about the nature of some other (usually primary) data. They describe properties such as location, format, updating, availability, usage restrictions, and so forth. Correspondingly, *metainforma-tion* is information about the nature of information. "'The battery is flat' is encoded in English" is a simple example.

δ4 *Operational data.* These are data regarding the operations of the whole data system and the systems performance. Correspondingly, *operational informa-tion* is information about the dynamics of an information system. Suppose the car has a yellow light that, when flashing, indicates that the car check-ing system is malfunctioning. The fact that the light is on may indicate that the low battery indicator is not working properly, thus undermining the hypothesis that the battery is flat.

δ5 *Derivative data.* These are data that can be extracted from some data when-ever the latter are used as indirect sources in search of patterns, clues or inferential evidence about other things than those directly addressed by the data themselves, e.g. for comparative and quantitative analyses (*ideome-try*). As it is difficult to define this category precisely, a familiar example may be helpful to convey the point. Credit cards notoriously leave a trail of derivative information. From someones credit card bill, concerning e.g. the purchase of petrol in a certain petrol station, one may derive the informa-tion of her whereabouts at a given time. Again, derivative information is not something new. Hume provides a beautiful example in these days of global warming. In the *Essays oral, Political, and Literary* (Part II, Essay 11. Of the Populousness of Ancient Nations, Para. 155/186 mp. 448 gp. 432, see now Hume (1987)) he reports that

> It is an observation of LAbbe du Bos, that Italy is warmer at present than it was in ancient times. *The annals of Rome* tell us, says he, that in the year 480 ab U.C. the winter was so severe that it destroyed the trees. [...] Many passages of Horace suppose the streets of Rome full of snow and ice. We should have more certainty with regard to this point, had the ancients known the use of ther-mometers: But their writers, without intending it, give us informa-tion, sufficient to convince us, that the winters are now much more temperate at Rome than formerly.

Hume has just extracted some derivative information from some primary information provided by LAbbe du Bos.

Let us now return to our question: can there be dataless information? GDI does not specify which types of data constitute information. This *typological neutrality* is justified by the fact that, when the apparent absence of data is not reducible to the occurrence of *negative* primary data, what becomes available and qualifies as information is some further non-primary information μ about σ constituted by some non-primary data δ.2-δ.5. For example, if a database query provides an answer, it will provide at least a *negative* answer, e.g. "no documents found". This is primary negative information. However, if the database provides no answer, either it fails to provide any data at all, in which case no specific information σ is available so the rule "no information without data" still applies or it can provide some data δ to establish, for example, that it is running in a loop. Likewise, silence, this time as a reply to a question, could represent negative primary information, e.g. as implicit assent or denial, or it could carry some non-primary information μ, e.g. about the fact that the person has not heard the question, or about the amount of noise in the room.

2.6 Ontological Neutrality

By rejecting the possibility of dataless information, GDI also endorses the following modest thesis of ontological neutrality:

ON) no information without data representation.
Following Landauer and Bennett (1985, and Landauer (1987), (1991), (1996), ON is often interpreted materialistically, as advocating the impossibility of physically disembodied information, through the equation "representation = physical implementation", that is:

ON.1) no information without physical implementation.
ON.1 is an inevitable assumption, when working on the physics of computation, since computer science must necessarily take into account the physical properties and limits of the data carriers. Thus, the debate on ON.1 has flourished especially in the context of the philosophy of quantum information and computing (see Deutsch (1985);(1997) and Di Vincenzo and Loss (1998); Steane (1998) provides a review). ON.1 is also the ontological assumption behind the Physical Symbol System Hypothesis in AI and Cognitive Science (Newell and Simon (1976). But ON, and hence GDI, does not specify whether, ultimately, the occurrence of every discrete state necessarily requires a *material* implementation of the data representations. Arguably, environments in which all entities, properties and processes are ultimately noetic (e.g. Berkeley, Spinoza), or in which the material or extended universe has a noetic or non-extended matrix as its ontological foundation (e.g. Pythagoras, Plato, Descartes, Leibniz, Fichte, Hegel), seem perfectly capable of upholding ON without necessarily embracing ON.1. The relata in Dd could be *dedomena*, such as Leibnizian monads, for example. Indeed, the classic realism debate on the ultimate nature of "being" can be reconstructed in terms of the possible interpretations of ON.

All this explains why GDI is also consistent with two other popular slogans, this time favourable to the proto-physical nature of information and hence completely antithetic to ON.1:

ON.2) "*It from bit*". Otherwise put, every "it" every particle, every field of force, even the space-time continuum itself derives its function, its meaning, its very existence (even if in some contexts indirectly) from the apparatus-elicited answers to yes-or-no questions, binary choices, *bits*. "It from bit" symbolizes the idea that every item of the physical world has at bottom a very deep bottom, in most instances an immaterial source and explanation; that which we call reality arises in the last analysis from the posing of yes-no questions and the registering of equipment-evoked responses; in short, that all things physical are information-theoretic in origin and that this is a *participatory universe* (Wheeler (1990), 5);

ON.3)   [information is] a name for the content of what is exchanged with the outer world as we adjust to it, and make our adjustment felt upon it. (Wiener (1954), 17).

Information is information, not matter or energy. No materialism which does not admit this can survive at the present day (Wiener (1961, 132).

ON.2 endorses an information-theoretic, metaphysical monism: the universe's essential nature is digital, being fundamentally composed of information as data/dedomena instead of matter or energy, with material objects as a complex secondary manifestation (a similar position has been defended more recently in physics by Frieden (1998), whose work is based on a loosely Platonist perspective). ON.2 may but does not have to endorse a computational view of information processes. ON.3 advocates a more pluralistic approach along similar lines. Both are compatible with GDI.

A final comment concerning GDI.3 can be introduced by discussing a fourth slogan:

ON.4)   In fact, what we mean by information - the elementary unit of information - is a difference which makes a difference. (Bateson (1973), 428).

ON.4 is one of the earliest and most popular formulations of GDI (see for example Franklin (1995), 34 and Chalmers (1996), 281). The formulation in Mackay (1969) - that is "information is a *distinction* that makes a difference" - predates Batesons but it is slightly different from it in that, by speaking of "distinction" instead of "difference", it has an epistemological rather than an ontological twist. A "difference" (a "distinction") is just a discrete state, namely a datum, and "making a difference" simply means that the datum is "meaningful", at least potentially.

2.7   Genetic Neutrality

Finally, let us consider the semantic nature of the data. How data can come to have an assigned meaning and function in a semiotic system in the first place is one of the hardest problems in semantics. Luckily, the point in question here is not *how* but *whether* data constituting information as semantic content can be meaningful independently of an informee. The *genetic neutrality* (GeN) supported by GDI states that:

GeN) δ can have a semantics *independently* of any informee.

Before the discovery of the Rosetta Stone, Egyptian hieroglyphics were already regarded as information, even if their semantics was beyond the comprehension of any interpreter. The discovery of an interface between Greek and Egyptian did not affect the semantics of the hieroglyphics but only its accessibility. This is the weak, conditional-counterfactual sense in which GDI.3 speaks of meaningful data being embedded in information-carriers informee-independently. GeN supports the possibility of *information without an informed subject*, to adapt a Popperian phrase. Meaning is not (at least not only) in the mind of the user. GeN is to be distinguished from the stronger, realist thesis, supported for example by Dretske (1981), according to which data could also have their own semantics independently of an intelligent *producer/informer*. This is also known as *environmental information*, a concept sufficiently important to deserve a brief presentation before we close this first part.

Environmental information. One of the most often cited example of environmental information is the series of concentric rings visible in the wood of a cut tree trunk, which may be used to estimate its age. Yet "environmental" information does not need to be *natural*. Going back to our example, when you turned the ignition key, the red light of the low battery indicator flashed. This signal too can be interpreted as an instance of environmental information.

Environmental information is defined relative to an observer (an information agent), who is supposed to have no direct access to pure data in themselves. It requires two systems a and b to be coupled in such a way that a's being (of type, or in state) F is correlated to b being (of type, or in state) G, thus carrying for the observer the information that b is G (this analysis is adapted from Barwise and Seligman (1997), who improve on a similar account by Dretske (1981)).

Environmental information: two systems *a* and *b* are coupled in such a way that *a*'s being (of type, or in state) *F* is correlated to *b* being (of type, or in state) *G*, thus carrying for the information agent the information that *b* is *G*.

Fig. 5. Environmental information

The correlation in Fig. 5 is usually *nomic* (it follows some law). It may be engineered as in the case of the low battery indicator (a) whose flashing (F) is triggered by, and hence it is informative about, the battery (b) being flat (G). Or it may be natural, as when litmus - a natural colouring matter from lichens - is used as an acid-alkali indicator because it turns red in acid solutions and blue in alkaline solutions. Other typical examples include the correlation between fingerprints and personal identification.

One may be so used to see the low battery indicator flashing as carrying the information that the battery is flat to find it hard to distinguish, with sufficient clarity, between environmental and semantic information. However, it is important to stress that environmental information may require or involve no semantics at all. It may consist of (networks or patterns of) correlated data understood as mere differences or constraining affordances. Plants (e.g., a sunflower), animals (e.g., an amoeba) and mechanisms (e.g., a photocell) are certainly capable of making practical use of environmental information even in the absence of any (semantic processing of) *meaningful* data.

2.8   Summary of the First Part

To summarise, GDI defines information, broadly understood, as syntactically well-formed and meaningful data. Its four types of neutrality (TyN, TaN, ON and GeN) represent an obvious advantage, as they make GDI perfectly scalable to more complex cases and reasonably flexible in terms of applicability and compatibility. Indeed, philosophers have variously interpreted and tuned these four neutralities according to their theoretical needs.

Our next step is to check whether GDI is satisfactory when discussing the most important type of semantic information, namely factual information. Before addressing this issue, however, we need to pause and look at the mathematical theory of communication (MTC).

MTC is not the only successful mathematical approach to the concept of information. *Fisher information* (Frieden (1998) and the *algorithmic information theory* (Chaitin (1987)) provide two other important examples. However, MTC is certainly the most widely known among philosophers. As such, it has had a profound impact on philosophical analyses of semantic information, to which it has provided both the technical vocabulary and at least the initial conceptual frame of reference. One needs to grasp its main gist if one wishes to make sense of the issuing philosophical debate.

3   Information as Data Communication

Some features of information are intuitive. We are used to information being *encoded, transmitted* and *stored.* One also expects it to be additive (information a + information b = information $a + b$) and *non-negative*, like other things in life, such as probabilities and interest rates. If you ask a question, the worst scenario is that you receive no answer or a wrong answer, which will leave you with zero new information.

Similar properties of information are quantifiable. They are investigated by the *mathematical theory of communication* (MTC) with the primary aim of devising efficient ways of encoding and transferring data.

The name for this branch of probability theory comes from Shannon's seminal work (Shannon and Weaver (1949)). Shannon pioneered this field and obtained many of its principal results, but he acknowledged the importance of previous work done by other researchers and colleagues at Bell laboratories, most notably Nyquist and Hartley (see Cherry (1978) and Mabon (1975)). After Shannon, MTC became known as *information theory*, an appealing but unfortunate label, which continues to cause endless misunderstandings. Shannon came to regret its widespread popularity, and we shall avoid using it in this context.

This second part of the article outlines some of the key ideas behind MTC, with the aim of understanding the relation between MTC and some philosophical theories of semantic information. The reader with no taste for mathematical formulae may wish to go directly to section 3.2, where some conceptual implications of MTC are outlined. The reader interested in knowing more may start by reading Weaver (1949), Pierce (1980), Shannon and Weaver (1949), then Jones

(1979), and finally Cover and Thomas (1991). The latter two are technical texts. Floridi (2003a) provides a simplified analysis oriented to philosophy students.

3.1   The Mathematical Theory of Communication

MTC has its origin in the field of electrical engineering, as the study of communication limits. It develops a quantitative approach to information as a means to answer two fundamental problems: the ultimate level of data compression (how small can a message be, given the same amount of information to be encoded?) and the ultimate rate of data transmission (how fast can data be transmitted over a channel?). The two solutions are the entropy H in equation [9] (see below) and the channel capacity C. The rest of this section illustrates how to get from the problems to the solutions.

To have an intuitive sense of the approach, let us return to our example. Recall the telephone conversation with the mechanic. In Fig. 6, the wife is the *informer*, the mechanic is the *informee*, "the battery is flat" is the (semantic) message (the *informant*), there is a coding and decoding procedure through a natural language (English), a channel of communication (the telephone system) and some possible noise. Informer and informee share the same background knowledge about the collection of usable symbols (technically known as the *alphabet*; in the example this is English).

MTC is concerned with the efficient use of the resources indicated in Fig. 6. Now, the conversation with the mechanic is fairly realistic and hence

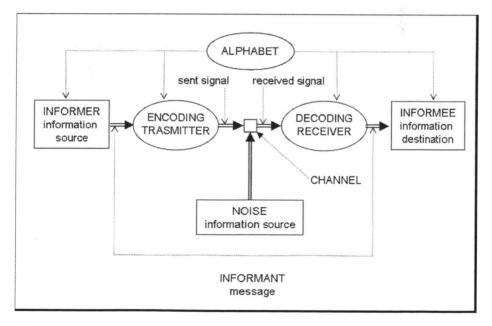

Fig. 6. Communication model (adapted from Shannon and Weaver [1949 rep. 1998])

more difficult to model than a simplified case. We shall return to it later but, in order to introduce MTC, imagine instead a very boring device that can produce only one symbol. Edgar Alan Poe wrote a short story in which a raven can answer only "nevermore" to any question. Poe's raven is called a *unary device*. Imagine you ring the garage and your call is answered by Poe's raven. Even at this elementary level, Shannons simple model of communication still applies. It is obvious that the raven (a unary device) produces zero amount of information. Simplifying, we already know the outcome of the communication exchange, so our ignorance (expressed by our question) cannot be decreased. Whatever the informational state of the system is, asking appropriate questions (e.g. "will I be able to make the car start?", "can you come to fix the car?") of the raven does not make any difference. Note that, interestingly enough, this is the basis of Platos famous argument in the *Phaedrus* against the value of semantic information provided by written texts:

> [Socrates]: Writing, Phaedrus, has this strange quality, and is very like painting; for the creatures of painting stand like living beings, but if one asks them a question, they preserve a solemn silence. And so it is with written words; you might think they spoke as if they had intelligence, but if you question them, wishing to know about their sayings, they always say only one and the same thing [they are unary devices, in our terminology]. And every word, when [275e] once it is written, is bandied about, alike among those who understand and those who have no interest in it, and it knows not to whom to speak or not to speak; when ill-treated or unjustly reviled it always needs its father to help it; for it has no power to protect or help itself.

As Plato well realises a unary source answers every question all the time with only one message, not with silence or message, since silence counts as a message, as we saw in 2.5, when discussing the nature of secondary information. It follows that a completely silent source also qualifies as a unary source. And if silencing a source (censorship) may be a nasty way of making a source uninformative, it is well known that crying wolf is a classic case in which an informative source degrades to the role of uninformative unary device.

Consider now a binary device that can produce two symbols, like a fair coin A with its two equiprobable symbols $\{h, t\}$; or, as Matthew 5:37 suggests, "Let your communication be Yea, yea; Nay, nay: for whatsoever is more than these cometh of evil". Before the coin is tossed, the informee (for example a computer) is in a state of data deficit greater than zero: the informee does not "know" which symbol the device will actually produce. Shannon used the technical term "uncertainty" to refer to *data deficit*. In a non-mathematical context this can be a very misleading term because of the strong epistemological connotations of this term. Remember that the informee can be a simple machine, and psychological, mental or doxastic states are clearly irrelevant.

Once the coin has been tossed, the system produces an amount of information that is a function of the possible outputs, in this case 2 equiprobable symbols, and equal to the data deficit that it removes.

Let us now build a slightly more complex system, made of two fair coins A and B. The AB system can produce 4 ordered outputs: $\langle h, h \rangle, \langle h, t \rangle, \langle t, h \rangle, \langle t, t \rangle$ It generates a data deficit of 4 units, each couple counting as a symbol in the source alphabet. In the AB system, the occurrence of each symbol $\langle -, - \rangle$ removes a higher data deficit than the occurrence of a symbol in the A system. In other words, each symbol provides more information. Adding an extra coin would produce a 8 units of data deficit, further increasing the amount of information carried by each symbol in the ABC system, and so on.

We are now ready to generalise the examples. Call the number of possible symbols N. For N = 1, the amount of information produced by a unary device is 0. For N = 2, by producing an equiprobable symbol, the device delivers 1 unit of information. And for N = 4, by producing an equiprobable symbol the device delivers the sum of the amount of information provided by a device producing one of two equiprobable symbols (coin A in the example above) plus the amount of information provided by another device producing one of two equiprobable symbols (coin B), that is, 2 units of information, although the total number of symbols is obtained by multiplying As symbols by Bs symbols. Now, our information measure should be a continuous and monotonic function of the probability of the symbols. The most efficient way of satisfying these requirements is by using the logarithm to the base 2 of the number of possible symbols (the logarithm to the base 2 of a number n is the power to which 2 must be raised to give the number n, for example $log_2 = 3$, since $2^3 = 8$). Logarithms have the useful property of turning multiplication of symbols into addition of information units. By taking the logarithm to the base 2 (henceforth log simply means log_2) we have the further advantage of expressing the units in bits. The base is partly a matter of convention, like using centimetres instead of inches, partly a matter of convenience, since it is useful when dealing with digital devices that use binary codes to represent data.

Given an alphabet of N equiprobable symbols, we can now rephrase some examples more precisely (Fig. 7) by using equation [1]:

average informativeness per symbol (or "uncertainty")

$$= log_2(N) \text{bits of information per symbol} \tag{1}$$

The basic idea is all in equation [1]. Information can be quantified in terms of decrease in data deficit (Shannon's "uncertainty"). Unfortunately, real coins are always biased. To calculate how much information they produce one must rely on the frequency of the occurrences of symbols in a finite series of tosses, or on their probabilities, if the tosses are supposed to go on indefinitely. Compared to a fair coin, a slightly biased coin must produce less than 1 bit of information, but still more than 0. The raven produced no information at all because the occurrence of a string S of "nevermore" was not *informative* (not *surprising*, to use Shannon's more intuitive, but psychologistic vocabulary), and that is because the *probability* of the occurrence of "nevermore" was maximum, so overly predictable. Likewise, the amount of information produced by the biased coin depends on the average *informativeness* (also known as average *surprisal*, another unfortunate term to

Device	Alphabet	Bits of information per symbol
Poe's raven (unary)	1 symbol	$\log(1) = 0$
1 coin (binary)	2 equiprobable symbols	$\log(2) = 1$
2 coins	4 equiprobable symbols	$\log(4) = 2$
1 die	6 equiprobable symbols	$\log(6) = 2.58$
3 coins	8 equiprobable symbols	$\log(8) = 3$

Fig. 7. Examples of communication devices and their information power

refer to the average statistical rarity) of the string S of h and t produced by the coin. The average informativeness of the resulting string S depends on the *probability* of the occurrence of each symbol. The higher the frequency of a symbol in S, the less information is being produced by the coin, up to the point when the coin is so biased to produce always the same symbol and stops being informative at all, behaving like the raven or the boy who cries wolf.

So, to calculate the average informativeness of S we need to know how to calculate S and the informativeness of the ith symbol in general. This requires understanding what the probability of the ith symbol (Pi) to occur is.

The probability P_i of the ith symbol can be "extracted" from equation [1], where it is embedded in log(N), a special case in which the symbols are equiprobable. Using some elementary properties of the logarithmic function, we have:

$$log\,(N) = -log\left(N^{-1}\right) = -log\left(\frac{1}{N}\right) = -log\,(P) \qquad (2)$$

The value of $\frac{1}{N} = P$ can range from 0 to 1. If Poe's raven is our source, the probability of it saying "good morning" is 0. In the case of the coin, $P(h)+P(t) = 1$, no matter how biased the coin is. Probability is like a cake that gets sliced more and more thinly depending on the number of guests, but never grows beyond its original size and, in the worst case scenario, can at most be equal to zero, but never become "negative". More formally, this means:

$$\sum_{i=1}^{N} P_i = 1 \qquad (3)$$

The sigma notation in [3] is simply a shortcut that indicates that if we add all probabilities values from i = 1 to i = N their sum is equal to 1.

We can now be precise about the raven: "nevermore" is not informative at all because $P_{nevermore} = 1$. Clearly, the lower the probability of occurrence of a symbol, the higher is the informativeness of its actual occurrence. The informativeness u of the ith symbol can be expressed by analogy with $-log\,(P)$ in equation [2]:

$$u_i = -log\left(P_i\right) \tag{4}$$

Next, we need to calculate the length of a general string S. Suppose that the biased coin, tossed 10 times, produces the string: $\langle h, h, t, h, h, t, t, h, h, t \rangle$. The (length of the) string S (in our case equal to 10) is equal to the number of times the h type of symbol occurs added to the numbers of times the t type of symbol occurs.

Generalising for i types of symbols:

$$S = \sum_{i=1}^{N} S_i \tag{5}$$

Putting together equations [4] and [5] we see that the average informativeness for a string of S symbols is the sum of the informativeness of each symbol divided by the sum of all symbols:

$$\frac{\sum_{i=1}^{N} S_i u_i}{\sum_{i=1}^{N} S_i} \tag{6}$$

Term [6] can be simplified thus:

$$\sum_{i=1}^{N} \frac{S}{S_i} u_i \tag{7}$$

Now $\frac{S_i}{S}$ is the frequency with which the ith symbol occurs in S when S is finite. If the length of S is left undetermined (as long as one wishes), then the frequency of the ith symbol becomes its probability P_i. So, further generalising term [7], we have:

$$\sum_{i=1}^{N} P_i u_i \tag{8}$$

Finally, by using equation [4] we can substitute for u_i and obtain

$$H = -\sum_{i=1}^{N} P_i log P_i \text{ (bits per symbol)} \tag{9}$$

Equation [9] is Shannon's formula for $H = $ uncertainty, which we have called *data deficit* (actually, Shannon's original formula includes a positive constant K which amounts to a choice of a unit of measure, bits in our case; apparently, Shannon used the letter H because of R.V.L. Hartley's previous work).

Equation [9] indicates that the quantity of information produced by a device corresponds to the amount of data deficit erased. It is a function of the average informativeness of the (potentially unlimited) string of symbols produced by the device. It is easy to prove that, if symbols are equiprobable, [9] reduces to [1] and that the highest quantity of information is produced by a system whose symbols are equiprobable (compare the fair coin to the biased one).

To arrive at [9] we have used some very simple examples: a raven and a hand-ful of coins. Things in life are far more complex, witness our Monday morning accident. For example, we have assumed that the strings of symbols are *ergodic*: the probability distribution for the occurrences of each symbol is assumed to be stable through time and independently of the selection of a certain string. Our raven and coins are *discrete* and *zero-memory sources*. The successive symbols they produce are statistically independent. But in real life occurrences of sym-bols are often interdependent. Sources can be non-ergodic and have a memory. Symbols can be continuous, and the occurrence of one symbol may depend upon a finite number n of preceding symbols, in which case the string is known as a Markov chain and the source an n-*th* order Markov source. Consider for ex-ample the probability of hearing "n" (followed by the string "ing") after having received the string of letters "Good mor-" over the phone, when you called the garage. And consider the same example through time, in the case of a child (the son of the mechanic) who is learning how to answer the phone instead of his father. In brief, MTC develops the previous analysis to cover a whole variety of more complex cases. We shall stop here, however, because in the rest of this section we need to concentrate on other central aspects of MTC.

The quantitative approach just sketched plays a fundamental role in coding theory (hence in cryptography) and in data storage and transmission techniques. MTC is primarily a study of the properties of a channel of communication and of codes that can efficiently encipher data into recordable and transmittable sig-nals. Since data can be distributed either in terms of here/there or now/then, diachronic communication and synchronic analysis of a memory can be based on the same principles and concepts (our coin becomes a bistable circuit or flip-flop, for example). Two concepts that play a pivotal role both in communication anal-ysis and in memory management are so important to deserve a brief explanation: *redundancy* and *noise*.

Consider our AB system. Each symbol occurs with 0.25 probability. A simple way of encoding its symbols is to associate each of them with two digits, as in Fig. 8:

| <h, h>= 00 | <h, t>= 01 | <t, h>= 10 | <t, t>= 11 |

Fig. 8. Code 1

Call this Code 1. In Code 1 a message conveys 2 bits of information, as expected. Do not confuse *bits* as *bi*-nary units of information (recall that we decided to use log_2 also as a matter of convenience) with *bits* as *bi*-nary digi*ts*, which is what a 2-symbols system like a CD-ROM uses to encode a message. Suppose now that the AB system is biased, and that the four symbols occur with the following probabilities (Fig. 9):

| <h, h>= 0.5 | <h, t>= 0.25 | <t, h>= 0.125 | <t, t>= 0.125 |

Fig. 9. A biased system

This biased system produces less information, so by using Code 1 we would be wasting resources. A more efficient Code 2 (Fig. 10) should take into account the symbols probabilities, with the following outcomes:

<h, h>= 0	0.5 × 1 binary digit = .5
<h, t>= 10	0.25 × 2 binary digits = .5
<t, h>= 110	0.125 × 3 binary digits = .375
<t, t>= 111	0.125 × 3 binary digits = .375

Fig. 10. Code 2 (Fano Code)

In Code 2, known as Fano Code, a message conveys 1.75 bits of information. One can prove that, given that probability distribution, no other coding system will do better than Fano Code.

In real life, a good codification is also modestly redundant. *Redundancy* refers to the difference between the physical representation of a message and the mathematical representation of the same message that uses no more bits than necessary. *Compression* procedures work by reducing data redundancy, but redundancy is not always a bad thing, for it can help to counteract *equivocation* (data sent but never received) and *noise* (data received but unwanted). A message + noise contains more data than the original message by itself, but the aim of a communication process is *fidelity*, the accurate transfer of the original message from sender to receiver, not data increase. We are more likely to reconstruct a message correctly at the end of the transmission if some degree of redundancy counterbalances the inevitable noise and equivocation introduced by the physical process of communication and the environment. Noise extends the informee's freedom of choice in selecting a message, but it is an undesirable freedom and some redundancy can help to limit it. That is why the manual of your car includes both verbal explanations and pictures to convey (slightly redundantly) the same information.

We are now ready to understand Shannons two fundamental theorems. Suppose the 2-coins biased system AB produces the following message:

$$\langle t, h \rangle \ \langle h, h \rangle \ \langle t, t \rangle \ \langle h, t \rangle \ \langle h, t \rangle$$

Using Fano Code we obtain: 11001111010. The next step is to send this string through a channel. Channels have different transmission rates (C), calculated in terms of bits per second (bps). Shannons fundamental theorem of the noiseless channel states that there is no method of encoding which gives an equivocation less than H C, as explained in Fig. 11.

Let a source have entropy H (bits per symbol) and a channel have a capacity C (bits per second). Then it is possible to encode the output of the source in such a way as to transmit at the average rate of $C/H - \varepsilon$ symbols per second over the channel where ε is arbitrarily small. It is not possible to transmit at an average rate greater than C/H.
Shannon and Weaver (1949 rep. 1998), 59.

Fig. 11. Shannon's fundamental theorem of the noiseless channel

> Let a discrete channel have the capacity C and a discrete source the entropy per second H. If $H \leq C$ there exists a coding system such that the output of the source can be transmitted over the channel with an arbitrarily small frequency of errors (or an arbitrarily small equivocation). If $H > C$ it is possible to encode the source so that the equivocation is less than $H - C + \varepsilon$ where ε is arbitrarily small. There is no method of encoding which gives an equivocation less than $H - C$.
>
> Shannon and Weaver (1949 rep. 1998), 71.

Fig. 12. Shannon's fundamental theorem for a discrete channel

In other words, if you devise a good code you can transmit symbols over a noiseless channel at an average rate as close to C/H as one may wish but, no matter how clever the coding is, that average can never exceed C/H. We have already seen that the task is made more difficult by the inevitable presence of noise. However, the fundamental theorem for a discrete channel with noise comes to our rescue, as explained in Fig. 12.

Roughly, if the channel can transmit as much or more information than the source can produce, then one can devise an efficient way to code and transmit messages with as small an error probability as desired. These two fundamental theorems are among Shannon's greatest achievements. They are limiting results in information theory that constrain any conceptual analysis of semantic information. They are thus comparable to Gdel's Turing's and Church's theorems in logic and computation. With our message finally sent, we may close this section and return to a more philosophical approach.

3.2   Conceptual Implications of the Mathematical Theory of Communication

For the mathematical theory of communication (MTC), information is only a selection of one symbol from a set of possible symbols, so a simple way of grasping how MTC quantifies information is by considering the number of yes/no questions required to determine what the source is communicating. One question is sufficient to determine the output of a fair coin, which therefore is said to produce 1 bit of information. A 2-fair-coins system produces 4 ordered outputs: $\langle h, h \rangle$, $\langle h, t \rangle$, $\langle t, h \rangle$, $\langle t, t \rangle$, and therefore requires at least two questions, each output containing 2 bits of information, and so on. This *erotetic* (the Greek word for "question") analysis clarifies two important points.

First, MTC is not a theory of information in the ordinary sense of the word. In MTC, information has an entirely technical meaning. Consider some examples. According to MTC, two equiprobable "yes"'s contain the same quantity of information, no matter whether their corresponding questions are "have the lights of your car been left switched on for too long, without recharging the battery?" or "would you marry me?". If we knew that a device could send us, with equal probabilities, either this article or the whole *Stanford Encyclopedia of Philosophy*, by receiving one or the other we would receive very different amounts of bytes of data but actually only one bit of information in the MTC sense of the word. On June 1

1944, the BBC broadcasted a line from Verlaine's *Song of Autumn*: "Les sanglots longs des violons de Autumne". The message contained almost 1 bit of information, an increasingly likely "yes" to the question whether the D-Day invasion was imminent. The BBC then broadcasted the second line "Blessent mon coeur d'une longueur monotone". Another almost meaningless string of letters, but almost another bit of information, since it was the other long-expected "yes" to the question whether the invasion was to take place immediately. German intelligence knew about the code, intercepted those messages and even notified Berlin, but the high command failed to alert the Seventh Army Corps stationed in Normandy. Hitler had all the information in Shannons sense of the word, but failed to understand (or believe in) the crucial importance of those two small bits of data. As for ourselves, we were not surprised to conclude in the previous section that the maximum amount of information (again, in the MTC sense of the word) is produced by a text where each character is equally distributed, that is by a perfectly random sequence. According to MTC, the classic monkey randomly pressing typewriter keys is indeed producing a lot of information.

Second, since MTC is a theory of information without meaning (not in the sense of meaningless, but in the sense of not yet meaningful), and since we have seen that [information - meaning = data], "mathematical theory of data communication" is a far more appropriate description of this branch of probability theory than "information theory". This is not a mere question of labels. Information, as semantic content (more on this shortly), can also be described erotetically as *data + queries*. Imagine a piece of (propositional) information such as "the earth has only one moon". It is easy to polarise almost all its semantic content by transforming it into a [query + binary answer], such as [does the earth have only one moon? + yes]. Subtract the "yes" - which is at most 1 bit of information, in the equiprobable case of a yes or no answer - and you are left with virtually all the semantic content, fully de-alethicised (from *aletheia*, the Greek word for truth; the query is neither true nor false). To use a Fregean expression, *semantic content* is *unsaturated information*, where the latter is semantic information that has been "eroteticised" and from which a quantity of information has been subtracted equal to $log P(yes)$, with P being the probability of the yes-answer.

The datum "yes" works as a key to unlock the information contained in the query. MTC studies the codification and transmission of information by treating it as data keys, that is, as the amount of details in a signal or message or memory space necessary to saturate the informees unsaturated information. As Weaver (1949) remarked

> the word information relates not so much to what you do say, as to what you could say. The mathematical theory of communication deals with the carriers of information, symbols and signals, not with information itself. That is, information is the measure of your freedom of choice when you select a message (p.12).

Since MTC deals not with semantic information itself but with the data that constitute it, that is, with messages comprising uninterpreted symbols encoded in well-formed strings of signals, it is commonly described as a study of information at the *syntactic* level. MTC can be successfully applied in ICT (information and communication technologies) because computers are syntactical devices. What remains to be clarified is how H in equation [9] should be interpreted.

H is also known in MTC as *entropy*. It seems we owe this confusing label to John von Newman, who recommend it to Shannon:

> You should call it entropy for two reasons: first, the function is already in use in thermodynamics under the same name; second, and more importantly, most people don't know what entropy really is, and if you use the word *entropy* in an argument you will win every time (quoted by Golan (2002)).

Von Newman proved to be right on both accounts, unfortunately.

Assuming the ideal case of a noiseless channel of communication, H is a measure of three equivalent quantities:

a) the average amount of information per symbol produced by the informer, or
b) the corresponding average amount of data deficit (Shannons uncertainty) that the informee has before the inspection of the output of the informer, or
c) the corresponding informational potentiality of the same source, that is, its *informational entropy*.

H can equally indicate (a) or (b) because, by selecting a particular alphabet, the informer automatically creates a data deficit (uncertainty) in the informee, which then can be satisfied (resolved) in various degrees by the *informant*. Recall the erotetic game. If you use a single fair coin, I immediately find myself in a 1 bit deficit predicament: I do not know whether it is head or tail. Use two fair coins and my deficit doubles, but use the raven, and my deficit becomes null. My empty glass (point (b) above) is an exact measure of your capacity to fill it (point (a) above). Of course, it makes sense to talk of information as quantified by H only if one can specify the probability distribution.

Regarding (c), MTC treats information like a physical quantity, such as mass or energy, and the closeness between equation [9] and the formulation of the concept of entropy in statistical mechanics was already discussed by Shannon. The informational and the thermodynamic concept of entropy are related through the concepts of probability and *randomness* ("randomness" is better than "disorder" since the former is a syntactical concept whereas the latter has a strongly semantic value, that is, it is easily associated to interpretations, as I used to try to explain to my parents when I was young). Entropy is a measure of the amount of "mixedupness" in processes and systems bearing energy or information. Entropy can also be seen as an indicator of reversibility: if there is no change of entropy then the process is reversible. A highly structured, perfectly organised message contains a lower degree of entropy or randomness, less information in Shannon sense, and hence it causes a smaller data deficit, which can be close to zero (remember the raven). By contrast, the higher the potential randomness of the symbols in the alphabet, the more bits of information can be produced by the device. Entropy assumes its maximum value in the extreme case of uniform distribution, which is to say that a glass of water with a cube of ice contains less entropy than the glass of water once the cube has melted, and a biased coin has less entropy than a fair coin. In thermodynamics, we know that the greater the entropy, the less available the energy. This means that high entropy corresponds to high energy deficit, but so does entropy in MTC: higher values of H correspond to higher quantities of data deficit.

4   Information as Semantic Content

We have seen that, when data are well-formed and meaningful, the result is also known as *semantic content* (Bar-Hillel and Carnap (1953; Bar-Hillel (1964). Information, understood as semantic content, comes in two main varieties: factual and instructional. In our example, one may translate the red light flashing into semantic content in two senses:

a) as a piece of factual information, representing the fact that the battery is flat; and

b) as a piece of instructional information, conveying the need for a specific action, e.g. the re-charging or replacing of the flat battery.

In this third part of the article we shall be concerned primarily with (a), so it is better to clear the ground by considering (b) first. It is the last detour in our journey.

4.1   Instructional Information

Instructional information is a type of semantic content. An instruction booklet, for example, provides instructional information, either imperatively - in the form of a recipe: first do this, then do that - or conditionally, in the form of some inferential procedure: if such and such is the case do this, otherwise do that.

Instructional information is not about a situation, a fact, or a state of affairs w and does not model, or describe or represent w. Rather, it is meant to (help to) bring about w. For example, when the mechanic tells one over the phone to connect a charged battery to the flat battery of ones car, the information one receives is not factual, but instructional.

There are many plausible contexts in which a stipulation ("let the value of x = 3" or "suppose we discover the bones of a unicorn"), an invitation ("you are cordially invited to the college party"), an order ("close the window!"), an instruction ("to open the box turn the key"), a game move ("1.e2-e4 c7-c5" at the beginning of a chess game) may be correctly qualified as kinds of instructional information. The printed score of a musical composition or the digital files of a program may also be counted as typical cases of instructional information.

All these instances of information have a semantic side: they have to be at least potentially meaningful (interpretable) to count as information. Moreover, instructional information may be related to factual (descriptive) information in performative contexts, such as christening - e.g. "this ship is now called *HMS The Informer*" - or programming - e.g. as when deciding the type of a variable. The two types of semantic information (instructional and factual) may also come together in magic spells, where semantic representations of x may be (wrongly) supposed to provide some instructional power and control over x. Nevertheless, as a test, one should remember that instructional information does not qualify alethically (cannot be correctly qualified as true or false). In the example, it would be silly to ask whether the information "only use batteries with the same rated voltage" is true. Stipulations, invitations, orders, instructions, game moves, and software cannot be true or false. As Wittgenstein remarks "The way music speaks. Do not forget that a poem, even though it is composed in the language of information, is not used in the language-game of giving information." (Zettel, §160, see Wittgenstein (1981).

4.2   Factual Information

In the language game that Wittgenstein seems to have in mind, the notion of "semantic information" is intended in a declarative or factual mode. Factual information may be true or untrue (false, in case one adopts a binary logic). *True semantic content* is the most common sense in which information seems to be understood (Floridi (2004b)). It is also one of the most important, since information as true semantic content is a necessary condition for knowledge. Some elaboration is in order, and in the following sub-sections we shall briefly look at the concept of data as constraining affordances, at the role played by levels of abstraction in the transformation of constraining affordances into factual information, and finally at the relation between factual information and truth.

4.2.1   Constraining Affordances

The data that constitute factual information allow or invite certain constructs (they are *affordances* for the information agent that can take advantage of them) and resist or impede some others (they are *constraints* for the same agent), depending on the interaction with, and the nature of, the information agent that processes them. For example, the red light flashing repetitively and the engine not starting allow you (or any other information agent like you) to construct the information that (a) the battery is flat, while making it more difficult to you (or any other information agent like you) to construct the information that (b) there is a short circuit affecting the proper functioning of the low battery indicator, where the engine fails to start because there is no petrol in the tank, a fact not reported by the relevant indicator which is affected by the same short circuit. This is the sense in which data are *constraining affordances* for (an information agent responsible for) the elaboration of factual information.

4.2.2   Levels of Abstraction

In section 2.3, we saw that the concept of pure data in themselves (dedomena) is an abstraction, like Kant's noumena or Locke's substance. The point made was that data are never accessed and elaborated (by an information agent) independently of a *level of abstraction* (LoA). The time has come to clarify what a LoA is.

A LoA is a specific set of typed variables, intuitively representable as an interface, which establishes the scope and type of data that will be available as a resource for the generation of information (Floridi and Sanders (forthcoming). This concept of LoA is purely epistemological, and it should not be confused with other forms of "levellism" that are more or less explicitly based on an ontological commitment concerning the intrinsic architecture, syntax or structure of the system discussed (Dennett (1971), Marr (1982) , Newell (1982), Simon (1969), see now Simon (1996) ; Poli (2001) provides a reconstruction of ontological levellism; more recently, Craver (2004) has analysed ontological levellism, especially in biology and cognitive science, see also Craver (forthcoming)). Ontological levellism has come under increasing attack. Heil (2003) and Schaffer (2003) have seriously and convincingly questioned its plausibility. However, epistemological levellism is flourishing, especially in computer science (Roever et al. (1998), Hoare and Jifeng (1998)), where it is regularly used to satisfy the requirement that systems constructed in levels (in order to tame their complexity) function correctly.

Through a LoA, an information agent (the observer) accesses a physical or conceptual environment, the system. LoAs are not necessarily hierarchical and they are comparable. They are interfaces that mediate the epistemic relation between the observed and the observer. Consider, for example, a motion detector (Fig. 13). In the past, motion detectors caused an alarm whenever a movement was registered within the range of the sensor, including the swinging of a tree branch (object a in Fig. 13). The old LoA1 consisted of a single typed variable, which may be labelled MOVEMENT. Nowadays, when a PIR (passive infrared) motion detector registers some movement, it also monitors the presence of an infrared signal, so the entity detected has to be something that also emits infrared radiation usually perceived as heat before the sensor activates the alarm. The new LoA2 consists of two typed variables: MOVEMENT and INFRARED RADIATION. Clearly, your car (object b in Fig. 13) leaving your house is present for both LoAs; but for the new LoA2, which is more finely grained, the branch of the tree swinging in the garden is absent. Likewise, a stone in the garden (object c in Fig. 13) is absent for both the new and the old LoA, since it satisfies no typed variable of either one.

The method of LoA is an efficient way of making explicit and managing the ontological commitment of a theory. In our case, "the battery is what provides electricity to the car" is a typical example of information elaborated at a drivers LoA. An engineers LoA may output something like "12-volt lead-acid battery is made up of six cells, each cell producing approximately 2.1 volts", and an economists LoA may suggest that "a good quality car battery will cost between $50 and $100 and, if properly maintained, it should last five years or more".

Data as constraining affordances - answers waiting for the relevant questions - are transformed into factual information by being processed semantically at a given LoA (alternatively: the relevant question is associated to the right answer

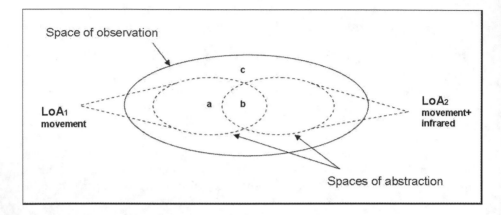

Fig. 13. An example of Levels of Abstraction

at a given LoA). Once data as constraining affordances have been elaborated into factual information at a given LoA, the next question is whether truth values supervene on factual information.

4.2.3   Information and Truth

Does some factual content qualify as information only if it is true? Defenders of the alethic neutrality of semantic information (Fetzer (2004) and Dodig-Crnkovic (2005), who criticises Floridi (2004b) ; Colburn (2000) , Fox (1983), and, among, situation theorists, Devlin (1991)) argue that meaningful and well-formed data already qualify as information, no matter whether they represent or convey a truth or a falsehood or indeed have no alethic value at all. Opponents, on the other hand, object that "[...] *false* information and *mis*-information are not kinds of information - any more than decoy ducks and rubber ducks are kinds of ducks" (Dretske (1981), 45) and that "false information is not an inferior kind of information; it just is not information" (Grice (1989), 371; other philosophers who accept a truth-based definition of semantic information are Barwise and Seligman (1997) and Graham (1999)). The result is a definition of factual semantic information as well-formed, meaningful and truthful data (defended in Floridi (2004b), (2005b)), where "truthful" is only a stylistic choice to be preferred to "true" because it enables one to say that a map conveys factual information insofar as it is truthful.

Once again, the debate is not about a mere definition, but concerns the possible consequences of the alethic neutrality thesis, three of which can be outlined here, whereas a fourth requires a longer analysis and will be discussed in section 5.1.

If the thesis "meaningful and well-formed data already qualify as information" is correct then

i) false information (including contradictions) would count as a genuine type of semantic information, not as pseudo-information;

ii) all necessary truths (including tautologies) would qualify as information (on this issue see Bremer (2003)); and

iii) "it is true that p" - where p is a variable that can be replaced by any instance of genuine semantic information - would not be a redundant expression; for example, "it is true" in the conjunction "'the earth is round' qualifies as information and it is true" could not be eliminated without semantic loss.

All these new issues are grafted to some old branches of the philosophical tree.

Whether false information is a genuine type of information has important repercussions on any philosophy and pragmatics of communication.

The question about the informative nature (or lack thereof) of necessary truths, tautologies, equations or identity statements is an old one, as it runs through Hume, Kant, Frege and Wittgenstein. The latter, for example, interestingly remarked:

> Another expression akin to those we have just considered is this: 'Here it is; take it or leave it!' And this again is akin to a kind of introductory statement which we sometimes make before remarking on certain alternatives, as when we say: 'It either rains or it doesn't rain; if it rains we'll stay in my room, if it doesn't ...'. The first part of this sentence is no piece of information (just as 'Take it or leave it' is no order). Instead of, 'It either rains or it doesn't rain' we could have said, 'Consider the two cases ...'. Our expression underlines these cases, presents them to your attention (*The Blue and Brown Books*, The Brown Book, II, p. 161, see Wittgenstein (1960)).

The solution of the problem of hyperintensionality (how one can draw a semantic distinction between expressions that are supposed to have the same meaning according to a particular theory of meaning that is usually model-theoretic or modal in character) depends on how one can make sense of the relation between truth and informativeness in the case of logically equivalent expressions.

Finally, the possibly redundant qualification of information as true is also linked with the critique of the deflationary theories of truth (DTT), since one could accept a deflationary T-schema as perfectly correct, while rejecting the explanatory adequacy of DTT. "It is true that" in "it is true that p" could be redundant in view of the fact that there cannot be factual information that is not true, but DTT could mistake this linguistic or conceptual redundancy for unqualified dispensability. "It is true that" could be redundant because, strictly speaking, information is not a truth-bearer but already encapsulates truth as truthfulness. Thus, DTT may be satisfactory as theories of truth-ascriptions while being inadequate as theories of truthfulness.

Once information is available, knowledge can be built in terms of *justifiable* or *explainable semantic information*. An information agent knows that the battery is flat not by merely guessing rightly, but because e.g. it perceives that the red light of the low battery indicator flashing and/or that the engine does not start. In this sense, information provides the basis of any further scientific investigation. Note, however, that the fact that data may count as *resources* for (i.e. inputs an agent can use to construct) information, and hence for knowledge, rather than *sources*, may lead to constructionist arguments against mimetic theories that interpret information as some sort of picture of the world. The point requires some elaboration.

Whether empirical or conceptual, data make possible only a certain range of information constructs, and not all constructs are made possible equally easily. An analogy may help here. Suppose one has to build a shelter. The design and complexity of the shelter may vary, but there is a limited range of "realistic" possibilities, determined by the nature of the available resources and constraints (size, building materials, location, weather, physical and biological environment, working force, technical skills, purposes, security, time constraints, etc.). Not any shelter can be built. And the type of shelter that will be built more often will be the one that is more likely to take close-to-optimal advantage of the available resources and constraints. The same applies to data. Data are at the same time the resources and constraints that make possible the construction of information. The best information is that better tuned to the constraining affordances available. Thus informational coherence and adequacy do not necessarily entail nor support nave or direct realism, or a correspondence theory of truth as this is ordinarily presented. Ultimately, information is the result of a process of data modelling; it does not have to represent or photograph or portray or photocopy, or map or show or uncover or monitor or ... the intrinsic nature of the system analysed, no more than an igloo describes the intrinsic nature of snow or the Parthenon indicates the real properties of stones.

When *semantic content is false*, this is a case of *misinformation* (Fox (1983)). And if the source of misinformation is aware of its nature, one may speak of *disinformation*, as when one says to the mechanic "my husband forgot to turn the lights on". Disinformation and misinformation are ethically censurable but may be successful in achieving their purpose: tell the mechanic that your husband left the lights on last night, and he will still be able to provide you with the right advice. Likewise, information may still fail to be successful; just imagine telling the mechanic that your car is out of order.

5   Philosophical Approaches to Semantic Information

What is the relation between MTC and the sort of semantic information that we have called factual? The mathematical theory of communication approaches information as a physical phenomenon. Its central question is whether and how much uninterpreted data can be encoded and transmitted efficiently by means of a given alphabet and through a given channel. MTC is not interested in the meaning, "aboutness", relevance, reliability, usefulness or interpretation of information, but only in the level of detail and frequency in the uninterpreted data, being these symbols, signals or messages. Philosophical approaches differ from MTC in two main respects.

First, they seek to give an account of information as *semantic* content, investigating questions like "how can something count as information? and why?", "how can something carry information about something else?", "how can semantic information be generated and flow?", "how is information related to error,

truth and knowledge?", "when is information useful?". Wittgenstein, for example, remarks that "One is inclined to say: 'Either it is raining, or it isn't - how
I know, how the information has reached me, is another matter.' But then let
us put the question like this: What do I call information that it is raining'?
(Or have I only information of this information too?) And what gives this 'information' the character of information about something? Doesn't the form of
our expression mislead us here? For isn't it a misleading metaphor to say: "My
eyes give me the information that there is a chair over there"? (*Philosophical
Investigations*, I. §356, see now Wittgenstein (2001)).

Second, philosophical theories of semantic information also seek to connect it
to other relevant concepts of information and more complex forms of epistemic,
mental and doxastic phenomena. For instance, Dretske (1981) and Barwise and
Seligman (1997) attempt to ground information, understood as factual semantic contents, on environmental information. The approach is also known as the
naturalization of information. A similar point can be made about Putnams twin
earths argument, the externalization of semantics and teleosemantics.

Philosophical analyses usually adopt a propositional orientation and an epistemic outlook, endorsing, often implicitly, the prevalence or centrality of factual
information within the map outlined in Fig. 1. They tend to base their analyses
on cases such as "Paris is the capital of France" or "The Bodleian Library is in
Oxford". How relevant is MTC to similar researches?

In the past, some research programs tried to elaborate information theories
alternative to MTC, with the aim of incorporating the semantic dimension. Donald M. Mackay (1969) proposed a quantitative theory of qualitative information
that has interesting connections with *situation logic* (see below). According to
MacKay, information is linked to an increase in knowledge on the receiver's side:

> Suppose we begin by asking ourselves what we mean by information.
> Roughly speaking, we say that we have gained information when we
> know something now that we didn't know before; when 'what we know'
> has changed. (Mackay (1969), p. 10).

Around the same years, Doede Nauta (1972) developed a semiotic-cybernetic
approach. Nowadays, few philosophers follow these lines of research. The majority agrees that MTC provides a rigorous constraint to any further theorising
on all the semantic and pragmatic aspects of information. The disagreement
concerns the crucial issue of the *strength* of the constraint.

At one extreme of the spectrum, any philosophical theory of semantic-factual
information is supposed to be *very strongly* constrained, perhaps even overdeter-
mined, by MTC, somewhat as mechanical engineering is by Newtonian physics.
Weavers optimistic interpretation of Shannons work is a typical example.

At the other extreme, any philosophical theory of semantic-factual information is supposed to be *only weakly* constrained, perhaps even completely underdetermined, by MTC, somewhat as tennis is constrained by Newtonian physics,
that is in the most uninteresting, inconsequential and hence disregardable sense
(see for example Sloman (1978) and Thagard (1990)).

The emergence of MTC in the 1950s generated earlier philosophical enthusiasm that has gradually cooled down through the decades. Historically, philosophical theories of semantic-factual information have moved from "very strongly constrained" to "only weakly constrained". Recently, we find positions that carefully appreciate MTC for what it can provide in terms of a robust and well-developed statistical theory of correlations between states of different systems (the sender and the receiver) according to their probabilities. This can have important consequences in mathematically-friendly contexts, such as some approaches to naturalised epistemology (Harms (1998)) or scientific explanation (Badino (2004)).

Although the philosophy of semantic information has become increasingly autonomous from MTC, two important connections have remained stable between MTC and even the most recent philosophical accounts:

1. the communication model, explained in section 3.1 (see Fig. 6); and
2. what Barwise labelled the "Inverse Relationship Principle" (IRP).

The communication model has remained virtually unchallenged, even if nowadays theoretical accounts are more likely to consider as basic cases multiagent and distributed systems interacting in parallel, rather than individual agents related by simple, sequential channels of communication. In this respect, the philosophy of information (Floridi (2002);(2004a)) is less Cartesian than "social".

IRP refers to the inverse relation between the probability of p - which may range over sentences of a given language (as in Bar-Hillel and Carnap) or events, situations or possible worlds (as in Dretske) - and the amount of semantic information carried by p (recall that Poe's raven, as a unary source provides no information because its answers are entirely predictable). It states that information goes hand in hand with unpredictability. Popper (1935) is often credited as the first philosopher to have advocated IRP explicitly. However, systematic attempts to develop a formal calculus involving it were made only after Shannons breakthrough.

We have seen that MTC defines information in terms of probability space distribution. Along similar lines, the *probabilistic approach* to semantic information defines the semantic information in p in terms of logical probability space and the inverse relation between information and the probability of p. This approach was initially suggested by Bar-Hillel and Carnap (1953) (see also Bar-Hillel (1964)) and further developed by Kemeny (1953), Smokler (1966), Hintikka and Suppes (1970) and Dretske (1981). The details are complex but the original idea is simple. The semantic content (CONT) in p is measured as the complement of the a priori probability of p:

$$CONT(p) = 1 - P(p) \tag{10}$$

CONT does not satisfy the two requirements of additivity and conditionalization, which are satisfied by another measure, the informativeness (INF) of p, which is calculated, following equations [9] and [10], as the reciprocal of $P(p)$, expressed in bits, where $P(p) = 1 - CONT(p)$:

$$INF(p) = log\frac{1}{1 - CONT(p)} = -logP(p) \qquad (11)$$

Things are complicated by the fact that the concept of probability employed in equations [10] and [11] is subject to different interpretations. In Bar-Hillel and Carnap (1953), the probability distribution is the outcome of a logical construction of atomic statements according to a chosen formal language. This introduces a problematic reliance on a strict correspondence between observational and formal language. In Dretske, the solution is to make probability values refer to the observed states of affairs (s), that is:

$$I(s) = -logP(s) \qquad (12)$$

The *modal approach* further modifies the probabilistic approach by defining semantic information in terms of modal space and in/consistency. The information conveyed by p becomes the set of all possible worlds, or (more cautiously) the set of all the descriptions of the relevant possible states of the universe, that are excluded by *p*.

The *systemic approach*, developed especially in situation logic (Barwise and Perry 1983, Israel and Perry 1990, Devlin (1991); Barwise and Seligman (1997) provide a foundation for a general theory of information flow) also defines information in terms of states space and consistency. However, it is less ontologically demanding than the modal approach, since it assumes a clearly limited domain of application. It is also compatible with Dretske's probabilistic approach, although it does not require a probability measure on sets of states. The informational content of p is not determined a priori, through a calculus of possible states allowed by a representational language, but in terms of factual content that p carries with respect to a given situation. Information tracks possible transitions in a system's states space under normal conditions. Both Dretske and situation theorists require some presence of information already immanent in the environment (*environmental information*), as nomic regularities or constraints. This "semantic externalism" can be controversial.

The *inferential approach* defines information in terms of entailment space: information depends on valid inference relative to an information agent's theory or epistemic state.

Each of the previous extensionalist approaches can be given an intentionalist interpretation by considering the relevant space as a doxastic space, in which information is seen as a reduction in the degree of personal uncertainty, given a state of knowledge of the informee. Wittgenstein addressed this distinction in his *Remarks on the Philosophy of Psychology* I §817 (Wittgenstein (1980))

> The important insight is that there is a language-game [Wittgenstein seems to have in mind here the information game we have already encountered above] in which I produce information automatically, information which can be treated by other people quite as they treat non-automatic information only here there will be no question of any 'lying' - information which I myself may receive like that of a third person. The

'automatic' statement, report etc. might also be called an 'oracle'. - But of course that means that the oracle must not avail itself of the words 'I believe ... '.

5.1   The Bar-Hillel-Carnap Paradox

Insofar as they subscribe to the Inverse Relationship Principle, the extensionalist approaches outlined in the previous section can be affected by what has been defined, with a little hyperbole, as the *Bar-Hillel-Carnap Paradox* (BCP, Floridi (2004b)).

In a nutshell, we have seen that, following IRP, the less probable or possible p is the more semantic information p is assumed to be carrying. This explains why most philosophers agree that tautologies convey no information at all, for their probability or possibility is 1. But it also leads one to consider contradictions - which describe impossible states, or whose probability is 0 - as the sort of messages that contain the highest amount of semantic information. It is a slippery slope. Make a statement less and less likely and you gradually increase its informational content, but at certain point the statement "implodes" (in the quotation below, it becomes "too informative to be true").

Bar-Hillel and Carnap (1953) were among the first to make explicit this prima facie counterintuitive inequality. Note how their careful wording betrays the desire to defuse the problem:

BCP)

It might perhaps, at first, seem strange that a self-contradictory sentence, hence one which no ideal receiver would accept, is regarded as carrying with it the most inclusive information. It should, however, be emphasized that semantic information is here not meant as implying truth. A false sentence which happens to say much is thereby highly informative in our sense. Whether the information it carries is true or false, scientifically valuable or not, and so forth, does not concern us. A self-contradictory sentence asserts too much; it is too informative to be true (p. 229).

Since its formulation, BCP has been recognised as an unfortunate, yet perfectly correct and logicall inevitable consequence of any quantitative *theory of weakly semantic information* (TWSI; "weakly" because truth values play no role in it). As a consequence, the problem has often been either ignored or tolerated (Bar-Hillel and Carnap (1953)) as the price of an otherwise valuable approach. Sometimes, however, attempts have been made to circumscribe its counterintuitive consequences. This has happened especially in Information Systems Theory (Winder et al. (1997)) - where consistency is an essential constraint that must remain satisfied for a database to preserve data integrity - and in Decision Theory, where inconsistent information is obviously of no use to a decision maker.

In these cases, whenever there are no possible models that satisfy a statement or a theory, instead of assigning to it the maximum quantity of semantic information, three strategies have been suggested:

1. assigning to all inconsistent cases the same, infinite information value (Lozin-skii (1994)). This is in line with an economic approach, which defines x as impossible if and only if x has an infinite price;
2. eliminating all inconsistent cases a priori from consideration, as impossible outcomes in decision-making (Jeffrey (1990)). This is in line with the syntactic approach developed by MTC;
3. assigning to all inconsistent cases the same zero information value (Mingers (1997), Aisbett and Gibbon (1999).

The latter approach is close to the *strongly semantic approach*, to which we shall now turn.

5.2   The Strongly Semantic Approach to Information

The general hypothesis is that BCP indicates that something has gone essentially amiss with TWSI. TWSI is based on a semantic principle that is too weak, namely that truth-values are independent of semantic information. A semantically stronger approach, according to which information encapsulates truth, can avoid the paradox and is more in line with the ordinary conception of what generally counts as factual information, as we have seen in section 4.2.3. MTC already provides some initial reassurance. MTC identifies the quantity of information associated with, or generated by, the occurrence of a signal (an event or the realisation of a state of affairs) with the elimination of possibilities (reduction in uncertainty) represented by that signal (event or state of affairs). In MTC, no counterintuitive inequality comparable to BCP occurs, and the line of argument is that, as in the case of MTC, a *theory of strongly semantic information* (TSSI), based on alethic and discrepancy values rather than probabilities, can also successfully avoid BCP (Floridi (2004b); (2005b), see Bremer and Cohnitz (2004) chap. 2 for an overview). The idea is to define semantic-factual information in terms of data space, as well-formed, meaningful and truthful data. This constrains the probabilistic approach introduced above, by requiring first a qualification of the content as truthful. Once the content is so qualified, the quantity of semantic information in p is calculated in terms of distance of p from the situation/resource w that p is supposed to model. Total distance is equivalent to a p true in all cases (all possible worlds or probability 1), including w and hence minimally informative, whereas maximum closeness is equivalent to the precise modelling of w at the agreed level of abstraction.

Suppose there will be exactly three guests for dinner tonight. This is our situation w. Imagine we are told that

T) there may or may not be some guests for dinner tonight; or
V) there will be some guests tonight; or
P) there will be three guests tonight.

The *degree of informativeness* of T is zero because, as a tautology, T applies both to w and to $\neg\ w$. V performs better, and P has the maximum degree of informativeness because, as a fully accurate, precise and contingent truth, it "zeros in" on its target w. Generalising, the more distant some semantic-factual information σ is from its target w, the larger is the number of situations to which it applies, the lower its degree of informativeness becomes. A tautology is a true σ that is most "distant" from the world.

Let us now use the letter ϑ to refer to the distance between a true σ and w. Using the more precise vocabulary of situation logic, ϑ indicates the degree of support offered by w to σ. We can now map on the x axis of a Cartesian diagram the values of ϑ given a specific σ and a corresponding target w. In our example, we know that $\vartheta(T) = 1$ and $\vartheta(P) = 0$. For the sake of simplicity, let us assume that $\vartheta(V) = 0.25$ (see Floridi (2004b) on how to calculate ϑ values). We now need a formula to calculate the *degree of informativeness* ι of σ in relation to $\vartheta(\sigma)$. It can be shown that the most elegant solution is provided by the complement of the square value of $\vartheta(\sigma)$, that is $y = 1 - x^2$. Using the symbols just introduced, we have:

$$\iota(\sigma) = 1 - \vartheta(\sigma)^2 \tag{13}$$

Fig. 14 shows the graph generated by equation [13] when we include also negative values of distance for false σ (ϑ ranges from $-1 =$ contradiction to $1 =$ tautology).

If σ has a very high degree of informativeness ι (very low ϑ) we want to be able to say that it contains a large quantity of semantic information and, vice

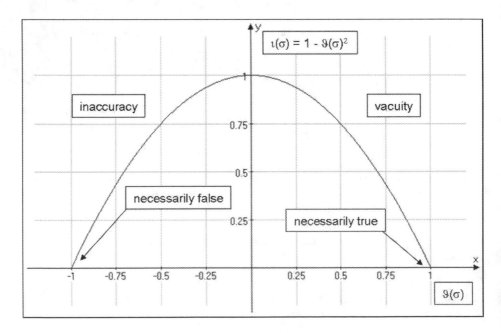

Fig. 14. Degree of informativeness

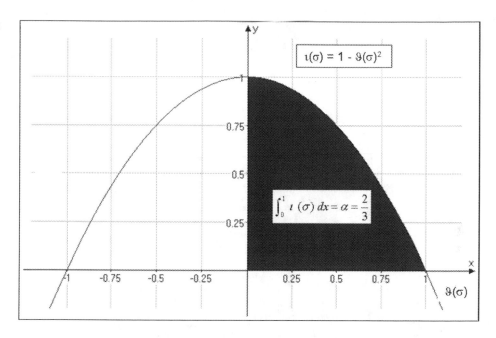

Fig. 15. Maximum amount of semantic information α carried by σ

versa, the lower the degree of informativeness of σ is, the smaller the quantity of semantic information conveyed by σ should be. To calculate the quantity of semantic information contained in σ relative to $\iota(\sigma)$ we need to calculate the area delimited by equation [13], that is, the definite integral of the function $\iota(\sigma)$ on the interval [0, 1]. As we know, the maximum quantity of semantic information (call it α) is carried by P, whose $\vartheta = 0$. This is equivalent to the whole area delimited by the curve. Generalising to σ we have:

$$\int_0^1 \iota(\sigma)\, dx = \alpha = \frac{2}{3} \tag{14}$$

Fig. 15 shows the graph generated by equation [14]. The shaded area is the maximum amount of semantic information α carried by σ.

Consider now V, "there will be some guests tonight". V can be analysed as a (reasonably finite) string of disjunctions, that is V = ["there will be one guest tonight" or "there will be two guests tonight" or ... "there will be n guests tonight"], where n is the reasonable limit we wish to consider (things are more complex than this, but here we only need to grasp the general principle). Only one of the descriptions in V will be fully accurate. This means that V also contains some (perhaps much) information that is simply irrelevant or redundant. We shall refer to this "informational waste" in V as vacuous information in V. The amount of vacuous information (call it β) in V is also a function of the distance ϑ of V from w, or more generally

$$\int_0^{\vartheta} \iota\left(\sigma\right) dx = \beta \tag{15}$$

Since $\vartheta(\mathrm{V}) = 0.25$, we have

$$\int_0^{0.25} \iota\left(\sigma\right) dx = 0.24479 \tag{16}$$

Fig. 16 shows the graph generated by equation [16]. The shaded area is the amount of vacuous information β in V. Clearly, the amount of semantic information in V is simply the difference between α (the maximum amount of information that can be carried in principle by σ) and β (the amount of vacuous information actually carried by σ), that is the clear area in the graph of Fig. 16. More generally, and expressed in bits, the amount of semantic information γ in σ is:

$$\gamma\left(\sigma\right) = log\left(\alpha - \beta\right) \tag{17}$$

Note the similarity between [14] and [15]. When $\vartheta\left(\sigma\right) = 1$, that is, when the distance between σ and w is maximum, then $\alpha = \beta$ and $\gamma\left(\sigma\right) = 0$. This is what happens when we consider T. T is so distant from w as to contain only vacuous information. In other words, T contains as much vacuous information as P contains relevant information.

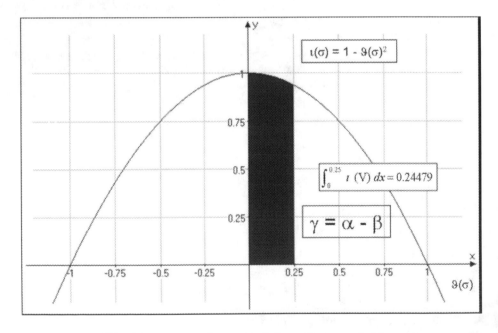

Fig. 16. Amount of semantic information γ carried by σ

6   Conclusion

Philosophical theories of semantic information have recently contributed to a new area of research in itself, the philosophy of information (Adams (2003), Floridi (2002), (2003b), (2004a)). The two special issue volumes of *Minds and Machines* on the philosophy of information (Floridi (2003c)) provide an overview of the scope and depth of current work in the field. Information seems to have become a key concept to unlock several philosophical problems. "The most valuable commodity I know of is information", boldly declares Gordon Gekko in Oliver Stones *Wall Street* (1987). Euphranor would probably have concurred. The problem is that we still have to agree about what information is exactly.

Acknowledgements

This article is based on Floridi (2005a) and I am grateful to Ed Zalta and the Editorial Board of the *Stanford Encyclopedia of Philosophy* for permission to reproduce the original text. I benefited enormously from many insightful editorial comments by Fred Kroon, Jerry Seligman and Ed Zalta on previous drafts. I am also very grateful to several colleagues and friends for their helpful suggestions and conversations on previous drafts and past papers on which this article is based. They are responsible only for the improvements not for any remaining mistake: Frederick R. Adams, Mark Bedau, John Collier, Ian C. Dengler, Michael Dunn, Roger Brownsword, Timothy Colburn, James Fetzer, Phil Fraundorf, Gian Maria Greco, Ken Herold, Bernard Katz, Philipp Keller, Gianluca Paronitti, Jeff Sanders, Sebastian Sequoiah-Grayson, Janet D. Sisson, Giovanni Sommaruga, Ernest Sosa, J. L. Speranza and Matteo Turilli. Finally, I would like to thank all the participants to the Workshop on information theories, and the organizers, Juerg Kohlas and Giovanni Sommaruga.

References

Adams, F.: The Informational Turn in Philosophy Minds and Machines 13(4), 471–501 (2003)

Aisbett, J., Gibbon, G.: A practical measure of the information in a logical theory. Journal of Experimental and Theoretical Artificial Intelligence 11(2), 201–218 (1999)

Armstrong, D.M.: A Materialist Theory of the Mind. London Routledge and Kegan Paul (1968)

Armstrong, D.M.: A Materialist Theory of the Mind, 2nd edn. London Routledge (1993)

Badino, M.: An Application of Information Theory to the Problem of the Scientific Experiment Synthese 140, 355–389 (2004)

Bar-Hillel, Y.: Language and Information: Selected Essays on Their Theory and Application. Addison-Wesley, Reading (1964)

Bar-Hillel, Y., Carnap, R.: An Outline of a Theory of Semantic Information (1953); repr. in Bar-Hillel, 221–274 (1964)

Barwise, J., Seligman, J.: Information Flow: The Logic of Distributed Systems. Cambridge University Press, Cambridge (1997)

Bateson, G.: Steps to an Ecology of Mind, Frogmore, St. Albans, Paladin (1973)

Berkeley, G.: Alciphron, Or the Minute Philosopher, 1732, Edinburgh 1948-57 Thomas Nelson

Braman, S.: Defining Information Telecommunications Policy 13, 233–242 (1989)

Bremer, M.E.: Do Logical Truths Carry Information? Minds and Machines 13(4), 567–575 (2003)

Bremer, M., Cohnitz, D.: Information and Information Flow - an Introduction. Ontos Verlag, Frankfurt, Lancaster (2004)

Chaitin, G.J.: Algorithmic Information Theory. Cambridge University Press, Cambridge (1987)

Chalmers, D.J.: The Conscious Mind: In Search of a Fundamental Theory. Oxford Univ. Press, New York (1996)

Cherry, C.: On Human Communication: A Review, a Survey, and a Criticism, 3rd edn. MIT Press, Cambridge (1978)

Colburn, T.R.: Philosophy and Computer Science. M.E. Sharpe, Armonk (2000)

Cover, T.M., Thomas, J.A.: Elements of Information Theory. Wiley, Chichester (1991)

Craver, C.F.: A Field Guide to Levels Proceedings and Addresses of the American Philosophical Association 77(3) (2004)

Craver, C.F.: Explaining the Brain: A Mechanist's Approach (forthcoming)

Debons, A., Cameron, W.J. (eds.): Perspectives in Information Science: Proceedings of the Nato Advanced Study Institute on Perspectives in Information Science, Aberystwyth, Wales, Uk, August 13-24, 1973. Leiden, Noordhoff (1975)

Dennett, D.C.: Content and Consciousness. Routledge and Kegan Paul, London (1969)

Dennett, D.C.: Intentional Systems. The Journal of Philosophy 68, 87–106 (1971)

Dennett, D.C.: Content and Consciousness, 2nd edn. Routledge and Kegan Paul, London (1986)

Deutsch, D.: Quantum Theory, the Church-Turing Principle and the Universal Quantum Computer Proceedings of the Royal Society. 400, 97–117 (1985)

Deutsch, D.: The Fabric of Reality. Penguin, London (1997)

Devlin, K.J.: Logic and Information. Cambridge University Press, Cambridge (1991)

Di Vincenzo, D.P., Loss, D.: Quantum Information Is Physical, Superlattices and Microstructures - Special issue on the occasion of Rolf Landauer's 70th birthday 23, 419–432 (1998)

Dodig-Crnkovic, G.: System Modeling and Information Semantics. In: Bubenko, J., Eriksson, O., Fernlund, H., Lind, M. (eds.) Proceedings of the Fifth Promote IT Conference, Borlänge, Sweden, Studentlitteratur, Lund (2005)

Dretske, F.I.: Knowledge and the Flow of Information. Blackwell, Oxford (1981); CSLI Publications, Stanford (reprinted, 1999)

Dunn, J.M.: The Concept of Information and the Development of Non-Classical Logics Non-Classical Approaches in the Transition from Traditional to Modern Logic, Stelzner, W. (ed.): pp. 423–448. de Gruyter, Berlin

Fetzer, J.H.: Information, Misinformation, and Disinformation Minds and Machines 14(2), 223–229 (2004)

Floridi, L.: Philosophy and Computing: An Introduction. Routledge, London (1999)

Floridi, L.: What Is the Philosophy of Information? Metaphilosophy 33(1-2), 123–145 (2002)

Floridi, L. (ed.): Information in The Blackwell Guide to the Philosophy of Computing and Information, pp. 40–61. Blackwell, Oxford (2003a)

Floridi, L.: Two Approaches to the Philosophy of Information. Minds and Machines 13(4), 459–469 (2003)

Floridi, L.: Open Problems in the Philosophy of Information. Metaphilosophy 35(4), 554–582 (2004)

Floridi, L.: Outline of a Theory of Strongly Semantic Information. Minds and Machines 14(2), 197–222 (2004)

Floridi, L.: Information, Semantic Conceptions Of, Stanford Encyclopedia of Philosophy, Zalta, E.N. (ed.) (2005),
http://plato.stanford.edu/entries/information-semantic/

Floridi, L.: Is Information Meaningful Data? Philosophy and Phenomenological Research 70(2), 351–370 (2005)

Floridi, L. (ed.): Minds and Machines (Special issue of the journal on The Philosophy of Information) (2003)

Floridi, L., Sanders, J.W.: Levellism and the Method of Abstraction (fortcoming)

Fox, C.J.: Information and Misinformation: An Investigation of the Notions of Information, Misinformation, Informing, and Misinforming. Greenwood Press, Westport (1983)

Franklin, S.: Artificial Minds. MIT Press, Cambridge (1995)

Frieden, B.R.: Physics from Fisher Information: A Unification. Cambridge University Press, Cambridge (1998)

Frieden, B.R.: Science from Fisher Information: A Unification, 2nd edn. Cambridge University Press, Cambridge (2004)

Golan, A.: Information and Entropy Econometrics - Editor's View. Journal of Econometrics 107(1-2), 1–15 (2002)

Graham, G.: The Internet: A Philosophical Inquiry. Routledge, London (1999)

Grice, H.P.: Studies in the Way of Words. Harvard University Press, Cambridge (1989)

Hanson, P.P. (ed.): Information, language, and cognition. University of British Columbia Press, Vancouver (1990)

Harms, W.F.: The Use of Information Theory in Epistemology. Philosophy of Science 65(3), 472–501 (1998)

Heil, J.: Levels of Reality, Ratio 16(3), 205–221 (2003)

Hintikka, J., Suppes, P. (eds.): Information and Inference. Reidel, Dordrecht (1970)

Hoare, C.A.R., Jifeng, H.: Unifying Theories of Programming. Prentice-Hall, London (1998)

Hume, D.: Essays, Moral, Political, and Literary 1987, Indianapolis, Liberty Classics. Edited and with a foreword, notes, and glossary by Eugene F. Miller; with an apparatus of variant readings from the, edition by T.H. Green and T.H. Grose. Based on the 1777 edn. originally published as v. 1 of Essays and treatises on several subjects (1987)

Jeffrey, R.C.: The Logic of Decision, 2nd edn. University of Chicago Press, Chicago (1990)

Jones, D.S.: Elementary Information Theory. Clarendon Press, Oxford (1979)

Kemeny, J.: A Logical Measure Function. Journal of Symbolic Logic 18, 289–308 (1953)

Landauer, R.: Computation: A Fundamental Physical View. Physica Scripta 35, 88–95 (1987)

Landauer, R.: Information Is Physical. Physics Today 44, 23–29 (1991)

Landauer, R.: The Physical Nature of Information. Physics Letter A 217, 188 (1996)

Landauer, R., Bennett, C.H.: The Fundamental Physical Limits of Computation. Scientific American, 48–56 (July 1985)

Larson, A.G., Debons, A.: Information Science in Action: System Design. In: Proceedings of the Nato Advanced Study Institute on Information Science, Crete, Greece, August 1-11, M. Nijhoff, The Hague (1983)

Losee, R.M.: A Discipline Independent Definition of Information. Journal of the American Society for Information Science 48(3), 254–269 (1997)

Lozinskii, E.: Information and evidence in logic systems. Journal of Experimental and Theoretical Artificial Intelligence 6, 163–193 (1994)

Mabon, P.C.: Mission Communications: The Story of Bell Laboratories. Bell Telephone Laboratories, Murray Hill (1975)

Machlup, F., Mansfield, U. (eds.): The Study of Information: Interdisciplinary Messages. Wiley, Chichester (1983)

MacKay, D.M.: Information, Mechanism and Meaning. MIT Press, Cambridge (1969)

Marr, D.: Vision: A Computational Investigation into the Human Representation and Processing of Visual Information. W.H. Freeman, San Francisco (1982)

Mingers, J.: The Nature of Information and Its Relationship to Meaning. In: Winder, R.L., et al. (eds.) Philosophical Aspects of Information Systems, pp. 73–84. Taylor and Francis, London (1997)

Nauta, D.: The Meaning of Information. The Hague, Mouton (1972)

Newell, A.: The Knowledge Level. Artificial Intelligence 18, 87–127 (1982)

Newell, A., Simon, H.A.: Computer Science as Empirical Inquiry: Symbols and Search. Communications of the ACM 19, 113–126 (1976)

Pierce, J.R.: An Introduction to Information Theory: Symbols, Signals and Noise, 2nd edn. Dover Publications, New York (1980)

Poli, R.: The Basic Problem of the Theory of Levels of Reality. Axiomathes 12, 261–283 (2001)

Popper, K.R.: Logik Der Forschung: Zur Erkenntnistheorie Der Modernen Naturwissenschaft, 1935, Wien, J. Springer. Eng. tr. The Logic of Scientific Discovery, London, Hutchinson (1959)

Roever, W.P.d., Engelhardt, K., Buth, K.-H.: Data Refinement: Model-Oriented Proof Methods and Their Comparison. Cambridge University Press, Cambridge (1998)

Sayre, K.M.: Cybernetics and the Philosophy of Mind. Routledge and Kegan Paul, London (1976)

Schaffer, J.: Is There a Fundamental Level? Nous 37(3), 498–517 (2003)

Shannon, C.E.: Collected Papers. In: Sloane, N.J.A., Wyner, A.D. (eds.). IEEE Press, New York (1993)

Shannon, C. E., Weaver, W.: The Mathematical Theory of Communication, 1949 rep., Urbana, University of Illinois Press. Foreword by Richard E. Blahut and Bruce Hajek (1998)

Simon, H.A.: The Sciences of the Artificial, 1st edn. MIT Press, Cambridge (1969)

Simon, H.A.: The Sciences of the Artificial, 3rd edn. MIT Press, Cambridge (1996)

Sloman, A.: The Computer Revolution in Philosophy: Philosophy, Science and Models of Mind, Hassocks, Harvester (1978)

Smokler, H.: Informational Content: A Problem of Definition. The Journal of Philosophy 63(8), 201–211 (1966)

Steane, A.M.: Quantum Computing. Reports on Progress in Physics 61, 117–173 (1998)

Thagard, P.R.: Comment: Concepts of Information in 1990

Weaver, W.: The Mathematics of Communication. Scientific American 181(1), 11–15 (1949)

Wheeler, J.A.: Information, Physics, Quantum: The Search for Links, Complexity, Entropy, and the Physics of Information. In: Zureck, W.H. (ed.). Addison Wesley, Redwood City (1990)

Wiener, N.: The Human Use of Human Beings: Cybernetics and Society, 1st edn. (1954); 2nd edn., London, reissued with a new introduction by Steve J. Heims, London, Free Association (1989)

Wiener, N.: Cybernetics or Control and Communication in the Animal and the Machine, 2nd edn. MIT Press, Cambridge (1961)

Winder, R.L., Probert, S.K., Beeson, I.A.: Philosophical Aspects of Information Systems. Taylor & Francis, London (1997)

Wittgenstein, L.: Preliminary Studies for the Philosophical Investigations: Generally Known as the Blue and Brown Books, 2nd edn. Basil Blackwell, Oxford (1960)

Wittgenstein, L.: Remarks on the Philosophy of Psychology, vol. 2, University of Chicago Press, Chicago. Basil Blackwell, Oxford (1980); Edited by G. E. M. Anscombe and G. H. von Wright; translated by G. E. M. Anscombe; vol. 2 edited by G.H. von Wright and H. Nyman; translated by C.G. Luckhardt and A.E. Aue

Wittgenstein, L.: Zettel, 2nd edn. Blackwell, Oxford (1981); Edited by G.E.M. Anscombe and G.H. von Wright; translated by G.E.M. Anscombe

Wittgenstein, L.: Philosophical Investigations: The German Text with a Revised English Translation, 3rd edn. Blackwell, Oxford (2001); Translated by G.E.M. Anscombe. Incorporates final revisions made by Elizabeth Anscombe to her English edition. Some typesetting errors have been corrected, and the text has been repaginated

Information Theory, Relative Entropy and Statistics

François Bavaud

University of Lausanne
Switzerland

1 Introduction: The Relative Entropy as an Epistemological Functional

Shannon's Information Theory (IT) (1948) definitely established the purely *mathematical* nature of entropy and relative entropy, in contrast to the previous identification by Boltzmann (1872) of his "*H*-functional" as the *physical* entropy of earlier thermodynamicians (Carnot, Clausius, Kelvin). The following recounting is attributed to Shannon (Tribus and McIrvine 1971):

> *My greatest concern was what to call it. I thought of calling it "information", but the word was overly used, so I decided to call it "uncertainty". When I discussed it with John von Neumann, he had a better idea. Von Neumann told me, "You should call it entropy, for two reasons. In the first place your uncertainty function has been used in statistical mechanics under that name, so it already has a name. In the second place, and more important, nobody knows what entropy really is, so in a debate you will always have the advantage."*

In IT, the entropy of a message limits its minimum coding length, in the same way that, more generally, the complexity of the message determines its compressibility in the Kolmogorov-Chaitin-Solomonov algorithmic information theory (see e.g. Li and Vitanyi (1997)).

Besides coding and compressibility interpretations, the relative entropy also turns out to possess a direct probabilistic meaning, as demonstrated by the *asymptotic rate formula* (4). This circumstance enables a complete exposition of classical inferential statistics (hypothesis testing, maximum likelihood, maximum entropy, exponential and log-linear models, EM algorithm, etc.) under the guise of a discussion of the properties of the relative entropy.

In a nutshell, the relative entropy $K(f\|g)$ has two arguments f and g, which both are probability distributions belonging to the same simplex. Despite formally similar, the arguments are epistemologically contrasted: f represents the observations, the data, what *we see*, while g represents the expectations, the models, what *we believe*. $K(f\|g)$ is an asymmetrical measure of dissimilarity between empirical and theoretical distributions, able to capture the various aspects of the confrontation between models and data, that is the art of classical statistical inference, including Popper's refutationism as a particulary case. Here lies the dialectic charm of $K(f\|g)$, which emerges in that respect as an *epistemological functional*.

G. Sommaruga (Ed.): Formal Theories of Information, LNCS 5363, pp. 54–78, 2009.

We have here attempted to emphasize and synthetize the conceptual significance of the theory, rather than insisting on its mathematical rigor, the latter
being thoroughly developped in a broad and widely available litterature (see e.g.
Cover and Thomas (1991) and references therein). Most of the illustrations bear
on independent and identically distributed (i.i.d.) finitely valued observations,
that is on *dice models*. This convenient restriction is not really limiting, and can
be extended to Markov chains of finite order, as illustrated in the last part on
textual data with presumably original applications, such as heating and cooling
texts, or additive and multiplicative text mixtures.

2   The Asymptotic Rate Formula

2.1   Model and Empirical Distributions

$D = (x_1 x_2 \ldots x_n)$ denotes the data, consisting of n observations, and M denotes
a possible model for those data. The corresponding probability is $P(D|M)$, with

$$P(D|M) \geq 0 \qquad\qquad \sum_D P(D|M) = 1.$$

Assume (dice models) that each observation can take on m discrete values, each
observation X being i.i.d. distributed as

$$f_j^M := P(X = j) \qquad\qquad j = 1, \ldots, m.$$

f^M is the *model distribution*. The *empirical distribution*, also called *type* (Csiszár
and Körner 1980) in the IT framework, is

$$f_j^D := \frac{n_j}{n} \qquad\qquad j = 1, \ldots, m$$

where n_j counts the occurences of the j-th category and $n = \sum_{j=1}^m n_j$ is the
sample size.

Both f^M and f^D are discrete distributions with m modalities. Their collection
form the *simplex* S_m (figure 1)

$$S \equiv S_m := \{f \mid f_j \geq 0 \quad \text{and} \quad \sum_{j=1}^m f_j = 1\}.$$

2.2   Entropy and Relative Entropy: Definitions and Properties

Let $f, g \in S_m$. The *entropy* $H(f)$ of f and the *relative entropy* $K(f||g)$ between
f and g are defined (in nats) as

$$H(f) := -\sum_{j=1}^m f_j \ln f_j = \text{entropy of } f$$

$$K(f||g) := \sum_{j=1}^m f_j \ln \frac{f_j}{g_j} = \text{relative entropy of } f \text{ with respect to } g.$$

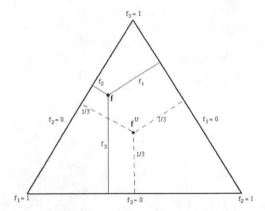

Fig. 1. The simplex S_3, where $f^U = (\frac{1}{3}, \frac{1}{3}, \frac{1}{3})$ denotes the uniform distribution. In the interior of S_m, a distribution f can be varied along $m-1$ independant directions, that is $\dim(S_m) = m - 1$.

$H(f)$ is concave in f, and constitutes a measure of the *uncertainty of the outcome* among m possible outcomes (proofs are standard):

$$0 \leq H(f) \leq \ln m$$

where

- $H(f) = 0$ iff f is a *deterministic* distribution concentrated on a single modality (minimum uncertainty)
- $H(f) = \ln m$ iff f is the *uniform* distribution (of the form $f_j = 1/m$) (maximum uncertainty).

$K(f\|g)$, also known as the *Kullback-Leibler divergence*, is convex in both arguments, and constitutes a *non-symmetric* measure of the *dissimilarity* between the distributions f and g, with

$$0 \leq K(f\|g) \leq \infty$$

where

- $K(f\|g) = 0$ iff $f \equiv g$
- $K(f\|g) < \infty$ iff f is absolutely continuous with respect to g, that is if $g_j = 0$ implies $f_j = 0$.

Let the categories $j = 1, \ldots, m$ be *coarse-grained*, that is aggregated into groups of super-categories $J = 1, \ldots, M < m$. Define

$$F_J := \sum_{j \in J} f_j \qquad G_J := \sum_{j \in J} g_j .$$

Then

$$H(F) \leq H(f) \qquad K(F\|G) \leq K(f\|g) . \tag{1}$$

2.3 Derivation of the Asymptotic Rate (i.i.d. Models)

On one hand, straightforward algebra yields

$$P(D|f^M) := \quad P(D|M) \quad = P(x_1 x_2 \ldots x_n|M)$$
$$= \prod_{i=1}^{n} (f_j^M)^{n_j} = \exp[-nK(f^D||f^M) - nH(f^D)]. \tag{2}$$

On the other hand, each permutation of the data $D = (x_1, \ldots, x_n)$ yields the same f^D. Stirling's approximation $n! \cong n^n \exp(-n)$ (where $a_n \cong b_n$ means $\lim_{n\to\infty} \frac{1}{n} \ln(a_n/b_n) = 0$) shows that

$$P(f^D|M) = \frac{n!}{n_1! \cdots n_m!} P(D|M) \cong \exp(nH(f^D)) P(D|M). \tag{3}$$

(2) and (3) imply the **asymptotic rate formula**:

$$P(f^D|f^M) \cong \exp(-n\,K(f^D||f^M)) \qquad\qquad \text{asymptotic rate formula . } \tag{4}$$

Hence, $K(f^D||f^M)$ is the asymptotic rate of the quantity $P(f^D|f^M)$, the *probability* of the empirical distribution f^D for a given model f^M, or equivalently the *likelihood* of the model f^M for the data f^D. Without additional constraints, the model $\hat{f}^M$ maximizing the likelihood is simply $\hat{f}^M = f^D$ (section 3). Also, without further information, the most probable empirical distribution $\tilde{f}^D$ is simply $\tilde{f}^D = f^M$ (section 4).

2.4 Asymmetry of the Relative Entropy and Hard Falsificationism

$K(f||g)$ as a dissimilarity measure between f and g is *proper* (that is $K(f||g) = 0$ implies $f \equiv g$) but *not symmetric* ($K(f||g) \neq K(g||f)$ in general). Symmetrized dissimilarities such as $J(f||g) := \frac{1}{2}(K(f||g) + K(g||f))$ or $L(f||g) := K(f||\frac{1}{2}(f + g)) + K(g||\frac{1}{2}(f + g))$ have often been proposed in the literature.

The conceptual significance of such functionals can indeed be questioned: from equation (4), the first argument f of $K(f||g)$ should be an empirical distribution, and the second argument g a model distribution. Furthermore, *the asymmetry of the relative entropy does not constitute a defect, but perfectly matches the asymmetry between data and models.* Indeed

- if $f_j^M = 0$ and $f_j^D > 0$, then $K(f^D||f^M) = \infty$ and, from (4), $P(f^D|f^M) = 0$ and, unless the veracity of the data f^D is questioned, the model distribution f^M should be strictly rejected
- if on the contrary $f_j^M > 0$ and $f_j^D = 0$, then $K(f^D||f^M) < \infty$ and $P(f^D|f^M) > 0$ in general: f^M should not be rejected, at least for small samples.

Thus the theory "All crows are black" is refuted by the single observation of a white crow, while the theory "Some crows are black" is not refuted by the observation of a thousand white crows. In this spirit, Popper's falsificationist mechanisms (Popper 1963) are captured by the properties of the relative entropy, and can in the present IT framework be further extended to probabilistic or "soft falsificationist" situations, beyond the purely logical true/false context (see section 3.1).

2.5   The Chi-Square Approximation

Most of the properties of the relative entropy are shared by another functional, historically anterior and well-known to statisticians, namely the *chi-square* $\chi^2(f\|g) := n \sum_j (f_j - g_j)^2/g_j$. As a matter of fact, the relative entropy and the chi-square (divided by $2n$) are identical up to the third order:

$$2K(f\|g) = \sum_{j=1}^{m} \frac{(f_j - g_j)^2}{g_j} + O(\sum_j \frac{(f_j - g_j)^3}{g_j^2}) = \frac{1}{n}\chi^2(f\|g) + O(\|f - g\|^3) \quad (5)$$

Example: coin ($m = 2$). The values of the relative entropy and the chi-square read, for various f^M and f^D, as :

	f^M	f^D	$K(f^D\|f^M)$	$\chi^2(f^D\|f^M)/2n$
a)	$(0.5, 0.5)$	$(0.5, 0.5)$	0	0
b)	$(0.5, 0.5)$	$(0.7, 0.3)$	0.0823	0.08
c)	$(0.7, 0.3)$	$(0.5, 0.5)$	0.0822	0.095
d)	$(0.7, 0.3)$	$(0.7, 0.3)$	0	0
e)	$(0.5, 0.5)$	$(1, 0)$	0.69	0.5
f)	$(1, 0)$	$(0.99, 0.01)$	∞	∞

3   Maximum Likelihood and Hypothesis Testing

3.1   Testing a Single Hypothesis (Fisher)

As shown by (4), the higher $K(f^D\|f^M)$, the lower the likelihood $P(f^D|f^M)$. This circumstance permits to test the single hypothesis H_0 : "*the model distribution is f^M*". If H_0 were true, f^D should fluctuate around its expected value f^M, and fluctuations of too large amplitude, with occurrence probability less than α (the significance level), should lead to the rejection of f^M. Well-known results on the chi-square distribution (see e.g. Cramer (1946) or Saporta (1990)) together with approximation (5) shows $2nK(f^D\|f^M)$ to be distributed, under H_0 and for n large, as $\chi^2[\text{df}]$ with df $= \dim(S_m) = m - 1$ degrees of freedom.

Therefore, the test consists in *rejecting H_0 at level α* if

$$2\,n\,K(f^D\|f^M) \geq \chi^2_{1-\alpha}[m-1]. \quad (6)$$

In that respect, Fisher's classical hypothesis testing appears as a *soft falsificationist* strategy, yielding the rejection of a theory f^M for large values of $K(f^D\|f^M)$. It generalizes Popper's (hard) falsificationism which is limited to situations of strict refutation as expressed by $K(f^D\|f^M) = \infty$.

3.2   Testing a Family of Models

Very often, the hypothesis to be tested is *composite*, that is of the form H_0 : "$f^M \in \mathcal{M}$", where $\mathcal{M} \subset S = S_m$ constitutes a family of models containing a number $\dim(\mathcal{M})$ of free, non-redundant parameters.

If the observed distribution itself satisfies $f^D \in \mathcal{M}$, then there is obviously no reason to reject H_0. But $f^D \notin \mathcal{M}$ in general, and hence

$$\min_{f \in \mathcal{M}} K(f^D || f) = K(f^D || \hat{f}^M) \text{ is strictly positive, where } \hat{f}^M := \arg \min_{f \in \mathcal{M}} K(f^D || f).$$

$\hat{f}^M$ is known as the *maximum likelihood estimate* of the model, and depends on both f^D and $\mathcal{M}$. We assume $\hat{f}^M$ to be unique, which is e.g. the case if $\mathcal{M}$ is convex.

If $f^M \in \mathcal{M}$, $2nK(f^D || \hat{f}^M)$ follows a chi-square distribution with $\dim(\mathcal{S})$ − $\dim(\mathcal{M})$ degrees of freedom. Hence, one rejects H_0 at level α if

$$2nK(f^D || \hat{f}^M) \geq \chi^2_{1-\alpha}[\dim(S) - \dim(\mathcal{M})] . \tag{7}$$

If $\mathcal{M}$ reduces to a unique distribution f^M, then $\dim(\mathcal{M}) = 0$ and (7) reduces to (6). In the opposite direction, $\mathcal{M} = S$ defines the *saturated* model, in which case (7) yields the undefined inequality $0 \geq \chi^2_{1-\alpha}[0]$.

Example: coarse grained model specifications. Let f^M be a dice model, with categories $j = 1, \ldots, m$. Let $J = 1, \ldots, M < m$ denote *groups of categories*, and suppose that the model specifications are coarse-grained (see (1)), that is

$$\mathcal{M} = \{ f^M \mid \sum_{j \in J} f_j^M \overset{!}{=} F_J^M \quad J = 1, \ldots, M \}$$

where F^M is fixed. Let $J(j)$ denote the group to which j belongs. Then the maximum likelihood (ML) estimate is simply

$$\hat{f}_j^M = f_j^D \frac{F_{J(j)}^M}{F_{J(j)}^D} , \text{ where } F_J^D := \sum_{j \in J} f_j^D \text{ and } K(f^D || \hat{f}^M) = K(F^D || F^M). \tag{8}$$

Example: independence. Let X and Y two categorical variables with modalities $j = 1, \ldots, m_1$ and $k = 1, \ldots, m_2$. Let f_{jk} denote the joint distribution of (X, Y). The distribution of X alone (respectively Y alone) obtains as the marginal $f_{j\bullet} := \sum_k f_{jk}$ (respectively $f_{\bullet k} := \sum_j f_{jk}$). Let $\mathcal{M}$ denote the set of *independent distributions*, i.e.

$$\mathcal{M} = \{ f \in S \mid f_{jk} = a_j b_k \} .$$

The corresponding ML estimate $\hat{f}^M \in \mathcal{M}$ is

$$\hat{f}_{jk}^M = f_{j\bullet}^D f_{\bullet k}^D \quad \text{where} \quad f_{j\bullet}^D := \sum_k f_{jk}^D \quad \text{and} \quad f_{\bullet k}^D := \sum_j f_{jk}^D$$

with the well-known property (where $H_D(X)$ denotes the entropy associated to the empirical distribution of the variable X)

$$K(f^D \| \hat{f}^{\mathcal{M}}) = I(X:Y) = \frac{1}{2} \sum_{jk} \frac{(f_{jk}^D - \hat{f}_{jk}^{\mathcal{M}})^2}{\hat{f}_{jk}^{\mathcal{M}}} + 0(\|f^D - \hat{f}^{\mathcal{M}}\|^3) \qquad (9)$$

where the *mutual information* $I(X:Y) := H_D(X) + H_D(Y) - H_D(X,Y)$ is the information-theoretical measure of dependence between X and Y. Inequality $H_D(X,Y) \le H_D(X) + H_D(Y)$ insures its non-negativity. By (9), the corresponding test reduces to the usual chi-square test of independence, with $\dim(S) - \dim(\mathcal{M}) = (m_1 m_2 - 1) - (m_1 + m_2 - 2) = (m_1 - 1)(m_2 - 1)$ degrees of freedom.

3.3   Testing between Two Hypotheses (Neyman-Pearson)

Consider the two hypotheses $H_0 :$ " $f^M = f^0$ " and $H_1 :$ " $f^M = f^1$ ", where f^0 and f^1 constitute two distinct distributions in S. Let $W \subset S$ denote the *rejection region* for f^0, that is such that H_1 is accepted if $f^D \in W$, and H_0 is accepted if $f^D \in W^c := S \setminus W$. The errors of first, respectively second kind are

$$\alpha := P(f^D \in W \mid f^0) \qquad\qquad \beta := P(f^D \in W^c \mid f^1).$$

For n large, Sanov's theorem (18) below shows that

$$\begin{aligned} \alpha &\cong \exp(-nK(\tilde{f}^0 \| f^0)) & \tilde{f}^0 &:= \arg\min_{f \in W} K(f \| f^0) \\ \beta &\cong \exp(-nK(\tilde{f}^1 \| f^1)) & \tilde{f}^1 &:= \arg\min_{f \in W^c} K(f \| f^1). \end{aligned} \qquad (10)$$

The rejection region W is said to be *optimal* if there is no other region $W' \subset S$ with $\alpha(W') < \alpha(W)$ and $\beta(W') < \beta(W)$. The celebrated Neyman-Pearson lemma, together with the asymptotic rate formula (4), states that W is optimal iff it is of the form

$$W = \{f \mid \frac{P(f|f^1)}{P(f|f^0)} \ge T\} = \{f \mid K(f\|f^0) - K(f\|f^1) \ge \frac{1}{n}\ln T := \tau\}. \qquad (11)$$

One can demonstrate (see e.g. Cover and Thomas (1991) p.309) that the distributions (10) governing the asymptotic error rates coincide when W is optimal, and are given by the *multiplicative mixture*

$$\tilde{f}_j^0 = \tilde{f}_j^1 = f_j(\mu) := \frac{(f_j^0)^\mu (f_j^1)^{1-\mu}}{\sum_k (f_k^0)^\mu (f_k^1)^{1-\mu}} \qquad (12)$$

where μ is the value insuring $K(f(\mu)\|f^0) - K(f(\mu)\|f^1) = \tau$. Finally, the overall probability of error, that is the probability of occurrence of an error of first *or* second kind, is minimum for $\tau = 0$, with rate equal to

$$K(f(\mu^*)\|f^0) = K(f(\mu^*)\|f^1) = -\min_{0 \le \mu \le 1} \ln(\sum_k (f_k^0)^\mu (f_k^1)^{1-\mu}) =: C(f^0, f^1)$$

where μ^* is the value minimising the third term. The quantity $C(f^0, f^1) \ge 0$, known as *Chernoff information*, constitutes a symmetric dissimilarity between the distributions f^0 and f^1, and measures how easily f^0 and f^1 can be discriminated from each other. In particular, $C(f^0, f^1) = 0$ iff $f^0 = f^1$.

Example 2.5, continued: coins. Let $f := (0.5, 0.5)$, $g := (0.7, 0.3)$, $h := (0.9, 0.1)$ and $r := (1, 0)$. Numerical estimates yield (in nats) $C(f, g) = 0.02$, $C(f, h) = 0.11$, $C(g, h) = 0.03$ and $C(f, r) = \ln 2 = 0.69$.

3.4   Testing a Family within Another

Let $\mathcal{M}_0$ and $\mathcal{M}_1$ be two families of models, with $\mathcal{M}_0 \subset \mathcal{M}_1$ and $\dim(\mathcal{M}_0) < \dim(\mathcal{M}_1)$. Consider the test of H_0 *within* H_1, opposing $H_0 : $ "$f^M \in \mathcal{M}_0$" against $H_1 : $ "$f^M \in \mathcal{M}_1$".

By construction, $K(f^D || \hat{f}^{\mathcal{M}_0}) \geq K(f^D || \hat{f}^{\mathcal{M}_1})$ since $\mathcal{M}_1$ is a more general model than $\mathcal{M}_0$. Under H_1, their difference can be shown to follow asymptotically a chi-square distribution. Precisely, the *nested test* of H_0 within H_1 reads: "*under the assumption that H_1 holds, rejects H_0 if*

$$2n \left[K(f^D || \hat{f}^{\mathcal{M}_0}) - K(f^D || \hat{f}^{\mathcal{M}_1}) \right] \geq \chi^2_{1-\alpha}[\dim(\mathcal{M}_1) - \dim(\mathcal{M}_0)] \text{''}. \tag{13}$$

Example: quasi-symmetry, symmetry and marginal homogeneity. Flows can be represented by a square matrix $f_{jk} \geq 0$ such that $\sum_{j=1}^m \sum_{k=1}^m f_{jk} = 1$, with the representation "f_{jk} = *proportion of units located at place j at some time and at place k some fixed time later*".

A popular model for flows is the *quasi-symmetric class* QS (Caussinus 1966), known as the *Gravity model* in Geography (Bavaud 2002a)

$$\text{QS} = \{ f \mid f_{jk} = \alpha_j \beta_k \gamma_{jk} \qquad \text{with } \gamma_{jk} = \gamma_{kj} \}$$

where α_j quantifies the "push effect", β_k the "pull effect" and γ_{jk} the "distance deterrence function".

Symmetric and *marginally homogeneous* models constitute two popular alternative families, defined as

$$\text{S} = \{ f \mid f_{jk} = f_{kj} \} \qquad\qquad \text{MH} = \{ f \mid f_{j\bullet} = f_{\bullet j} \} .$$

Symmetric and quasi-symmetric ML estimates satisfy (see e.g. Bishop and al. (1975) or Bavaud (2002a))

$$\hat{f}^{\text{S}}_{jk} = \frac{1}{2}(f^D_{jk} + f^D_{kj}) \qquad \hat{f}^{\text{QS}}_{jk} + \hat{f}^{\text{QS}}_{kj} = f^D_{jk} + f^D_{kj} \qquad \hat{f}^{\text{QS}}_{j\bullet} = f^D_{j\bullet} \qquad \hat{f}^{\text{QS}}_{\bullet k} = f^D_{\bullet k}$$

from which the values of $\hat{f}^{\text{QS}}$ can be obtained iteratively. A similar yet more involved procedure permits to obtain the marginal homogeneous estimates $\hat{f}^{\text{MH}}$.

By construction, $\text{S} \subset \text{QS}$, and the test (13) consists in rejecting S (under the assumption that QS holds) if

$$2n \left[K(f^D || \hat{f}^{\text{S}}) - K(f^D || \hat{f}^{\text{QS}}) \right] \geq \chi^2_{1-\alpha}[m-1] . \tag{14}$$

Noting that $\text{S} = \text{QS} \cap \text{MH}$, (14) actually constitutes an alternative testing procedure for QS, avoiding the necessity of computing $\hat{f}^{\text{MH}}$ (Caussinus 1996).

Example 3.4 continued: inter-regional migrations. Relative entropies associated to Swiss inter-regional migrations flows 1985-1990 ($m = 26$ cantons; see Bavaud (2002a)) are $K(f^D||\hat{f}^S) = .00115$ (with df $= 325$) and $K(f^D||\hat{f}^{QS}) = .00044$ (with df $= 300$). The difference is $.00071$ (with df $= 25$ only) and indicates that flows asymmetry is mainly produced by the violation of marginal homogeneity (unbalanced flows) rather than the violation of quasi-symmetry. However, the sheer size of the sample ($n = 6'039'313$) leads, at conventional significance levels, to reject all three models S , MH and QS.

3.5   Competition between Simple Hypotheses: Bayesian Selection

Consider the set of q simple hypotheses "$H_a : f^M = g^a$ ", where $g^a \in S_m$ for $a = 1, \ldots, q$. In a Bayesian setting, denote by $P(H_a) = P(g^a) > 0$ the prior probability of hypothesis H_a, with $\sum_{a=1}^q P(H_a) = 1$. The posterior probability $P(H_a|D)$ obtains from Bayes rule as

$$P(H_a|D) = \frac{P(H_a)\,P(D|H_a)}{P(D)} \qquad \text{with} \quad P(D) = \sum_{a=1}^q P(H_a)\,P(D|H_a)\,.$$

Direct application of the asymptotic rate formula (4) then yields

$$P(g^a|f^D) \cong \frac{P(g^a)\,\exp(-n\,K(f^D||g^a))}{P(f^D)} \qquad \text{(Bayesian selection formula)} \quad (15)$$

which shows, for $n \to \infty$, the posterior probability to be concentrated on the (supposedly unique) solution of

$$\hat{g} = \arg\min_{g^a} K(f^*||g^a) \qquad \text{where} \quad f^* := \lim_{n \to \infty} f^D \,.$$

In other words, the asymptotically surviving model g^a minimises the relative entropy $K(g^a||f^*)$ with respect to the long-run empirical distribution f^*, in accordance with the ML principle.

For finite n, the relevant functional is $K(f^D||g^a)) - \frac{1}{n}\ln P(g^a)$, where the second term represents a prior penalty attached to hypothesis H_a. Attempts to generalize this framework to *families* of models $\mathcal{M}_a$ ($a = 1, \ldots, q$) lie at the heart of the so-called *model selection procedures*, with the introduction of penalties (as in the AIC, BIC, DIC, ICOMP, etc. approaches) increasing with the number of free parameters dim($\mathcal{M}_a$) (see e.g. Robert (2001)). In the alternative *minimum description length* (MDL) and *algorithmic complexity theory* approaches (see e.g. MacKay (2003) or Li and Vitanyi (1997)), richer models necessitate a longer description and should be penalised accordingly. All those procedures, together with Vapnik's *Structural Risk Minimization* (SRM) principle (1995), aim at controlling the problem of over-parametrization in statistical modelling. We shall not pursue any further those matters, whose conceptual and methodological unification remains yet to accomplish.

Example: Dirichlet priors. Consider the continuous *Dirichlet prior* $g \sim \mathcal{D}(\underline{\alpha})$, with density $\rho(g|\underline{\alpha}) = \frac{\Gamma(\alpha)}{\prod_j \Gamma(\alpha_j)} \prod_j g_j^{\alpha_j - 1}$, normalised to unity in S_m, where $\underline{\alpha} = (\alpha_1, \ldots, \alpha_m)$ is a vector of parameters with $\alpha_j > 0$ and $\alpha := \sum_j \alpha_j$. Setting $\pi_j := \alpha_j / \alpha = E(g_j|\underline{\alpha})$, Stirling approximation yields $\rho(g|\underline{\alpha}) \cong \exp(-\alpha K(\pi||g))$ for α large.

Alfter observing the data $\underline{n} = (n_1, \ldots, n_m)$, the posterior distribution is well-known to be $\mathcal{D}(\underline{\alpha} + \underline{n})$. Using $f_j^D = n_j/n$, one gets $\rho(g|\underline{\alpha} + \underline{n})/\rho(g|\underline{\alpha}) \cong \exp(-nK(f^D||g))$ for n large, as it must from (15). Hence

$$\rho(g|\underline{\alpha} + \underline{n}) \cong \exp[-\alpha K(\pi||g) - nK(f^D||g)] \cong \exp[-(\alpha + n)K(f^{\text{post}}||g)] \quad (16)$$

$$\text{where} \quad f_j^{\text{post}} = E(g_j|\underline{\alpha} + \underline{n}) = \lambda \pi_j + (1 - \lambda)f_j^D \quad \text{with} \quad \lambda := \frac{\alpha}{\alpha + n}. \quad (17)$$

(16) and (17) show the parameter α to measure the *strength of belief in the prior guess*, measured in units of the sample size (Ferguson 1974).

4 Maximum Entropy

4.1 Large Deviations: Sanov's Theorem

Suppose data to be *incompletely observed*, i.e. one only knows that $f^D \in \mathcal{D}$, where $\mathcal{D} \subset S$ is a subset of the simplex S, the set of all possible distributions with m modalites. Then, for an i.i.d. process, a theorem due to Sanov (1957) says that, for sufficiently regular $\mathcal{D}$, the asymptotic rate of the probability that $f^D \in \mathcal{D}$ under model f^M decreases exponentially as

$$P(f^D \in \mathcal{D}|f^M) \cong \exp(-n K(\tilde{f}^{\mathcal{D}}||f^M)) \quad \text{where} \quad \tilde{f}^{\mathcal{D}} := \arg\min_{f \in \mathcal{D}} K(f||f^M). (18)$$

$\tilde{f}^{\mathcal{D}}$ is the so-called *maximum entropy* (ME) solution, that is the most probable empirical distribution under the prior model f^M and the knowledge that $f^D \in \mathcal{D}$. Of course, $\tilde{f}^{\mathcal{D}} = f^M$ if $f^M \in \mathcal{D}$.

4.2 On the Nature of the Maximum Entropy Solution

When the prior is uniform $(f_j^M = 1/m)$, then

$$K(f^D||f^M) = \ln m - H(f^D)$$

and *minimising* (over $f \in \mathcal{D}$) the relative entropy $K(f||f^M)$ amounts in *maximising* the entropy $H(f^D)$ (over $f \in \mathcal{D}$).

For decades (ca. 1950-1990), the "maximum entropy" principle, also called "minimum discrimination information (MDI) principle" by Kullback (1959), has largely been used in science and engineering as a *first-principle*, "maximally non-informative" method of generating *models*, maximising our ignorance (as represented by the entropy) under our available knowledge ($f \in \mathcal{D}$) (see in particular Jaynes (1957), (1978)).

However, (18) shows the maximum entropy construction to be justified from Sanov's theorem, and to result form the minimisation of the *first* argument of the relative entropy, which points towards the *empirical* (rather than theoretical) nature of the latter. In the present setting, $\tilde{f}^{\mathcal{D}}$ appears as the most likely data reconstruction under the prior model and the incomplete observations (see also section 5.3).

Example: unobserved category. Let f^M be given and suppose one knows that a category, say $j = 1$, has *not* occured. Then

$$\tilde{f}_j^{\mathcal{D}} = \begin{cases} 0 & \text{for } j = 1 \\ \frac{f_j^M}{1 - f_1^M} & \text{for } j > 1 \end{cases} \qquad \text{and} \quad K(\tilde{f}^{\mathcal{D}} \| f^M) = -\ln(1 - f_1^M),$$

whose finiteness (for $f_1^M < 1$) contrasts the behavior $K(f^M \| \tilde{f}^{\mathcal{D}}) = \infty$ (for $f_1^M > 0$). See example 2.5 f).

Example: coarse grained observations. Let f^M be a given distribution with categories $j = 1, \ldots, m$. Let $J = 1, \ldots, M < m$ denote *groups of categories*, and suppose that observations are aggregated or coarse-grained, i.e. of the form

$$\mathcal{D} = \{ f^D \mid \sum_{j \in J} f_j^D \overset{!}{=} F_J^D \quad J = 1, \ldots, M \}.$$

Let $J(j)$ denote the group to which j belongs. The ME distribution then reads (see (8) and example 3.2)

$$\tilde{f}_j^{\mathcal{D}} = f_j^M \frac{F_{J(j)}^D}{F_{J(j)}^M} \text{ where } F_J^M := \sum_{j \in J} f_j^M \text{ and } K(\tilde{f}^{\mathcal{D}} \| f^M) = K(F^D \| F^M). \quad (19)$$

Example: symmetrical observations. Let f_{jk}^M be a given joint model for *square* distributions ($j, k = 1, \ldots, m$). Suppose one knows the data distribution to be symmetrical, i.e.

$$\mathcal{D} = \{ f \mid f_{jk}^D = f_{kj}^D \}.$$

$$\text{Then} \qquad \tilde{f}_{jk}^{\mathcal{D}} = \frac{\sqrt{f_{jk}^M f_{kj}^M}}{Z} \qquad \text{where} \qquad Z := \sum_{jk} \sqrt{f_{jk}^M f_{kj}^M}$$

which contrasts the result $\hat{f}_{jk}^{\mathcal{M}} = \frac{1}{2}(f_{jk}^D + f_{kj}^D)$ of example 3.4 (see section 5.1).

4.3   "Standard" Maximum Entropy: Linear Constraint

Let $\mathcal{D}$ be determined by a *linear constraint* of the form

$$\mathcal{D} = \{f \mid \sum_{j=1}^{m} f_j a_j = \bar{a} \,\} \qquad \text{with} \qquad \min_j a_j \leq \bar{a} \leq \max_j a_j \,.$$

In other words, one knows the empirical average of some quantity $\{a_j\}_{j=1}^{m}$ to be fixed to $\bar{a}$. Minimizing over $f \in S$ the functional

$$K(f\|f^M) + \theta A(f) \qquad\qquad A(f) := \sum_{j=1}^{m} f_j a_j \qquad\qquad (20)$$

yields $\qquad \tilde{f}_j^{\mathcal{D}} = \dfrac{f_j^M \, \exp(\theta a_j)}{Z(\theta)} \qquad\qquad Z(\theta) := \sum_{k=1}^{m} f_k^M \, \exp(\theta a_k) \qquad (21)$

where the Lagrange multiplier θ is determined by the constraint $\bar{a}(\theta) := \sum_j \tilde{f}_j^{\mathcal{D}}(\theta)\, a_j \overset{!}{=} \bar{a}$ (see figure 2).

Example: average value of a dice. Suppose one believes a dice to be fair ($f_j^M = 1/6$), and one is told that the empirical average of its face values is say $\bar{a} = \sum_j f_j^{\mathcal{D}}\, j = 4$, instead of $\bar{a} = 3.5$ as expected. The value of θ in (21) insuring $\sum_j \tilde{f}_j^{\mathcal{D}}\, j = 4$ turns out to be $\theta = 0.175$, insuring $\sum_j \tilde{f}_j^{\mathcal{D}}\, j = 4$, as well as $\tilde{f}_1^{\mathcal{D}} = 0.10$, $\tilde{f}_2^{\mathcal{D}} = 0.12$, $\tilde{f}_3^{\mathcal{D}} = 0.15$, $\tilde{f}_4^{\mathcal{D}} = 0.17$, $\tilde{f}_5^{\mathcal{D}} = 0.25$, $\tilde{f}_6^{\mathcal{D}} = 0.30$ (Cover and Thomas (1991) p. 295).

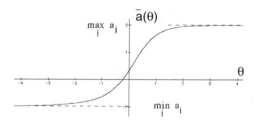

Fig. 2. Typical behaviour of $\bar{a}(\theta)$

Example: Statistical Mechanics. An interacting particle system can occupy $m \gg 1$ configurations $j = 1, \ldots, m$, a priori equiprobable ($f_j^M = 1/m$), with corresponding energy E_j. Knowing the average energy to be $\bar{E}$, the resulting ME solution (with $\beta := -\theta$) is the *Boltzmann-Gibbs distribution*

$$\tilde{f}_j^{\mathcal{D}} = \frac{\exp(-\beta E_j)}{Z(\beta)} \qquad\qquad Z(\beta) := \sum_{k=1}^{m} \exp(-\beta E_k) \qquad (22)$$

minimising the *free energy* $F(f) := E(f) - TH(f)$, obtained (up to a constant term) by multiplying the functional (20) by the *temperature* $T := 1/\beta = -1/\theta$. Temperature plays the role of an arbiter determining the trade-off between the contradictory objectives of energy minimisation and entropy maximisation:

- at high temperatures $T \to \infty$ (i.e. $\beta \to 0^+$), the Boltzmann-Gibbs distribution $\tilde{f}^{\mathcal{D}}$ becomes uniform and the entropy $H(\tilde{f}^{\mathcal{D}})$ maximum (fluid-like organisation of the matter).
- at low temperatures $T \to 0^+$ (i.e. $\beta \to \infty$), the Boltzmann-Gibbs distribution $\tilde{f}^{\mathcal{D}}$ becomes concentrated on the *ground states* $j_- := \arg\min_j E_j$, making the average energy $E(\tilde{f}^{\mathcal{D}})$ minimum (crystal-like organisation of the matter).

Example 3.4, continued: quasi-symmetry. ME approach to gravity modelling consists in considering flows constrained by q linear constraints of the form

$$\mathcal{D} = \{f \mid \sum_{j,k=1}^{m} f_{jk} a_{jk}^{\alpha} = \bar{a}^{\alpha} \qquad \alpha = 1, \dots, q\}$$

such that, typically

1) $a_{jk} := d_{jk} = d_{kj}$ (fixed average trip distance, cost or time d_{jk})
2) $a_{jk}^{\alpha} := \delta_{j\alpha}$ (fixed origin profiles, $\alpha = 1, \dots, m$)
3) $a_{jk}^{\alpha} := \delta_{\alpha k}$ (fixed destination profiles, $\alpha = 1, \dots, m$)
4) $a_{jk} := \delta_{jk}$ (fixed proportion of stayers)
5) $a_{jk} := \delta_{j\alpha} - \delta_{\alpha k}$ (balanced flows, $\alpha = 1, \dots, m$)

Constraints 1) to 5) (and linear combinations of them) yield all the "classical Gravity models" proposed in Geography, such as the *exponential decay model* (with $f_{jk}^M = a_j \, b_k$):

$$\tilde{f}_{jk}^{\mathcal{D}} = \alpha_j \beta_k \exp(-\beta d_{jk})$$

Moreover, if the prior f^M is quasi-symmetric, so is $\tilde{f}^{\mathcal{D}}$ under the above constraints (Bavaud 2002a).

5   Additive Decompositions

5.1   Convex and Exponential Families of Distributions

Definition: A family $\mathcal{F} \subset S$ of distributions is a **convex** family iff

$$f, g \in \mathcal{F} \Rightarrow \lambda f + (1 - \lambda)g \in \mathcal{F} \qquad \forall \, \lambda \in [0, 1]$$

Observations typically involve the identification of merged categories, and the corresponding empirical distributions are coarse grained, that is determined through aggregated values $F_J := \sum_{j \in J} f_j$ only. Such coarse grained distributions form a convex family (see table 1). More generally, linearly constrained distributions (section 4.3) are convex. Distributions (11) belonging to the optimal Neyman-Pearson regions W (or W^c), posterior distributions (17) as well as

Table 1. Some convex and/or exponential families

Family $\mathcal{F}$	characterization	remark	convex	exponential
deficient	$f_1 = 0$		yes	yes
deterministic	$f_1 = 1$		yes	yes
coarse grained	$\sum_{j \in J} f_j = F_J$		yes	no
mixture	$f_j = f_{(Jq)} = \rho_q h_J^q$	$\{h_J^q\}$ fixed	yes	yes
mixture	$f_j = f_{(Jq)} = \rho_q h_J^q$	$\{h_J^q\}$ adjustable	no	yes
independent	$f_{jk} = a_j b_k$		no	yes
marginally homogeneous	$f_{j\bullet} = f_{\bullet j}$	square tables	yes	no
symmetric	$f_{jk} = f_{kj}$	square tables	yes	yes
quasi-symmetric	$f_{jk} = a_j b_k c_{jk},\ c_{jk} = c_{kj}$	square tables	no	yes

marginally homogeneous distributions (example 3.4) provide other examples of convex families.

Definition: A family $\mathcal{F} \subset S$ of distributions is an **exponential** family iff

$$f, g \in \mathcal{F} \Rightarrow \frac{f^\mu g^{1-\mu}}{Z(\mu)} \in \mathcal{F} \quad \text{where} \quad Z(\mu) := \sum_{j=1}^{m} f_j^\mu g_j^{1-\mu} \quad \forall\, \mu \in [0, 1]$$

Exponential families are a favorite object of classical statistics. Most classical discrete or continuous probabilistic models (log-linear, multinomial, Poisson, Dirichlet, Normal, Gamma, etc.) constitute exponential families. Amari (1985) has developed a local parametric characterisation of exponential and convex families in a differential geometric framework.

5.2 Factor Analyses

Independence models are exponential but not convex (see table 1): the weighted sum of independent distributions is not independent in general. Conversely, non-independent distributions can be decomposed as a sum of (latent) independent terms through *factor analysis*. The spectral decomposition of the chi-square producing the *factorial correspondence analysis* of contingency tables turns out to be exactly applicable on mutual information (9) as well, yielding an "entropic" alternative to (categorical) factor analysis (Bavaud 2002b).

Independent component analysis (ICA) aims at determining the linear transformation of multivariate (continuous) data making them as independent as possible. In contrast to *principal component analysis*, limited to the second-order statistics associated to gaussian models, ICA attempts to take into account higher-order dependencies occurring in the mutual information between variables, and extensively relies on information-theoretic principles, as developed in Lee et al. (2000) or Cardoso (2003) and references therein.

5.3 Pythagorean Theorems

The following results, sometimes referred to as the *Pythagorean theorems of IT*, provide an exact additive decomposition of the relative entropy:

Decomposition theorem for convex families: if $\mathcal{D}$ is a convex family, then

$$K(f\|f^M) = K(f\|\tilde{f}^{\mathcal{D}}) + K(\tilde{f}^{\mathcal{D}}\|f^M) \qquad \text{for any } f \in \mathcal{D} \qquad (23)$$

where $\tilde{f}^{\mathcal{D}}$ is the ME distribution for $\mathcal{D}$ with prior f^M.

Decomposition theorem for exponential families: if $\mathcal{M}$ is an exponential family, then

$$K(f^D\|g) = K(f^D\|\hat{f}^{\mathcal{M}}) + K(\hat{f}^{\mathcal{M}}\|g) \qquad \text{for any } g \in \mathcal{M} \qquad (24)$$

where $\hat{f}^{\mathcal{M}}$ is the ML distribution for $\mathcal{M}$ with data f^D.

Sketch of the proof of (23) (see e.g. Simon 1973): if $\mathcal{D}$ is convex with $\dim(\mathcal{D}) = \dim(\mathcal{S}) - q$, its elements are of the form $\mathcal{D} = \{f \mid \sum_j f_j a_j^\alpha = a_0^\alpha \text{ for } \alpha = 1,\ldots,q\}$, which implies the maximum entropy solution to be of the form $\tilde{f}_j^{\mathcal{D}} = \exp(\sum_\alpha \lambda_\alpha a_j^\alpha) f_j^M / Z(\lambda)$. Substituting this expression and using $\sum_j f_j a_j^\alpha = \sum_j \tilde{f}_j^{\mathcal{D}} a_j^\alpha$ proves (23).

Sketch of the proof of (24) (see e.g. Simon 1973): if $\mathcal{M}$ is exponential with $\dim(\mathcal{M}) = r$, its elements are of the form $f_j = \rho_j \exp(\sum_{\alpha=1}^r \lambda_\alpha a_j^\alpha)/Z(\lambda)$ (where the *partition function* $Z(\lambda)$ insures the normalisation), containing r free non-redundant parameters $\lambda \in \mathbb{R}^r$. Substituting this expression and using the optimality condition $\sum_j \hat{f}_j^{\mathcal{M}} a_j^\alpha = \sum_j f_j^D a_j^\alpha$ for all $\alpha = 1,\ldots,r$ proves (24).

Equations (23) and (24) show that $\tilde{f}^{\mathcal{D}}$ and $\hat{f}^{\mathcal{M}}$ can both occur as *left* and *right* arguments of the relative entropy, underlining their somehow *hybrid* nature, *intermediate between data and models* (see section 4.2).

Example: nested tests. Consider two exponential families $\mathcal{M}$ and $\mathcal{N}$ with $\mathcal{M} \subset \mathcal{N}$. Twofold application of (24) demonstrates the identity

$$K(f^D\|\hat{f}^{\mathcal{M}}) - K(f^D\|\hat{f}^{\mathcal{N}}) = K(\hat{f}^{\mathcal{N}}\|\hat{f}^{\mathcal{M}})$$

occuring in nested tests such as (14).

Example: conditional independence in three-dimensional tables. Let $f_{ijk}^D := n_{ijk}/n$ with $n := n_{\bullet\bullet\bullet}$ be the empirical distribution associated to the $n_{ijk} =$ "*number of individuals in the category i of X, j of Y and k of Z* ". Consider the families of models

$$\mathcal{L} = \{f \in S \mid f_{ijk} = a_{ij}b_k\} = \{f \in S \mid \ln f_{ijk} = \lambda + \alpha_{ij} + \beta_k\}$$
$$\mathcal{M} = \{f \in S \mid f_{\bullet jk} = c_i d_k\} = \{f \in S \mid \ln f_{\bullet jk} = \mu + \gamma_j + \delta_k\}$$
$$\mathcal{N} = \{f \in S \mid f_{ijk} = e_{ij}h_{jk}\} = \{f \in S \mid \ln f_{ijk} = \nu + \epsilon_{ij} + \eta_{jk}\}.$$

Model $\mathcal{L}$ expresses that Z is independent from X and Y (denoted $Z \perp (X,Y)$). Model $\mathcal{M}$ expresses that Z and Y are independent ($Y \perp Z$). Model $\mathcal{N}$ expresses that, conditionally to Y, X and Z are independent ($X \perp Z|Y$). Models $\mathcal{L}$ and $\mathcal{N}$ are exponential (in S), and $\mathcal{M}$ is exponential in the space of joint distributions on (Y, Z). They constitute well-known examples of *log-linear models* (see e.g. Christensen (1990)).

Maximum likelihood estimates and associated relative entropies obtain as (see example 3.2)

$$\hat{f}^{\mathcal{L}}_{ijk} = f^D_{ij\bullet} f^D_{\bullet\bullet k} \quad \Rightarrow \quad K(f^D || \hat{f}^{\mathcal{L}}) = H_D(XY) + H_D(Z) - H_D(XYZ)$$

$$\hat{f}^{\mathcal{M}}_{ijk} = \frac{f^D_{ijk}}{f^D_{\bullet jk}} f^D_{\bullet j\bullet} f^D_{\bullet\bullet k} \quad \Rightarrow \quad K(f^D || \hat{f}^{\mathcal{M}}) = H_D(Y) + H_D(Z) - H_D(YZ)$$

$$\hat{f}^{\mathcal{N}}_{ijk} = \frac{f^D_{ij\bullet} f^D_{\bullet jk}}{f^D_{\bullet j\bullet}} \quad \Rightarrow \quad K(f^D || \hat{f}^{\mathcal{N}}) = H_D(XY) + H_D(YZ) - H_D(XYZ) - H_D(Y)$$

and permit to test the corresponding models as in (7). As a matter of fact, the present example illustrates another aspect of exact decomposition, namely

$$\mathcal{L} = \mathcal{M} \cap \mathcal{N} \qquad f^D_{ijk} \hat{f}^{\mathcal{L}}_{ijk} = \hat{f}^{\mathcal{M}}_{ijk} \hat{f}^{\mathcal{N}}_{ijk}$$

$$K(f^D || \hat{f}^{\mathcal{L}}) = K(f^D || \hat{f}^{\mathcal{M}}) + K(f^D || \hat{f}^{\mathcal{N}}) \qquad \mathrm{df}^{\mathcal{L}} = \mathrm{df}^{\mathcal{M}} + \mathrm{df}^{\mathcal{N}}$$

where df denotes the appropriate degrees of freedom for the chi-square test (7).

5.4 Alternating Minimisation and the EM Algorithm

Alternating minimisation. Maximum likelihood and maximum entropy are particular cases of the general problem

$$\min_{f \in \mathcal{F}} \min_{g \in \mathcal{G}} K(f || g) . \tag{25}$$

Alternating minimisation consists in defining recursively

$$f^{(n)} := \arg \min_{f \in \mathcal{F}} K(f || g^{(n)}) \tag{26}$$

$$g^{(n+1)} := \arg \min_{g \in \mathcal{G}} K(f^{(n)} || g) . \tag{27}$$

Starting with some $g^{(0)} \in \mathcal{G}$ (or some $f^{(0)} \in \mathcal{F}$), and for $\mathcal{F}$ and $\mathcal{G}$ convex, $K(f^{(n)} || g^{(n)})$ converges towards (25) (Csiszár (1975); Csiszár and Tusnády, 1984).

The EM algorithm. Problem (26) is easy to solve when $\mathcal{F}$ is the coarse grained family $\{ f \mid \sum_{j \in J} f_j = F_J \}$, with solution (19) $f^{(n)}_j = g^{(n)}_j F_{J(j)} / G^{(n)}_{J(j)}$ and the result $K(f^{(n)} || g^{(n)}) = K(F || G^{(n)})$ (see example 4.2).

The present situation describes *incompletely observed data*, in which F only (and not f) is known, with corresponding model $G(g)$ in $\mathcal{M} := \{ G \mid G_J = \sum_{j \in J} g_j$ and $g \in \mathcal{G} \}$. Also

$$\min_{G \in \mathcal{M}} K(F || G) = \min_{g \in \mathcal{G}} K(F || G(g)) = \min_{g \in \mathcal{G}} \min_{f \in \mathcal{F}} K(f || g)$$

$$= \lim_{n \to \infty} K(f^{(n)} || g^{(n)}) = \lim_{n \to \infty} K(F || G^{(n)})$$

which shows $G^{(\infty)}$ to be the solution of $\min_{G \in \mathcal{M}} K(F\|G)$. This particular version of the alternating minimisation procedure is known as the *EM algorithm* in the literature (Dempster et al. 1977), where (26) is referred to as the "*expectation step*" and (27) as the "*maximisation step*".

Of course, the above procedure is fully operational provided (27) can also be easily solved. This occurs for instance for *finite-mixture* models determined by c *fixed* distributions h_J^q (with $\sum_{J=1}^m h_J^q = 1$ for $q = 1, \ldots, c$), such that the categories $j = 1, \ldots, m$ read as product categories of the form $j = (J, q)$ with

$$g_j = g_{(Jq)} = \rho_q\, h_J^q \qquad \rho_q \geq 0 \qquad \sum_{q=1}^c \rho_q = 1 \qquad G_J = \sum_q \rho_q h_J^q$$

where the "mixing proportions" ρ_q are freely adjustable. Solving (27) yields

$$\rho_q^{(n+1)} = \sum_J f_{(Jq)}^{(n)} = \rho_q^{(n)} \sum_J \frac{h_J^q\, F_J}{\sum_r h_J^r\, \rho_r^{(n)}}$$

which converges towards the optimal mixing proportions $\rho_q^{(\infty)}$, unique since $\mathcal{G}$ is convex. Continuous versions of the algorithm (in which J represents a position in an Euclidean space) generate the so-called *soft clustering* algorithms, which can be further restricted to the hard clustering and K-means algorithms. However, the distributions h_J^q used in the latter cases generally contain additional *adjustable* parameters (typically the mean and the covariance matrix of normal distributions), which break down the convexity of $\mathcal{G}$ and cause the algorithm to converge towards *local minima*.

6   Beyond Independence: Markov Chain Models and Texts

As already proposed by Shannon (1948), the independence formalism can be extended to stationary dependent sequences, that is on categorical time series or "textual" data $D = x_1 x_2 \ldots x_n =: x_1^n$, such as

$D=$bbaabbaabbbaabbbaabbbaabbaabaabbaabbaabbaabbaabbaabbaabbaab
aabbaabbbaabaabaabbbaabbbaabbaabbaabbaabaabbbaabbbaabbaabaabaa
bbaabaabbaabbaabbbaabbaabaabaabbaabbbbaabbaabaabaabaabaabaabaa
bbaabbaabbaabbbbaab   .

In this context, each occurence x_i constitutes a *letter* taking values ω_j in a state space Ω, the *alphabet*, of cardinality $m = |\Omega|$. A sequence of r letters $\alpha := \omega_1 \ldots \omega_r \in \Omega^r$ is an *r-gram*. In our example, $n = 202$, $\Omega = \{a, b\}$, $m = 2$, $\Omega^2 = \{aa, ab, ba, bb\}$, etc.

6.1   Markov Chain Models

A *Markov chain model of order r* is specified by the *conditional* probabilities

$$f^M(\omega|\alpha) \geq 0 \qquad \omega \in \Omega \qquad \alpha \in \Omega^r \qquad \sum_{\omega \in \Omega} f^M(\omega|\alpha) = 1\,.$$

$f^M(\omega|\alpha)$ is the probability that the symbol following the r-gram α is ω. It obtains from the *stationary distributions* $f^M(\alpha\omega)$ and $f^M(\alpha)$ as

$$f^M(\omega|\alpha) = \frac{f^M(\alpha\omega)}{f^M(\alpha)}.$$

The set $\mathcal{M}_r$ of models of order r constitutes an exponential family, nested as $\mathcal{M}_r \subset \mathcal{M}_{r+1}$ for all $r \geq 0$. In particular, $\mathcal{M}_0$ denotes the independence models, and $\mathcal{M}_1$ the ordinary (first-order) Markov chains.

The corresponding empirical distributions $f^D(\alpha)$ give the relative proportion of r-grams $\alpha \in \Omega^r$ in the text D. They obtain as

$$f^D(\alpha) := \frac{n(\alpha)}{n - r + 1} \qquad \text{with} \qquad \sum_{\alpha \in \Omega^r} f^D(\alpha) = 1$$

where $n(\alpha)$ counts the number of occurrences of α in D. In the above example, the tetragrams counts are for instance:

α	$n(\alpha)$	α	$n(\alpha)$	α	$n(\alpha)$
aaaa	0	aaab	0	aaba	16
aabb	35	abaa	16	abab	0
abba	22	abbb	11	baaa	0
baab	51	baba	0	babb	0
bbaa	35	bbab	0	bbba	11
bbbb	2			total	199

6.2   Simulating a Sequence

Under the assumption that a text follows a r-order model $\mathcal{M}_r$, empirical distributions $f^D(\alpha)$ (with $\alpha \in \Omega^{r+1}$) converge for n large to $f^M(\alpha)$. The latter define in turn r-order transition probabilities, allowing the generation of new texts, started from the stationary distribution.

Example. The following sequences are generated form the empirical probability transitions of the *Universal declaration of Human Rights*, of length $n = 8'149$ with $m = 27$ states (the alphabet + the blank, without punctuation):

$r = 0$ (independent process)
```
    iahthire edr pynuecu d lae mrfa ssooueoilhnid nritshfssmo nise yye
noa it eosc e lrc jdnca tyopaooieoegasrors c hel niooaahettnoos rnei
s sosgnolaotd t atiet
```

$r = 1$ (first-order Markov chain)
```
    erionjuminek in l ar hat arequbjus st d ase scin ero tubied pmed
beetl equly shitoomandorio tathic wimof tal ats evash indimspre tel
sone aw onere pene e ed uaconcol mo atimered
```

$r = 2$ (second-order Markov chain)

```
  mingthe rint son of the frentery and com andepent the halons hal
to coupon efornitity the rit noratinsubject will the the in priente
hareeducaresull ch infor aself and evell
```

$r = 3$ (third-order Markov chain)

```
  law socience of social as the right or everyone held genuinely
available sament of his no one may be enties the right in the cons as
the right to equal co one soveryone
```

$r = 4$ (fourth-order Markov chain)

```
  are endowed with other means of full equality and to law no one is
the right to choose of the detent to arbitrarily in science with pay
for through freely choice work
```

$r = 9$ (ninth-order Markov chain)

```
  democratic society and is entitled without interference and to
seek receive and impartial tribunals for acts violating the
fundamental rights indispensable for his
```

Of course, empirical distributions are expected to accurately estimate model distributions for n large enough, or equivalently for r small enough, typically for

$$r < r_{\max} := \frac{1}{2} \frac{\ln n}{\ln m}.$$

Simulations with r above about $r_{\max}$ (here roughly equal to 2) are over-parameterized: the number of parameters to be estimated exceeds the sample abilities to do so, and simulations replicate fragments of the initial text rather than typical r-grams occurences of written English in general, providing a vivid illustration of the *curse of dimensionality* phenomenon.

6.3   Entropies and Entropy Rate

The *r-gram entropy* and the *conditional entropy of order r* associated to a (model or empirical) distribution f are defined by

$$H_r(f) := - \sum_{\alpha \in \Omega^r} f(\alpha) \ln f(\alpha) = H(X_1, \ldots, X_r)$$

$$h_{r+1}(f) := - \sum_{\alpha \in \Omega^r} f(\alpha) \sum_{\omega \in \Omega} f(\omega|\alpha) \ln f(\omega|\alpha) = H_{r+1}(f) - H_r(f) = H(X_{r+1}|X_1^r).$$

The quantity $h_r(f)$ is non-neagtive, and non-increasing in r. Its limit defines the *entropy rate*, measuring the conditional uncertainty on the next symbol knowing the totality of past occurrences:

$$h(f) := \lim_{r \to \infty} h_r(f) = \lim_{r \to \infty} \frac{H_r(f)}{r} \qquad \text{entropy rate.}$$

By construction, $0 \leq h(f) \leq \ln m$, and the so-called *redundancy* $R := 1 - (h/\ln m)$ satisfies $0 \leq R \leq 1$.

The entropy rate measures the randomness of the stationary process: $h(f) = \ln m$ (i.e. $R = 1$) characterizes a maximally random process is, that is a dice model with uniform distribution. The process is ultimately deterministic iff $h(f) = 0$ (i.e. $R = 0$).

Shannon's estimate of the entropy rate of the written English on $m = 27$ symbols is about $h = 1.3$ bits per letter, that is $h = 1.3 \times \ln 2 = 0.90$ nat, corresponding to $R = 0.73$: hundred pages of written English are in theory compressible without loss to $100 - 73 = 27$ pages. Equivalently, using an alphabet containing $\exp(0.90) = 2.46$ symbols only (and the same number of pages) is in principle sufficient to code the text without loss.

Example: entropy rates for ordinary Markov chains. For a regular Markov chain of order 1 with transition matrix $W = (w_{jk})$ and stationary distribution π_j, one gets

$$h_1 = -\sum_j \pi_j \ln \pi_j \geq h_2 = h_3 = \ldots = -\sum_j \pi_j \sum_k w_{jk} \ln w_{jk} = h \, .$$

Identity $h_1 = h$ holds iff $w_{jk} = \pi_k$, that is if the process is of order $r = 0$. Also, $h \to 0$ iff W tends to a permutation, that is iff the process becomes deterministic.

6.4   The Asymptotic Rate for Markov Chains

Under the assumption of a model f^M of order r, the probability to observe D is

$$P(D|f^M) \cong \prod_{i=1}^n P(x_{i+r}|x_i^{i+r-1}) \cong \prod_{\omega \in \Omega} \prod_{\alpha \in \Omega^r} f^M(\omega|\alpha)^{n(\alpha\omega)} \quad \sum_{\omega \in \Omega} \sum_{\alpha \in \Omega^r} n(\alpha\omega) = n$$

where finite "boundary effects", possibly involving the first or last r symbols of the sequence, are here neglected. Also, noting that a total of $n(\alpha)!/\prod_\omega n(\alpha\omega)!$ permutations of the sequence generate the same $f^D(\omega|\alpha)$, taking the logarithm and using Stirling approximation yields the asymptotic rate formula for Markov chains

$$P(f^D|f^M) \cong \exp(-n \, \kappa_{r+1}(f^D \| f^M)) \, , \quad \text{where}$$
$$\kappa_{r+1}(f\|g) := K_{r+1}(f\|g) - K_r(f\|g)) = \sum_{\alpha \in \Omega^r} f(\alpha) \sum_{\omega \in \Omega} f(\omega|\alpha) \ln \frac{f(\omega|\alpha)}{g(\omega|\alpha)}$$
$$\text{and} \quad K_r(f\|g)) := \sum_{\alpha \in \Omega^r} f(\alpha) \ln \frac{f(\alpha)}{g(\alpha)} \, .$$

$$(28)$$

Setting $r = 0$ returns the asymptotic formula (4) for independence models.

6.5   Testing the Order of an Empirical Sequence

For $s \leq r$, write $\alpha \in \Omega^r$ as $\alpha = (\beta\gamma)$ where $\beta \in \Omega^{r-s}$ and $\gamma \in \Omega^s$. Consider s-order models of the form $f^M(\omega|\beta\gamma) = f^M(\omega|\gamma)$. It is not difficult to prove the identity (see e.g. Billingsley (1961))

$$\min_{f^M \in \mathcal{M}_s} \kappa_{r+1}(f^D \| f^M) = -H_{r+1}(f^D) + H_r(f^D) + H_{s+1}(f^D) - H_s(f^D)$$
$$= h_{s+1}(f^D) - h_{r+1}(f^D) \geq 0 .$$

(29)

As an application, consider, as in section 3.4 , the log-likelihood nested test of H_0 within H_1, opposing H_0 : "$f^M \in \mathcal{M}_s$" against H_1 : "$f^M \in \mathcal{M}_r$". Identities (28) and (29) lead to the rejection of H_0 if

$$2n \left[h_{s+1}(f^D) - h_{r+1}(f^D) \right] \geq \chi^2_{1-\alpha}[(m-1)(m^r - m^s)] .$$

(30)

Example: test of independence. For $r = 1$ and $s = 0$, the test (30) amounts in testing independence, and the decision variable

$$h_1(f^D) - h_2(f^D) = H_1(f^D) + H_1(f^D) - H_2(f^D)$$
$$= H(X_1) + H(X_2) - H(X_1, X_2) = I(X_1 : X_2)$$

is (using stationarity) nothing but the mutual information between two consecutive symbols X_1 and X_2, as expected from example 3.2.

Example: sequential tests. For $r = 1$ and $s = r - 1$, inequality (30) implies that the model at least of order r. Setting $r = 1, 2, \ldots, r_{\max}$ (with df $= (m - 1)^2 m^{r-1}$) constitutes a sequential procedure permitting to detect the order of the model, if existing.

For instance, a binary Markov chain of order $r = 3$ and length $n = 1024$ in $\Omega = \{a, b\}$ can be simulated as $X_t := g(\frac{1}{4}(Z_t + Z_{t-1} + Z_{t-2} + Z_{t-3}))$, where Z_t are i.i.d. variables uniformly distributed as $\sim U(0, 1)$, and $g(z) := a$ if $z \geq \frac{1}{2}$ and $g(z) := b$ if $z < \frac{1}{2}$. Application of the procedure at significance level $\alpha = 0.05$ for $r = 1, \ldots 5 = r_{\max}$ is summarised in the following table, and shows to correctly detect the order of the model:

r	$h_r(f^D)$	$2n[h_r(f^D) - h_{r+1}(f^D)]$	df	$\chi^2_{0.95}[\text{df}]$
1	0.692	0.00	1	3.84
2	0.692	2.05	2	5.99
3	0.691	**110.59**	4	9.49
4	0.637	12.29	8	15.5
5	0.631	18.02	16	26.3

6.6   Heating and Cooling Texts

Let $f(\omega|\alpha)$ (with $\omega \in \Omega$ and $\alpha \in \Omega^r$) denote a conditional distribution of order r. In analogy to formula (22) of Statistical Mechanics, the distribution can be "heated" or "cooled" at relative temperature $T = 1/\beta$ to produce the so-called *annealed* distribution

$$f_\beta(\omega|\alpha) := \frac{f^\beta(\omega|\alpha)}{\sum_{\omega' \in \Omega} f^\beta(\omega'|\alpha)} .$$

Sequences generated with the annealed transitions hence simulate texts possessing a temperature T relatively to the original text.

Example: simulating hot and cold English texts. Conditional distributions of order 3, retaining tetragram structure, have been calibrated from Jane Austen's novel *Emma* (1816), containing $n = 868'945$ tokens belonging to $m = 29$ types (the alphabet, the blank, the hyphen and the apostrophe). A few annealed simulations are shown below, where the first trigram was sampled from the stationary distribution (Bavaud and Xanthos, 2002).

$\beta = 1$ (original process)
```
    feeliciousnest miss abbon hear jane is arer that isapple did ther
by the withour our the subject relevery that amile sament is laugh in
' emma rement on the come februptings he
```

$\beta = 0.1$ (10 times hotter)
```
    torables - hantly elterdays doin said just don't check comedina inglas
ratefusandinite his happerall bet had had habiticents' oh young most
brothey lostled wife favoicel let you cology
```

$\beta = 0.01$ (100 times hotter): any transition having occurred in the original text tends to occur again with uniform probability, making the heated text maximally unpredictable. However, most of the possible transitions did not occur initially, which explains the persistence of the English-like aspect.

```
    et-chaist-temseliving dwelf-ash eignansgranquick-gatefullied georgo
namissedeed fessnee th thusestnessful-timencurves - him duraguesdaird
vulgentroneousedatied yelaps isagacity in
```

$\beta = 2$ (2 times cooler) : conversely, frequent (rare) transitions become even more frequent (rare), making the text fairly predictable.

```
    's good of his compassure is a miss she was she come to the of his
and as it it was so look of it i do not you with her that i am superior
the in ther which of that the half - and the
```

$\beta = 4$ (4 times cooler): in the low temperature limit, dynamics is trapped in the most probable initial transitions and texts properly become crystal-like, as expected from Physics (see example 4.3):

```
    ll the was the was the was the was the was the was the was the was
the was the was the was the was the was the was the was the was the
was the was the was the was the was the was the
```

6.7 Additive and Multiplicative Text Mixtures

In the spirit of section 5.1, *additive* and *multiplicative* mixtures of two conditional distributions $f(\omega|\alpha)$ and $g(\omega|\alpha)$ of order r can be constructed as

$$h_\lambda(\omega|\alpha) := \lambda f(\omega|\alpha) + (1 - \lambda)g(\omega|\alpha) \quad h_\mu(\omega|\alpha) := \frac{f^\mu(\omega|\alpha)\, g^{(1-\mu)}(\omega|\alpha)}{\sum_{\omega' \in \Omega} f^\mu(\omega'|\alpha)\, g^{(1-\mu)}(\omega'|\alpha)}$$

where $0 < \lambda < 1$ and $0 < \mu < 1$. The resulting transition exists if it exists in at least one of the initial distributions (additive mixtures) or in both distributions (multiplicative mixtures).

Example: additive mixture of English and French. Let g denote the empirical distribution of order 3 of example (6.6), and define f as the corresponding distribution estimated on the $n = 725'001$ first symbols of the French novel *La bête humaine* from Emile Zola. Additive simulations with various values of λ read (Bavaud and Xanthos, 2002):

$\lambda = 0.17$

```
ll thin not alarly but alabouthould only to comethey had be the sepant
a was que lify you i bed at it see othe to had state cetter but of i
she done a la veil la preckone forma feel
```

$\lambda = 0.5$

```
daband shous ne findissouservait de sais comment do be certant she
cette l'ideed se point le fair somethen l'autres jeune suit onze muchait
satite a ponded was si je lui love toura
```

$\lambda = 0.83$

```
les appelleur voice the toodhould son as or que aprennel un revincontait
en at on du semblait juge yeux plait etait resoinsittairl on in and
my she comme elle ecreta-t-il avait autes foiser
```

showing, as expected, a gradual transformation from English- to French-likeness with increasing λ.

Example: multiplicative mixture of English and French. Applied now on multiplicative mixtures, the procedure described in example 6.7 yields (Bavaud and Xanthos, 2002)

$\mu = 0.17$

```
licatellence a promine agement ano ton becol car emm*** ever ans
touche-***i harriager gonistain ans tole elegards intellan enour bellion
genea***he succept wa***n instand instilliaristinutes
```

$\mu = 0.5$

```
n neignit innerable quit tole ballassure cause on an une grite chambe
ner martient infine disable prisages creat mellesselles dut***grange
accour les norance trop mise une les emm***
```

$\mu = 0.83$

```
es terine fille son mainternistonsidenter ing sile celles tout a
pard elevant poingerent une graver dant lesses jam***core son luxu***que
eles visagemensation lame cendance materroga***e
```

where the symbol *** indicates that the process is trapped in a trigram occuring in the English, but not in the French sample (or vice versa). Again, the

French-likeness of the texts increases with μ. Interestingly enough, some simulated subsequences are arguably evocative of Latin, whose lexicon contains an important part of the forms common to English and French.

From an inferential point of view, the multiplicative mixture is of the form (12), and hence lies at the boundary of the optimal Neyman-Pearson decision region, governing the asymptotic rate of errors of both kinds, namely confounding French with English or English with French.

References

Amari, S.-I.: Differential-Geometrical Methods in Statistics. Lecture Notes in Statistics, vol. 28. Springer, Heidelberg (1985)

Bavaud, F.: The Quasisymmetric Side of Gravity Modelling. Environment and Planning A 34, 61–79 (2002a)

Bavaud, F.: Quotient Dissimilarities, Euclidean Embeddability, and Huygens's Weak Principle. In: Jajuga, K., Solkolowski, A., Bock, H.-H. (eds.) Classification, Clustering and Data Analysis, pp. 195–202. Springer, Heidelberg (2002b)

Bavaud, F., Xanthos, A.: Thermodynamique et Statistique Textuelle: concepts et illustrations. In: Proceedings of JADT 2002 (6èmes Journées internationales d'Analyse statistique des Données Textuelles), St-Malo (2002)

Billingsley, P.: Statistical Inference for Markov Processes. University of Chicago Press, Chicago (1961)

Bishop, Y.M.M., Fienberg, S.E., Holland, P.W.: Discrete multivariate Analysis. The MIT Press, Cambridge (1975)

Boltzmann, L.: Weitere Studien über das Wärmegleichgewicht unter Gasmolekülen. Sitzungsberichte der Akademie der Wissenschaften 66, 275–370 (1872)

Cardoso, J.-F.: Dependence, Correlation and Gaussianity in Independent Component Analysis. Journal of Machine Learning Research 4, 1177–1203 (2003)

Caussinus, H.: Contribution à l'analyse statistique des tableaux de corrélation. Annales de la Faculté des Sciences de Toulouse 29, 77–183 (1966)

Christensen, R.: Log-Linear Models. Springer, Heidelberg (1990)

Cover, T.M., Thomas, J.A.: Elements of Information Theory. Wiley, Chichester (1991)

Cramer, H.: Mathematical Methods of Statistics. Princeton University Press, Princeton (1946)

Csiszár, I.: I-Divergence Geometry of Probability Distribution and Minimization Problems. The Annals of Probability 3, 146–158 (1975)

Csiszár, I., Körner, J.: Towards a general theory of source networks. IEEE Trans. Inform. Theory 26, 155–165 (1980)

Csiszár, I., Tusnády, G.: Information Geometry and Aternating Minimization Procedures. Statistics and Decisions (suppl. 1), 205–237 (1984)

Dempster, A.P., Laird, N.M., Rubin, D.B.: Maximum Likelihood from Incomplete Data via the EM Algorithm. J. Roy. Stat. Soc. B 39, 1–22 (1977)

Ferguson, T.S.: Prior Distributions on Spaces of Probability Measures. The Annals of Statistics 2, 615–629 (1974)

Jaynes, E.T.: Information theory and statistical mechanics. Physical Review 108, 171–190 (1957)

Jaynes, E.T.: Where do we stand on maximum entropy? In: Maximum Entropy Formalism Conference. MIT, Cambridge (1978)

Kullback, S.: Information Theory and Statistics. Wiley, Chichester (1959)

Lee, T.-W., Girolami, M., Bell, A.J., Sejnowski, T.J.: A unifying Information-Theoretic Framework for Independent Component Analysis. Computers and Mathematics with Applications 39, 1–21 (2000)

Li, M., Vitanyi, P.: An Introduction to Kolmogorov complexity and its applications. Springer, Heidelberg (1997)

MacKay, D.J.C.: Information Theory, Inference and Learning Algorithms. Cambridge University Press, Cambridge (2003)

Popper, K.: Conjectures and Refutations, Routledge (1963)

Robert, C.P.: The Bayesian Choice, 2nd edn. Springer, Heidelberg (2001)

Sanov, I.N.: On the probability of large deviations of random variables. Mat. Sbornik 42, 11–44 (1957); (English translation in Sel. Trans. Math. Statist. Probab., pp.213–244 (1961) (in Russian)

Saporta, G.: Probabilités, Analyse de Données et Statistique, Editions Technip, Paris (1990)

Simon, G.: Additivity of Information in Exponential Family Power Laws. Journal of the American Statistical Association 68, 478–482 (1973)

Shannon, C.E.: A mathematical theory of communication. Bell System Tech. J. 27, 379–423, 623-656 (1948)

Tribus, M., McIrvine, E.C.: Energy and Information. Scientific American 224, 178–184 (1971)

Vapnik, V.N.: The Nature of Statistical Learning Theory. Springer, Heidelberg (1995)

Information: The Algorithmic Paradigm

Cristian S. Calude

Department of Computer Science
University of Auckland, New Zealand
www.cs.auckland.ac.nz/~cristian

1 Introduction

Information has a diversity of meanings, from everyday usage to a variety of technical settings. There is no single theory of information, but several theories, Shannon's information theory [16, 27, 28], semantic theories [2], logic of information [18], information algebra [21], philosophy of information [19], information flow [3], quantum information theory [24], evolutionary information [30], algorithmic information theory [4, 15], to name just a few. Each theory focuses on some specific aspects of information, and overlaps are minimal. Information is context-sensitive and heavily dependent on the adopted coding.

In this paper we will present, through a sequence of examples, some ideas and results of the algorithmic approach to information. In this approach information is measured by counting bits encoding computations.

2 Counting Bits

Information, in a broad sense, can be measured in various units, from bits to dollars. In this paper we shall confine ourselves to bits. The bit, short for binary digit, was first used in 1946 by John Tukey. A single bit can hold only one of two values: 0 or 1. More information is obtained by combining consecutive bits into larger units, bit-strings (shortly, strings): 00, 01,10,11, 000, 001,..., 111, 0000,... Sometimes it is useful to consider the empty string denoted by λ.

Strings have length: the number of characters of a string. For example, the length of 0 is 1, the length of 1110111 is 7, the length of the empty string is 0. Strings can be concatenated: the concatenation of the strings x and y is xy. The length of xy is the sum of the lengths of x and y.

Bits can be very useful to measure information. The power of bits can be illustrated with the information which can be encoded in, say, 20 bits. With a simple strategy, twenty questions/answers elicit 20 bits of information, which correspond to a single choice among $2^{20} = 1,048,576$ equally probable alternatives. For example, with the information in a bit-string of length 20 one can identify any town in USA. The limit of this approach is most visible at the level of semantics: No meaning is captured! For example, translated in binary, 'happy birthday', 'ya dirthbppayh' have the same information content.

The following *guessing game* is a more interesting example illustrating the power of bits: one person chooses a (secret) natural number and another person tries to guess it. The person who guesses is only allowed to ask questions of the following form: "Is your number less than n?" for every natural $n \geq 0$; the other person truthfully answers *yes* or *no*. The aim is to guess the number as fast as possible, that is, with as few questions as possible.

As an example consider the following questions:

1. Is your number less than 1?
2. Is your number less than 2?
3. Is your number less than 3?
4. Is your number less than 4?
   . . .

and so on until the first *yes* comes out.

To guess the number 10 we need to ask the first 11 questions; in general, to guess the number n we have to ask the first $n + 1$ questions. This solution leads to an encoding of all naturals numbers: with a bit-string of length $n + 1$ we encode the number n.

Can we do it better? Certainly. For example, we start asking the questions "Is your number less than 2^i?" for $i = 1, 2, \ldots$ till we get the answer "yes". This will happen at a value i such that $2^{i-1} \leq n < 2^i$. Then, we continue by halving the length of the interval. For example, to guess the number 10 we need 8 questions (corresponding to $i = 1, 2, 4, 8, 16, 12, 9, 10$). In general, to guess the number n we have to ask the first $2 \log n + 1$ questions (here $\log n$ is the integer part of $\log_2 n$, the base-2 logarithm of n). Note that this approach is better than the first one for $n > 3$.

Still, can we do it better? This is possible if we consider large enough numbers n: we can design better and better solutions. Does there exist an optimal solution? An answer will be given in the next section.

3   The Halting Problem

The halting problem (for Turing machines) is the problem to decide whether an arbitrary Turing machine eventually halts on an arbitrary input:

Does there exist a Turing machine T_{halt} which given the code $code(T)$ of a Turing machine T, and the input x, eventually stops and produces 1 if $T(x)$ stops and 0 if $T(x)$ does not stop?

Turing's result states that *the halting problem cannot be solved by any Turing machine*, i.e. there is no such T_{halt}. Here is an information-theoretical proof, [4, 14]. Instead of Turing machines we will deal with the informal notion of "program". We assume that programs incorporate inputs—which are coded as natural numbers. So, a program may run forever (does not halt) or may eventually stop, in which case it prints a natural number.

Assume that there exists a *halting program* deciding whether an arbitrary program will ever halt. Construct the following program, called P:

1. read a natural N;
2. generate all programs up to N bits in size;
3. use the *halting program* to check for each generated program whether it halts;
4. simulate the running of the above generated programs, and
5. output a number different from each output produced by the above programs.

The program P halts for every natural N. How long is P? Answer: $\log N +$ constant bits. Reason: to code N we need about $\log_2 N$ bits and the rest of the program P is constant.

For large N, the program P belongs to the set of programs having less than N bits (because $\log N +$ constant $< N$). Accordingly, for such an N, the program P will be generated by itself at some stage of the computation. We have got a contradiction since P outputs a natural number different from the output produced by itself!

Consider now all programs of length at most n, i.e. $2^{n+1} - 1$ programs. Some programs halt, some do not halt. If we order lexicographically all programs of length n and ask, for each such program, whether it halts or not, we get a bit-string of length $2^{n+1} - 1$ encoding the whole information. Is it possible to encode the same amount of information with fewer bits?

The answer is affirmative and a solution will be presented in what follows. We need more technical details. We revert our discussion to Turing machines, but of a very special type, self-delimiting Turing machines. The domain of a Turing machine T—dom(T)—is the set of strings where T halts; if dom(T) is prefix-free, i.e. no string in dom(T) is a proper extension of another string in dom(T), then T is called a *self-delimiting Turing machine*. An important result is the *universality theorem*:

> *There effectively exists a self-delimiting Turing machine U, called universal, such that for every self-delimiting Turing machine T we can compute a constant c, depending only on U and T, satisfying the following property: if $T(x)$ stops, then $U(x') = T(x)$, for some string x' with length no longer than the length of x plus c.*

In the framework of self-delimiting Turing machines the above coding problem can be stated as follows: given a universal self-delimiting Turing machine U and an integer $n > 0$, find an encoding via a bit-string shorter than 2^n from which one can check which program x of length less than n stops on U. This encoding was discovered by Chaitin in 1975 who introduced the Omega number, see [4, 15]:

$$\Omega_U = \sum_{\{x \mid U(x)\,\text{halts}\}} 2^{-|x|}, \qquad (1)$$

where $|x|$ denotes the length of x. We consider the binary expansion of Ω_U

$$\Omega_U = 0.\omega_1\omega_2\cdots\omega_m\cdots \qquad (2)$$

Given the first n bits of the binary expansion (2) of Ω_U, $\omega_1\omega_2\cdots\omega_n$, we can decide which programs x of length less than n halt on U: we enumerate enough elements $p_1, p_2, \ldots, p_k$ in the domain of U till the sum $\sum_{i=1}^k 2^{-|p_i|}$ becomes larger than or equal to $0.\omega_1\omega_2\cdots\omega_n$. We have:

$$\{x \mid |x| \le n, U(x) \text{ halts}\} \subseteq \{p_1, p_2, \ldots, p_k\},$$

so the halting programs x with $|x| \le n$ can be obtained by eliminating from the set $\{p_1, p_2, \ldots, p_k\}$ all programs of length larger than n. Indeed,

$$\Omega_U < 0.\omega_1\omega_2\cdots\omega_n + 2^{-n},$$

and for every halting program $q \notin \{p_1, p_2, \ldots, p_k\}$ with $|q| \le n$ we then have:

$$\sum_{i=1}^k 2^{-|p_i|} + 2^{-|q|} \ge 0.\omega_1\omega_2\cdots\omega_n + 2^{-n} > \Omega_U,$$

a contradiction.

We have shown that the halting information for all programs of length less than or equal to n (a set containing $2^{n+1} - 1$ elements) can be compressed into a string of length n: $\omega_1\omega_2\cdots\omega_n$.[1]

We are now in the position to give an answer to the question posed at the end of the section 2. Recall, we are interested in constructing an infinite prefix-free set of bit-strings to code as efficiently as possible all non-negative integers. The domain of a universal self-delimiting Turing machine is such a code (cf. [1]):

Let A be a set of bit-strings. The following two conditions are equivalent:

a) The set A is the domain of a universal self-delimiting Turing machine.

b) For every computable one-one function $g : \{0, 1, 2, \ldots\} \to \Sigma^*$ having a prefix-free range, there exist a computable one-one function $f : \{0, 1, 2, \ldots\} \to \Sigma^*$ and a constant $k \ge 0$ such that

a. $f(\{0, 1, \ldots\}) \subseteq A$,

b. $|f(n)| \le |g(n)| + k$, for every $n \ge 0$.

The good news is that coding with programs in the domain of a universal self-delimiting Turing machine is optimal up to an additive constant (and one can show that a better coding does not exist). The bad news is that this coding is computably enumerable, but not computable.

Finally, we ask the question: which problems can be solved knowing finitely many bits of Ω_U (see 2)? To answer this question we will present an implementation of a specific universal self-delimiting Turing machine U based on a register machine program, see [6]. A register machine has a finite number of registers,

---

[1] The converse implication is not true: we may know exactly which programs of length less than n halt and still not know any bit of Ω_U, cf. [5].

each of which may contain an arbitrarily large non-negative binary integer. The register machine U (labelled) instructions are:

> L: EQ R1 R2 R3
> L: SET R1 R2
> L: ADD R1 R2
> L: READ R1
> L: HALT

The names of the above instructions are self-explanatory. For instance, the first instruction is the classical if-then-else condition. In all cases R2 denotes either a register or a binary constant of the form $1(0+1)^* + 0$, while R1 and R3 must be register variables.

A *register machine program* consists of a finite list of labelled instructions from the above list, with the restriction that the HALT instruction appears only once, as the last instruction of the list. The input data (a binary string) follows immediately after the HALT instruction. A program not reading the whole data or attempting to read past the last data-bit results in a runtime error. Some programs have no input data.

It is perhaps surprising that many problems in mathematics can be reformulated in terms of the halting/non-halting status of appropriately constructed self-delimiting Turing machines. For example, consider Fermat's Last Theorem, stating that there are no integers $x, y, z, n > 3$ such that $x^n + y^n = z^n$. We can construct a self-delimiting Turing machine T_{Fermat} which systematically enumerates all possible integers (for example, written in binary) $x, y, z, n > 3$, checks whether $x^n + y^n = z^n$, and stops if for some values x, y, z, n the relation is true (which would mean that the program has found a counter-example); otherwise, T generates a new 4-tuple x, y, z, n and repeats the above procedure. Fermat's Last Theorem is equivalent with the statement "T_{Fermat} never halts", hence knowing that Fermat's Last Theorem is true we know that T_{Fermat} never halts.

In this way we can measure the difficulty of Fermat's Last Theorem by the complexity of T_{Fermat}, for example, by the number of bits necessary to specify T_{Fermat} in some fixed formalism (say, U). Of course, there are many self-delimiting Turing machines equivalent to T_{Fermat}, so a natural way to evaluate the complexity is to consider the least complex such machine. And, of course, this extends to any problem Π for which we can construct a self-delimiting Turing machine T_Π such that Π is false if and only if $U(T_\Pi)$ halts (if such a program exists): the *difficulty* of such a problem Π is the minimal number of bits of Ω_U necessary to test whether C_Π stops on U.

Here are three important open questions that can be analysed with this method (cf. [6]):

- *Goldbach's Conjecture:*[2] the program $T_{Goldbach}$ has 135 instructions totalling 3,484 bits.

---

[2] The conjecture was tested up to 4×10^{17}, see [25].

- *Riemann Hypothesis*:[3] the program T_{Riemann} consists of 290 instructions totalling 7,780 bits.
- *Collatz' Conjecture*:[4] there is a *non-constructive* way to prove that there exists a program T_{Collatz} which never stops iff the Collatz' Conjecture is true.

Are the numbers specified above the exact *difficulties* of the corresponding problems? Definitely not, they are upper bounds! The bad news is that, as expected, the problem of computing the *difficulty* of a problem is not computable. The good news is one can work with upper bounds: changing U will result in a change of upper bounds, but the order of difficulty will be preserved, namely if Π_1 is more difficult than Π_2 for U, the same relation will be true for any other universal self-delimiting Turing machine U^*. Specifically, the above analysis shows that the Riemann Hypothesis is more difficult than the Goldbach's Conjecture. For Collatz' Conjecture we cannot even evaluate an upper bound for the difficulty as the proof is not constructive.

4   Can Computers Create Information?

Can computation produce new information? To answer this question we will introduce a measure of information based on counting bits encoding computations. The motivation for this complexity measure may be rooted in Leibniz's work (1686): "A *theory* must be simpler than the data is explains ". Hence, a bit-string for which there is no *theory* is "unexplainable", "incomprehensible" except as 'a thing in itself' (*Ding an sich* in Kant's terminology).

Bearing in mind these facts we say that if a self-delimiting Turing machine T with program p produces the bit-string x, then p generates x via T, and the amount of information T extracts from x is

$$H_T(x) = \min\{|p| \mid T(p) = x\}.$$

It is possible that $T(p) = x$ is false for any program p; in this case $H_T(x) = \infty$. This definition heavily depends on T, but using the universality theorem we can make H as independent as possible on the underlying Turing machine because H_U is optimal up to an additive constant in the class of all possible H_T:

For every self-delimiting Turing machine T there exists a constant c (depending on U and T) such that for all strings x:

$$H_U(x) \leq H_T(x) + c.$$

So, for now on we shall fix a universal self-delimiting Turing machine U and write H instead of H_U.

---

[3] One of the Clay Mathematical Institute Millennium Problems, see [17, 31].

[4] See more in [23]; the conjecture was tested and proved true up to $10 \cdot 2^{58}$, see [26].

If $H(x) > H(y)$, then the complexity of x is larger than the complexity of y, that is, x encodes more information than y. In this framework, to create information means to start with an input x and produce an output y which has more information than x, that is, $H(x) < H(y)$. Our initial question becomes: is there any computable process capable of producing infinitely many outputs each of which has more information than its corresponding input? The problem is trivial for finitely many inputs.

One possible way to answer the above question is to assume that we have a one-to-one self-delimiting Turing machine T that halts on infinitely many inputs x, each of which having $H(x) \leq |x| - c/2$, where c is a fixed constant. Is it possible that T produces infinitely many outputs with the property that $H(T(x)) \geq |T(x)| + c/2$, that is, T produces a fixed amount (c bits) of newly created information[5])? The answer is *negative: no T is capable of such performance.* Indeed, this is not possible because otherwise T would generate an infinite computably enumerable set of strings y with $H(y) \geq |y| + c/2$, an impossibility (because the set $\{z \mid H(z) \geq |z| + c\}$ is immune, see [4]).

The above result suggests that a computer cannot create too much new information. Then the next question is: how much information can we expect to be created by computation?

A "Gödelian theory" is a finitely-specified, arithmetically sound, consistent theory strong enough to formalize arithmetic. For example, ZFC—Zermelo-Fraenkel set theory with choice, the classical axiomatic system in which virtually all current mathematics can be formalised—is a Gödelian theory. Define a new complexity measure

$$\delta(x) = H(x) - |x|.$$

The motivation in working with δ instead of H is the following. The complexity measures H and δ are similar as δ is defined from H and a simple computable function; for example, both measures are uncomputable. But H and δ differ in an *essential* way: given a positive N, the set $\{x \mid H(x) \leq N\}$ is finite while the set $\{x \mid \delta(x) \leq N\}$ is infinite. A sentence with a large δ-complexity has also a large H-complexity, but the converse is not true. For example, the H-complexity of (true) sentences of the form "$1 + n = n + 1$" tends to infinity as $n \to \infty$; however, their δ-complexity is bounded.

We can now state the main result in [9]:

For every Gödelian theory there exists a constant N, such that the theory proves no statement x with $\delta(x) > N$.

Any Gödelian theory can be used to prove theorems which have a bit more information than the theory itself, but not too much: everything is "hardwired" into the theory, there is very little room for "creativity" to produce more information.

The above result is a form of Gödel's incompleteness:

Any statement x with $\delta(x) > N$ cannot be proved by the Gödelian theory.

---

[5] This is a very small increase in information.

Even more, the set of statement which cannot be either proved or disproved by the Gödelian theory is large.

5   The Algorithmic Coding Theorem

Shannon's coding theorem [27] says that the minimal average code string length is about equal to the entropy of the source string set. The coding theorem plays an important role in Shannon's information theory [16, 27, 28]. In what follows we will briefly present an algorithmic version of Shannon's coding theorem.

Self-delimiting Turing machines have a prefix-free domain. Prefix-free sets S satisfy Kraft's inequality, [16]:

$$\sum_{p \in S} 2^{-|p|} \leq 1.$$

The following (more general) converse result, known as *Kraft-Chaitin theorem* (see [4]), is frequently used to build self-delimiting Turing machines:

> *Given a computable list of "requirements" (n_i, s_i), $(s_i$ are strings, $n_i \geq 1)$ such that $\sum_i 2^{-n_i} \leq 1$, we can effectively construct a self-delimiting Turing machine T and a computable one-to-one enumeration $x_0, x_1, x_2, \ldots$ of strings x_i of length n_i such that $T(x_i) = s_i$, for all i, and $T(x)$ is undefined if $x \notin \{x_i \mid i \geq 1\}$.*

Let Σ^* be the set of all binary strings. A function $P : \Sigma^* \to [0,1]$ such that $\sum_x P(x) \leq 1$ is called a *semi-distribution* over the strings. In case $\sum_x P(x) = 1$, P is a *distribution*. A semi-distribution P is semi-computable from below (above) in case the set $\{(x,r) \mid x \in \Sigma^*, \ r \in \mathbf{Q}, \ P(x) > r\}$ ($\{(x,r) \mid x \in \Sigma^*, \ r \in \mathbf{Q}, \ P(x) < r\}$) is computably enumerable ($\mathbf{Q}$ is the set of rationals). A semi-distribution P is computable if it is semi-computable from below and from above. For example, the probability[6] that the self-delimiting Turing machine T produces the output x,

$$P_T(x) = \sum_{T(u)=x} 2^{-|u|},$$

is a semi-distribution semi-computable from below. The function $P(x) = 2^{-2|x|-1}$ is a computable distribution.

A *prefix-code* for strings is a one-to-one function $C : \Sigma^* \to \Sigma^*$ such that $C(\Sigma^*)$ is prefix-free. If $C(x) = u$, then u is a code for x. The injectivity of C implies unique decodability.

For every self-delimiting Turing machine T and string x such that $P_T(x) > 0$, we denote by

$$x_T^* = \min\{u \mid T(u) = x\},$$

where the minimum is taken according to the quasi-lexicographical ordering of strings ($\lambda < 0 < 1 < 00 < 01 < 10 < 11 < 000 < \cdots$); x_T^* is called the

---

[6] See more about the underlying probability space in [4].

minimal (canonical) program of x with respect to T. For every surjective self-delimiting Turing machine T, $C_T(x) = x_T^*$ is a prefix-code; universal machines are surjective.

The *average code-string length* of a prefix-code C with respect to a semi-distribution P is

$$L_{C,P} = \sum_x P(x) \cdot |C(x)|,$$

the *minimal average code-string length* with respect to a semi-distribution P is

$$L_P = \inf \{L_{C,P} \mid C \text{ prefix-code}\},$$

and the *entropy* of a semi-distribution P is

$$\mathcal{H}_P = -\sum_x P(x) \cdot \log P(x).$$

Shannon's classical *(noiseless) coding theorem* [16, 27] can be expressed in the language of semi-distributions as follows:

The following inequalities hold true for every semi-distribution P:

$$\mathcal{H}_P - 1 \leq \mathcal{H}_P + \left(\sum_x P(x)\right) \log \left(\sum_x P(x)\right) \leq L_P \leq \mathcal{H}_P + 1.$$

If P is a distribution, then $\log(\sum_x P(x)) = 0$, so we get the classical inequality $\mathcal{H}_P \geq L_P$, cf. [16]. However, this inequality is not true for every semi-distribution. For example, take $P(x) = 2^{-2|x|-3}$ and $C(x) = x_1 x_1 \ldots x_n x_n 01$. It follows that $L_P \leq L_{C,P} = \mathcal{H}_P - \frac{1}{4}$.

Under which conditions, given a semi-distribution P, can we find a (universal) self-delimiting Turing machine T such that $H_T(x)$ is equal, up to an additive constant, to $-\log P(x)$, i.e. the complexity is equal up to an additive constant to entropy? An answer is given by the following general result proved in [8]:

Assume that P is a semi-distribution such that $P(x) > 0$, for every x, and there exist a computably enumerable set $S \subset \Sigma^* \times \{0, 1, \ldots\}$ and a constant $c \geq 0$ such that the following two conditions are satisfied for every $x \in \Sigma^*$:

(i) $\sum_{(x,n) \in S} 2^{-n} \leq P(x),$

(ii) *if $P(x) > 2^{-n}$, then $(x, m) \in S$, for some $m \leq n + c$.*

Then, there exists a machine T (depending upon S) such that for all x,

$$-\log P(x) \leq H_T(x) \leq (1 + c) - \log P(x).$$

The above result makes no direct computability assumptions on P. To get sharper consequences we will introduce the halting probability of a self-delimiting Turing machine T, $\Omega_T{}^7$, and the minimal (canonical) programs with respect to T. First, in analogy with (1) we define

$$\Omega_T = \sum_{\{x | T(x)\,\text{halts}\}} 2^{-|x|}.$$

Specialising P we show that minimal programs are almost optimal for P. Minimal programs of universal machines are almost optimal for every semi-computable semi-distribution P:

Assume that P is a semi-distribution semi-computable from below. Then, there exists a machine T (depending upon P) such that for all x,

$$- \log P(x) \le H_T(x) \le 2 - \log P(x).$$

Consequently, minimal programs for T are almost optimal: the code C_T satisfies the inequalities:

$$0 \le L_{C_T,P} - \mathcal{H}_P \le 2.$$

When the semi-distribution P is given, an optimal prefix-code can be found for P. However, that code may be far from optimal for a different semi-distribution. For example, let C be a prefix-code such that $|C(x)| = 2^{|x|+2}$, for all x. Let $\alpha > 0$ and consider the distribution

$$P_\alpha(x) = (1 - 2^{-\alpha})\, 2^{-(\alpha+1)|x|}.$$

If $\alpha \le 1$, then

$$L_{C,P_\alpha} - \mathcal{H}_{P_\alpha} = \infty,$$

but if $\alpha > 1$, then

$$L_{C,P_\alpha} - \mathcal{H}_{P_\alpha} < \infty.$$

So, C is asymptotical optimal for every distribution P_α with $1 < \alpha$, but C is far away from optimality if $0 < \alpha \le 1$. Clearly, P_α is computable provided α is computable.

Minimal programs are asymptotical optimal for every semi-distribution semi-computable from below:

Let P be a semi-distribution semi-computable from below, and U a universal self-delimiting Turing machine. Then, there exists a constant c_P (depending upon P) such that

$$0 \le L_{C_U,P} - \mathcal{H}_P \le 1 + c_P.$$

---

[7] The reader may recall the number Ω_U introduce in section 1.

The next result establishes a tight relation between complexity (H_T) and entropy ($-\log P_T$):

Let T be a machine and $c \geq 0$. The following statements are equivalent:

(a) *for all x, $H_T(x) \leq (1 + c) - \log P_T(x)$,*

(b) *for all non-negative n, if $P_T(x) > 2^{-n}$, then $H_T(x) \leq n + c$.*

In particular we get the *algorithmic coding theorem* (Chaitin–Gács):

There exists a constant $c \geq 0$ such that for all strings x,

$$|H_U(x) + \log P_U(x)| \leq 1 + c.$$

6   Algorithmic Randomness and Incompleteness

Defining randomness is very tricky. There are many proposals, among them the algorithmic one which equates randomness with incompressibility, and then proves other natural properties of "algorithmic randomness": stochasticity, unpredictability, etc. Algorithmic randomness comes into two forms, for finite bit-strings and for infinite sequences.

An infinite sequence $\mathbf{x} = x_1 x_2 \ldots, x_n \ldots$ is *algorithmically random* if there exists a positive constant $c > 0$ such that $H(x_1 x_2 \ldots, x_n) \geq n - c$. Chaitin's theorem (see [13]) states

The sequence of bits of Ω_U (i.e. the sequence $\omega_1 \omega_2 \ldots \omega_n \ldots$ in (2)) is algorithmically random.

We say that the real Ω_U is algorithmically random.

Two questions come naturally: a) are there any other "natural" algorithmically random sequences?, b) Ω_U is not only algorithmically random, but also computably enumerable, that is, Ω_U is the limit of a computable increasing sequence of rationals; are there other computably enumerable and algorithmically random numbers?

The answer to the first question is affirmative while the second question has a negative answer.

Let bin : $\{1, 2, \ldots\} \to \Sigma^*$ be the bijection which associates to every $n \geq 1$ its binary expansion without the leading 1,

n	n_2	bin(n)	$|$bin$(n)|$
1	1	λ	0
2	10	0	1
3	11	1	1
4	100	00	2
$\vdots$	$\vdots$	$\vdots$	$\vdots$

If $A \subset \Sigma^*$, then we define $\Upsilon[A] = \{n \geq 1 \mid \mathrm{bin}(n) \in A\}$. In other terms, the binary expansion of n is $n_2 = 1\mathrm{bin}(n)$. The *zeta number* of the Turing machine M,[8] denoted ζ_M, is defined by

$$\zeta_M = \sum_{n \in \Upsilon[\mathrm{dom}(M)]} \frac{1}{n}.$$

In [10] one proves the following result:

The zeta number ζ_U of a universal self-delimiting Turing machine U is algorithmically random.

In fact, the above theorem is true for a larger class of Turing machines. A *convergent Turing machine* is a Turing machine V whose zeta number is finite, $\zeta_V < \infty$.[9] Every self-delimiting Turing machine is convergent, but the converse is not true. The universality theorem holds true for convergent Turing machines as well. We can now state a more general result, [10]:

The zeta number ζ_V of a universal convergent Turing machine V is algorithmically random.

The answer to the second question is provided by the following theorem (cf. [7, 22], see also [4]):

A real $\alpha \in (0,1)$ is computably enumerable and algorithmically random iff there exists a universal self-delimiting Turing machine U such that $\alpha = \Omega_U$.

Algorithmic randomness is intimately related to incompleteness in Gödel's sense. Here are two results:

Chaitin's theorem [13]: *Every Gödelian theory cannot determine more than finitely many digits of Ω_U.*

Solovay's theorem [29]: *Fix a Gödelian theory. We can construct universal self-delimiting Turing machines U such that the theory cannot determine any digit of Ω_U.*

Generalised Solovay's theorem [5]: *Fix a Gödelian theory and a universal self-delimiting Turing machine U. Assume that $\Omega_U = 0.11 \cdots 10 \cdots$. Then, we can effectively construct a universal self-delimiting Turing machine U' such that $\Omega_{U'} = \Omega_U$ and the theory can determine at most the digits of Ω_U before the first 0.*

---

[8] M is not necessarily self-delimiting; of course, ζ_M could be infinite.

[9] Clearly, $\Omega_V < \infty$ iff $\zeta_V < \infty$, so convergence can be equally defined in terms of zeta or Omega.

7   Algorithmic Randomness and Halting

In this section we answer the question: Can a program stop at an algorithmically random time?

First we introduce yet another complexity measure, the natural complexity, cf. [11]. The *natural complexity* of the string x (with respect to the Turing machine[10] M) is $\nabla_M(x) = \min\{n \geq 1 \mid M(\mathrm{bin}(n)) = x\}$. Using ∇ the universality theorem has the following form:

> *One can effectively construct a (universal) Turing machine V such that for every machine M, there is a constant $\varepsilon > 0$ (depending on V and M) such that $\nabla_V(x) \leq \varepsilon \cdot \nabla_M(x)$, for all strings x.*

We fix the universal Turing machine V and write ∇ instead of ∇_V. A binary string x is *algorithmically random* if $\nabla(x) \geq 2^{|x|}/|x|$.[11] One can prove (see [4, 15]) that algorithmically random strings have many properties one naturally associate with randomness, among them strong uncomputability:

> *No Turing machine is capable of enumerating an infinity of algorithmically random strings.*

Most binary strings of a given length n are algorithmically random because they have high density:

$$\mathrm{density}(n) = \#\{x \in \Sigma^* : |x| = n, \nabla(x) \geq 2^n/n\} \cdot 2^{-n} \geq 1 - 1/n,$$

hence

$$\lim_{n \to \infty} \mathrm{density}(n) = 1.$$

We are interested in the properties of the exact times programs stop. A time t will be called *algorithmically random* if $\mathrm{bin}(t)$ is algorithmically random. In [12] one proves the following result:

> *Let V be a universal Turing machine. One can effectively compute a constant c (depending on V) such that the following is true: if an N-bit program p has not stopped on V by the time $2^{2N+2c+1}$, where $N \geq 2$, then $V(p)$ cannot exactly stop at any algorithmically random time $t \geq 2^{2N+2c+1}$.*

In other words, given V and a program p of length N we can compute the time $\theta_{V,N} = 2^{2N+2c+1}$ with the following property: either $V(p)$ stops before the time $\theta_{V,N}$, or if it has not stopped by that time, then either $V(p)$ will never

---

[10] Not necessarily self-delimiting.

[11] In the language of the complexity H, the string x is algorithmically random if $H(x) \geq |x| - \log|x|$. Algorithmically randomness for strings is a matter of degree, so we can set various bounds on the complexity; see [4].

stop or $V(p)$ will stop at a non-algorithmically random time $t \geq \theta_{V,N}$. Because non-algorithmically random times have effectively zero density, "chances" that an N-bit program p that has not stopped on V by the time $\theta_{V,N}$ will eventually stop effectively approach zero:

> *For every length N, we can effectively compute a threshold time $\theta_{V,N}$ (which depends on V and N) such that if a program of length N runs for $\theta_{V,N}$ steps without halting, then the density of times greater than $\theta_{V,N}$ at which the program can stop has effective zero density. More precisely, if an N-bit program runs for $T > \max\{\theta_N, 2^{2+5 \cdot 2^k}\}$ steps, then the density of times at which the program can stop is less than 2^{-k}.*

8   Incompleteness and Uncertainty

Gödel's hostility to any suggestion regarding possible connections between his incompleteness theorem and physics, particularly, Heisenberg's uncertainty relation, is well-known: J. Wheeler was thrown out of Gödel's office for asking the question "Professor Gödel, what connection do you see between your incompleteness theorem and Heisenberg's uncertainty principle?"

Still, there is a huge interest in the relations between these two statements. For example, Hawking's view (see [20]) is that

> "a physical theory is self-referencing, like in Gödel's theorem ... Theories we have so far are both inconsistent and incomplete".[12]

In [12] a relation between incompleteness and uncertainty is established. To present it we will use the natural complexity $\nabla = \nabla_U$ induced by a universal self-delimiting Turing machine U; recall that $H = H_U$. One can see that

$$2^{H(x)} \leq \nabla(x) < 2^{H(x)+1},$$

hence, $\Delta(x) = 2^{H(x)}$, the uncertainty in the value $\nabla(x)$, is the difference between the upper and lower bounds given.

Finally let $\Delta_s = 2^{-s}$. The property of $\Omega = \Omega_U$ to be algorithmically random can be expressed in the following way:

$$\Delta_s \cdot \Delta(\omega_1 \ldots \omega_s) \geq 1, \tag{3}$$

In (3), an *uncertainty relation*, the complexity measures the uncertainty in the total information. One can prove that the relation (3) *implies* Chaitin's theorem (presented at the end of section 6), hence, Gödel's incompleteness.

Of course, this is a formal approach and much more is required to check its "physical" base (see more in [12]).

---

[12] It is worth noting that a theory which is inconsistent is not necessarily complete, although in many cases this is true.

References

1. Ambos-Spies, K., Calude, C.S., Merkle, W., Staiger, L.: On Universal Computably Enumerable Prefix Codes (in preparation)
2. Bar-Hillel, Y. (ed.): Language and Information: Selected Essays on Their Theory and Application. Addison-Wesley, Reading (1964)
3. Barwise, J., Seligman, J.: Information Flow: The Logic of Distributed Systems. Cambridge University Press, Cambridge (1997)
4. Calude, C.S.: Information and Randomness: An Algorithmic Perspective, 2nd edn. Springer, Berlin (2002)
5. Calude, C.S.: Chaitin Ω numbers, Solovay machines and incompleteness. Theoret. Comput. Sci. 284, 269–277 (2002)
6. Calude, C.S., Calude, E., Dinneen, M.J.: A new measure of the difficulty of problems. Journal for Multiple-Valued Logic and Soft Computing 10, 1–21 (2006)
7. Calude, C.S., Hertling, P., Khoussainov, B., Wang, Y.: Recursively enumerable reals and Chaitin Ω numbers. Theoret. Comput. Sci. 255, 125–149 (2001)
8. Calude, C.S., Ishihara, H., Yamaguchi, T.: Minimal programs are almost optimal. International Journal of Foundations of Computer Science 12(4), 479–489 (2001)
9. Calude, C.S., Jürgensen, H.: Is complexity a source of incompleteness? Advances in Applied Mathematics 35, 1–15 (2005)
10. Calude, C.S., Stay, M.A.: Natural halting probabilities, partial randomness, and Zeta functions. Information and Computation 204, 1718–1739 (2006)
11. Calude, C.S., Stay, M.A.: From Heisenberg to Gödel via Chaitin. International Journal of Theoretical Physics 44(7), 1053–1065 (2005)
12. Calude, C.S., Stay, M.A.: Most programs stop quickly or never halt. Advances in Applied Mathematics 40, 295–308 (2008)
13. Chaitin, G.J.: A theory of program size formally identical to information theory. J. Assoc. Comput. Mach. 22, 329–340 (1975)
14. Chaitin, G.J.: Information, Randomness and Incompleteness, Papers on Algorithmic Information Theory, 2nd edn. World Scientific, Singapore (1990)
15. Chaitin, G.J.: Algorithmic Information Theory. Cambridge University Press, Cambridge (1987)
16. Cover, T.M., Thomas, J.A.: Elements of Information Theory. Wiley, New York (1991)
17. Devlin, K.J.: The Millennium Problems. Basic Books, New York (2002)
18. Devlin, K.J.: Logic and Information. Cambridge University Press, Cambridge (1991)
19. Floridi, L.: What is the philosophy of information? Metaphilosophy 33(1-2), 123–145 (2002)
20. Hawking, S.W.: Gödel and the End of Physics. Dirac Centennial Celebration, Cambridge, UK (July 2002), http://www.damtp.cam.ac.uk/strtst/dirac/hawking/
21. Kohlas, J.: Information Algebras: Generic Structures for Inference. Springer, London (2003)
22. Kučera, A., Slaman, T.A.: Randomness and recursive enumerability. SIAM J. Comput. 31(1), 199–211 (2001)
23. Lagarias, J.C.: The $3x + 1$ problem and its generalizations. Amer. Math. Monthly 92, 3–23 (1985)
24. Nielsen, M.A., Chuang, I.L.: Quantum Computation and Quantum Information. Cambridge University Press, Cambridge (2000)

25. Oliveira e Silva, T.: Goldbach Conjecture verification, June 5 (2006),
    http://www.ieeta.pt/~tos/goldbach.html
26. Oliveira e Silva, T.: Computational verification of the 3x+1 conjecture, May 26
    (2006), http://www.ieeta.pt/~tos/3x+1.html
27. Shannon, C.E.: A mathematical theory of communication. Bell System Technical
    Journal 27, 379–423, 623–656 (1948)
28. Shannon, C.E., Weaver, W.: The Mathematical Theory of Communication. Uni-
    versity of Illinois Press, Urbana (1949) (paperback edition 1963; special fiftieth
    anniversary edition in 1999)
29. Solovay, R.M.: A version of Ω for which ZFC can not predict a single bit. In:
    Calude, C.S., Păun, G. (eds.) Finite Versus Infinite. Contributions to an Eternal
    Dilemma, pp. 323–334. Springer, London (2000)
30. Stonier, T.: Information and Meaning: An Evolutionary Perspective. Springer, Hei-
    delberg (1997)
31. http://www.claymath.org/millennium/Riemann_Hypothesis/

Information Algebra*

Jürg Kohlas and Cesar Schneuwly

University of Fribourg
Department of Informatics DIUF
Bd de Pérolles 90
1700 Fribourg, Switzerland
juerg.kohlas@unifr.ch, cesar.schneuwly@unifr.ch
http://diuf.unifr.ch/tcs/

1   Introduction

According to Shannon's classical information theory [19] information is measured
by the reduction of uncertainty and the latter is measured by *entropy*. This
theory is concerned with the transmission of symbols from a finite alphabet.
The uncertainty concerns the question which symbol is sent and the information
is given by a probabilistic model of the transmission channel and the symbol
observed at the output of the channel. This leads to a statistical communication
theory which is still the main subject of communication theory today.

There are some important elements in Shannon's approach that will be precked
up and reconsidered here, although in another direction and with other goals
than in Shannon's work. The first ingredient is that information relates to questions. In Shannon's case the question is fixed: what symbol is sent? In information
processing in general several questions, whole systems of interrelated questions,
will be considered. A piece of information may relate to a determined domain
and must then be focussed on the question or questions of interest. Further,
several pieces of information on related domains or questions may be available
and must be aggregated to get the overall picture. These elements introduce an
algebraic flavor into an extended information theory.

The theory proposed here can be sketched as follows: Questions can be represented by the possible answers they allow. There may be finer or coarser answers,
which corresponds to a finer or coarser *granularity* of questions. This can be captured by a *partial order* between questions or the *domains* of possible answers.
It will even be supposed that the system of questions or domains forms a *lattice*, such that two domains have a *supremum* or *join* representing the *combined
question*, i.e. the possible answers to both questions. Two domains have also
an *infimum* or *meet* representing the common part, the intersection, of both
questions. Associated with this lattice of domains is a system of information
consisting of pieces of information, each piece bearing on a determined domain
from the lattice. Within this system the operations of *combination* of information, representing aggregation, and of *projection* to a given domain, representing

* Research supported by grant No. 200020–109510 of the Swiss National Foundation
  for Research.

information extraction, are defined. This leads to a certain *two-sorted algebra* which is called an *information algebra* and which is the subject of this chapter.

First, in Section 2.1 the classical *relational algebra* associated with *relational databases* will be presented as a prototype of an information algebra. This serves as a motivation, since databases are surely depositories of information. In Section 2.2 the abstract axiomatic definition of information algebra is given. It is shown how information in this framework can be transported to arbitrary domains and thus relate to any question of the system considered. Further two equivalent variants of the algebraic structure are discussed: In Section 2.3 it is shown how focussing of information may in some cases be replaced by *variable elimination*. This positions information algebra in the context of *logics* and relates information extraction with *existential quantification*. The latter relation is elaborated in Section 3, especially in Subsections 3.2 and 3.3. The transport operation of information shows that the *same* piece of information may be represented equivalently relative to different domains. This leads to an equivalent *domain-free* version of the information algebra (Section 2.5). This variant may be better suited for some discussions than the original *labeled* version.

In Section 3 several instances or examples of information algebras are presented. They are mostly related to different systems of *logic* which provide besides databases a second basic form of representation of information. In particular the classical systems of *propositional* and *predicate* logic are presented as information algebras. This is clearly related to *algebraic theories* of logic as proposed for instance by [6, 7, 8]. The concept of *contexts* is proposed as a more general framework related to logic for obtaining information systems (Section 3.4). This concept is motivated by and related to *classifications* [1]. It is also connected to *concept analysis* [3]. Outside logic, a further example in Section 3.5 is linked to *fuzzy set theory* and *possibility theory*. These few examples should suffice to convince the reader about the justification and the interest of information algebras.

The last Section 4 establishes a first link of the theory of information algebras with Shannon's information measure, although it must be stressed that the algebraic theory so far is *not* a statistical theory. First we show how a natural *partial order* of information content arises from the algebra of information. It allows to compare information content both in an absolute way as well as with respect to a given question or domain. This order permits also to define particular algebras built form basic, finest information elements, called *atoms*. In those cases it will be possible to define a *quantitative information measure* using Hartley's measure (or entropy of uniform distributions) to quantify the *reduction of uncertainty* by an information element out of an information algebra. This measure is shown to respect the qualitative, partial order of information content. It is defined relative to any given domain, and there is also a *relative information measure* of a piece of information given another one. Several interesting properties of this measure are discussed. Again this is not a statistical theory of information, such that entropy displays not yet its full power. Motivated by relational algebra, dual information algebra and related measures in *Boolean information algebra* will also be considered in Section 4.2 and 4.4. Both measures have their proper interpretation and application.

Where do information algebras originate from? For Bayesian networks [14] proposed a so-called *local propagation algorithm*, which solved the dimensionality and efficiency problem of the naive solution problems. Based on this work [20] proposed a system of simple axioms which were sufficient for permitting local propagation and which were also sufficient for several formalisms of artificial intelligence. In [9] the algebraic theory of these, so-called *valuation algebras* was developed into some depth. And in particular, *information algebras* were proposed as valuation algebras which satisfy in addition the *idempotency property*. It is this property which allows the development of the information theory proposed here. So, whereas Shannon's theory is a theory of communication, resulting in efficient coding schemes, the theory of information algebra is a theory of *computation*, leading to efficient generic algorithms for important problems of *query processing*.

2   The Algebra of Information

2.1   A Prototype: Relational Algebra

Relational databases surely contain information. Therefore they may serve as a prototype example for the algebraic structure and theory we want to propose and discuss here. So let's summarize the basic elements of relational database theory.

Let $\mathcal{A}$ be a set of symbols, called *attributes*. For each $\alpha \in \mathcal{A}$ let D_α be a non-empty set, the set of possible values for attribute α. For example, if $\mathcal{A} = \{$name,age,income$\}$, then D_{name} could be the set of strings, whereas D_{age} and D_{income} are both the set of nonnegative integers.

Let $x \subseteq \mathcal{A}$. A x-*tuple* is a function f with domain x and values $f(\alpha) \in D_\alpha$ for each $\alpha \in x$. The set of all x-tuples is denoted by E_x. For any x-tuple f and a subset $y \subseteq x$ the restriction $f[y]$ is defined to be the y-tuple g such that $g(\alpha) = f(\alpha)$ for all $\alpha \in y$.

A *relation* R over x is a set of x-tuples, i.e. a subset of E_x. The set of attributes x is called the *domain* of R and denoted by $d(R)$. For $y \subseteq d(R)$ the *projection* of R onto y is defined as follows:

$$\pi_y(R) = \{f[y] : f \in R\}.$$

The *join* of a relation R over x and a relation S over y is defined by

$$R \bowtie S = \{f : f \in E_{x \cup y}, f[x] \in R, f[y] \in S\}.$$

It is easy to see that the relations satisfy the following properties:

1. The join is an *associative* and *commutative* operation, and E_x is a *neutral* element for relations over x, i.e. $R \bowtie E_x = R$ if $d(R) = x$,
2. $d(R \bowtie S) = d(R) \cup d(S)$,

3. If $x \subseteq d(R)$, then $d(\pi_x(R)) = x$,
4. If $x \subseteq y \subseteq d(R)$, then $\pi_x(\pi_y(R)) = \pi_x(R)$,
5. If $d(R) = x$ and $d(S) = y$, then $\pi_x(R \bowtie S) = R \bowtie \pi_{x \cap y}(S)$,
6. If $x \subseteq y$, then $\pi_x(E_y) = E_x$,
7. If $x \subseteq d(R)$, then $R \bowtie \pi_x(R) = R$.

In fact, this algebraic system is part of *relational algebra* as defined in relational database theory [15]. Besides join and projection there are further operations like complement, union and difference. Relational algebra is used for query processing in relational databases. The operations of join and projection, and especially property (5) above, play a particularly important role [2]. There is even a special term for the formula $R \bowtie \pi_{x \cap y}(S)$ occurring in (5). It is called a *semijoin*.

We propose in the next section to abstract an algebraic system from this example, which we claim covers important aspects of a general theory of information.

2.2   The Axioms

Relations R as defined in the previous section can be thought of as representing *pieces of information* indicating which tuples $f \in E_{d(R)}$ describe possible tuples of values of the attributes $\alpha \in d(R)$. A relation R with domain $d(R) = x$ answers the *question*, which of the elements of the cartesian space

$$D_x = \times_{\alpha \in d(R)} D_\alpha \tag{1}$$

represent the true values of the variables. A relation is however only a partial answer since it does not fix a unique, precise element as an answer. So, any piece of information R refers to a determined domain $d(R)$, which in turn represents a *question* related to the attributes in x, asking what are the possible elements of D_x. Further the *join* serves to combine or aggregate two pieces of information, represented by two relations R and S. The combined information, represented by the join $R \bowtie S$ refers to domain $d(R) \cup d(S)$, according to property (2) in the previous section. *Projection* serves to extract the information relative to a part $y \subseteq d(R)$ of the domain of an information R. It results in an information relative to domain y, see property (3) in the previous section.

Thus, in a general way, we assume a set D of elements which are called *domains* and which are thought to represent in an abstract sense questions. Domains may have different *granularity*, i.e. a domain $x \in D$ may be *coarser* than another domain $y \in D$, meaning that y represents a more precise question than x. This is modelled by a *partial order* in D. Thus, $x \leq y$ means that x is a coarser domain than y, or that domain y is finer than x. Moreover, given two domains x and y, there should be a coarsest domain, finer than both x and y, i.e. the *join* $x \vee y$ should exist within D. It represents the *combined question* composed of questions x and y. In the same way a finest domain coarser than both x and y should exist, i.e. the *meet* $x \wedge y$ should exist within D. This means that D is assumed to be a *lattice* [3].

In relational algebra the domains are represented by subsets x of the attribute set $\mathcal{A}$. The partial order is defined by set inclusion, $x \leq y$ if $x \subseteq y$. Join and meet of domains correspond to set union and intersection, i.e. $x \vee y = x \cup y$ and $x \wedge y = x \cap y$. This is a distributive lattice [3]. In many applications we will use subsets of attributes or variables as domains. We call this *multivariate domains*.

Alternatively, but equivalently, we could consider the domains D_x defined in equation (1). Then we have $D_x \leq D_y$ if $x \subseteq y$ and $D_x \vee D_y = D_{x \cup y}$ and $D_x \wedge D_y = D_{x \cap y}$. A cartesian product D_x induces a *partition* of the universe $D_{\mathcal{A}}$. In fact, another, more general and interesting class of domain lattices are given by lattices of partitions of a universe S [4]. We remark that such partition lattices are in general no more distributive.

Further we consider a set Φ of elements, called *pieces of information* whose generic elements we denote by $\phi, \psi, \dots$ etc. Each information ϕ concerns a certain domain $d(\phi) \in D$, which is attached to ϕ as a *label* or *mark*. The *combination* of information is defined by a binary operation $\Phi \times \Phi \to \Phi$, which will be denoted by $(\phi, \psi) \mapsto \phi \otimes \psi$. If x is a domain out of D and $\phi \in \Phi$ an information such that $x \leq d(\phi)$, then $\phi^{\downarrow x}$ denotes the part of information ϕ which concerns domain x. This operation of *projection* (sometimes also called *marginalization*) is defined as a partial mapping $\Phi \times D \to \Phi$.

Formally, we have thus a two-sorted algebra (Φ, D) with the following operations:

1. *Meet, Join:* $D \times D \to D$, $(x, y) \mapsto x \wedge y, x \vee y$,
2. *Combination:* $\Phi \times \Phi \to \Phi$, $(\phi, \psi) \mapsto \phi \otimes \psi$,
3. *Projection:* $\Phi \times D \to \Phi$, $(\phi, x) \mapsto \phi^{\downarrow x}$, defined for $x \leq d(\phi)$.

We impose the following axioms on this two-sorted algebra:

1. *Lattice:* D is a lattice with respect to the operations of meet and join.
2. *Semigroup:* Φ is associative and commutative under combination.
3. *Labeling:* $d(\phi \otimes \psi) = d(\phi) \vee d(\psi)$.
4. *Neutrality:* For all $x \in D$ there is a neutral element e_x such that $d(e_x) = x$ and for all $\phi \in \Phi$ with $d(\phi) = x$, $\phi \otimes e_x = \phi$; and for all $y \in D$, $x \geq y$, we have $e_x^{\downarrow y} = e_y$.
5. *Nullity:* For all $x \in D$ there is a null element z_x such that $d(z_x) = x$ and for all $\phi \in \Phi$ with $d(\phi) = x$, $\phi \otimes z_x = z_x$; and for all $y \in D$, $y \geq x$, we have $z_x \otimes e_y = z_y$.
6. *Projection:* If $\phi \in \Phi$, $x \in D$, $x \leq d(\phi)$, then $d(\phi^{\downarrow x}) = x$.
7. *Transitivity:* If $x \leq y \leq d(\phi)$, then $(\phi^{\downarrow y})^{\downarrow x} = \phi^{\downarrow x}$.
8. *Combination:* If $d(\phi) = x$, $d(\psi) = y$, then $(\phi \otimes \psi)^{\downarrow x} = \phi \otimes \psi^{\downarrow x \wedge y}$.
9. *Idempotency:* If $x \leq d(\phi)$, then $\phi \otimes \phi^{\downarrow x} = \phi$.

A two-sorted algebra (Φ, D) satisfying these axioms is called an *information algebra* [9]. That D is a lattice means that the operations of meet and join are both associative and commutative, idempotent (i.e. $a \wedge a = a \vee a = a$) and absorbing (i.e. $a \vee (a \wedge b) = a$ and $a \wedge (a \vee b) = a$). Axiom (2) says that Φ is a *commutative semigroup* under combination. The sequence of how pieces of information are combined does not matter. The labeling axiom (3) states that the combination of pieces of information relative to domains x and y relates to the combined question $x \vee y$. Axiom (4) establishes the existence of a neutral element, which represents *vacuous information*. It is stable, in the sense that projection vacuous information yields vacuous information. Similarly, axiom (5) establishes the existence of null elements, representing *contradiction*. Axiom (6) means that if the part relative to domain x is extracted from an information, then the resulting information relates to domain x. Transitivity (axiom (7)) says that projection can be done in steps. The combination axiom (8) tells us, that, in order to extract the part relative to domain x from a combined information on x and y, we can as well first extract the part relative to $x \wedge y$ from the information on y and then combine the two pieces of information. Finally, idempotency means that combining a piece of information with a part of it, gives nothing new. These seem reasonable properties to assume for an algebra of information. For relational algebra, these axioms correspond to the properties derived in the previous section. Relational algebra is thus an information algebra.

The next three assertions are immediate consequences of the axioms:

Lemma 1.   *1. If $d(\phi) = x$, then $\phi^{\downarrow x} = \phi$.*
   2. $\phi \otimes \phi = \phi$.
   3. $e_x \otimes e_y = e_{x \vee y}$.

Proof. (1) Let $x = d(\phi)$. Then, by the combination and stability axioms, we have $\phi^{\downarrow x} = (\phi \otimes e_x)^{\downarrow x} = \phi \otimes e_x^{\downarrow x} = \phi \otimes e_x = \phi$.

(2) Using (1) and idempotency, we obtain $\phi \otimes \phi = \phi \otimes \phi^{\downarrow x} = \phi$.

(3) By the labeling axiom, stability and idempotency we conclude that $e_x \otimes e_y = e_x \otimes e_y \otimes e_{x \vee y} = e_{x \vee y}^{\downarrow x} \otimes e_{x \vee y}^{\downarrow y} \otimes e_{x \vee y} = e_{x \vee y}$.   □

A central problem in applications can be formulated as follows: Given a number of pieces of information $\phi_1, \ldots, \phi_n$ with domains $d(\phi_i) = x_i$ and a goal domain x. The part relating to domain x of the total combined information is to be computed. Formally stated, we want to compute

$$(\phi_1 \otimes \cdots \otimes \phi_n)^{\downarrow x}.$$

This is the *projection problem*. If this is computed as written here, then by the labeling axiom, an information on the possibly very large domain $x_1 \vee \cdots \vee x_n$ has to be computed and then projected. This may be computationally infeasible. Instead, based in particular on the combination axiom, methods can be devised where ideally never information on larger domains than x_1 to x_n must be computed. These are called *local computation methods* [9, 12]. They were first proposed by [14] for *probabilistic networks*. Later [21] noted that these local computation methods can be used, if the elements satisfy some abstract axioms. The axioms of an information algebra are modelled after the Shenoy-Shafer system. In particular the idempotency axiom is added, which is not essential for local computation. But we shall see below that this axiom is essential for the theory of information presented here.

2.3   Variable Elimination

If we consider information algebras with *multivariate domains*, then an interesting variant of information algebras can be formed. Let V be a finite or countable set of variables, denoted by $X, Y, \ldots$ etc. Consider an information algebra (Φ, D), where D is the lattice of subsets of V. Using projection we define a new operation called *variable elimination* for $X \in d(\phi)$:

$$\phi^{-X} = \phi^{\downarrow d(\phi) - \{X\}}.$$

The following properties hold for variable elimination:

1. If $X \in d(\phi)$, then $d(\phi^{-X}) = d(\phi) - \{X\}$.
2. If $X, Y \in d(\phi)$, then $(\phi^{-X})^{-Y} = (\phi^{-Y})^{-X}$.
3. If $X \in d(\psi)$, $X \notin d(\phi)$, then $(\phi \otimes \psi)^{-X} = \phi \otimes \psi^{-X}$.
4. If $X \in d(\phi)$, then $\phi \otimes \phi^{-X} = \phi$.
5. If $X \subseteq z \in D$, then $e_z^{-X} = e_{z-\{X\}}$.

(1) follows immediately from the projection axiom, if the definition of variable elimination is used. Similarly, (2) follows directly from the transitivity axiom, (4) is the idempotency axiom and (5) follows from the neutrality axiom. Only (3) is a little bit more involved. We have $(\phi \otimes \psi)^{-X} = (\phi \otimes \psi)^{\downarrow (x \cup y) - \{X\}}$ if $d(\phi) = x$ and $d(\psi) = y$. Note that $x \subseteq z = (x \cup y) - \{X\} \subseteq x \cup y$. We claim that

$$(\phi \otimes \psi)^{\downarrow z} = \phi \otimes \psi^{\downarrow y \cap z}. \qquad (2)$$

Since $y \cap z = y \cap ((x \cup y) - \{X\}) = y - \{X\}$ because $X \in y$ and $X \notin x$, we have then $\phi \otimes \psi^{\downarrow y \cap z} = \phi \otimes \psi^{-X}$ which proves (3). In order to prove equation (2) we note that $z \cap (x \cup y) = z$. The labeling and combination axioms permit then to derive

$$(\phi \otimes \psi)^{\downarrow z} = (\phi \otimes \psi)^{\downarrow z} \otimes e_z$$
$$= (\phi \otimes \psi \otimes e_z)^{\downarrow z}$$
$$= (\phi \otimes e_z) \otimes \psi^{\downarrow y \cap z}$$
$$= (\phi \otimes \psi^{\downarrow y \cap z}) \otimes e_z.$$

The first term in this combination has domain $x \cup (y \cap z) = z$. This shows then that equation (2) holds indeed.

We may take properties (1) to (5) above for variable elimination as new axioms instead of axioms (4), (6), (7), (8) and (9) together with the remaining axioms (1), (2), (3) and (5). This gives a variant of an information algebra. In this system, property (2) above allows to define unambiguously the elimination of several variables $X_1, \ldots, X_n \in d(\phi)$ by

$$\phi^{-\{X_1, \ldots, X_n\}} = (\cdots ((\phi^{-X_1})^{-X_2}) \cdots)^{-X_n}.$$

According to property (2) the actual elimination sequences does not matter.

Variable elimination is only defined for *finite* sets of variables. Therefore, in general, it is less powerful than projection. If D in an information algebra (Φ, D)

is the lattice of *finite* subsets of a set of variables, then, for $x \in D$, projection may be defined in terms of variable elimination as follows:

$$\phi^{\downarrow x} = \phi^{-(d(\phi)-x)}.$$

It can easily be verified that, with this definition, the axioms of an information algebra are satisfied, if variable elimination satisfies properties (1) to (4) above. Thus, for multivariate systems with pieces of information always relating to finite sets of variables, the algebras with projection and variable elimination are *equivalent*.

2.4   Transport of Information

So far, information can only be projected to subdomains of its domain. However, transport of information from one domain to another one can be defined more generally. Let $(\varPhi, D)$ be an information algebra. Then, for $y \geq d(\phi)$, we define a new operation

$$\phi^{\uparrow y} = \phi \otimes e_y,$$

called the *vacuous extension* of ϕ to domain y. This term is justified, since, for $d(\phi) = x$,

$$(\phi^{\uparrow y})^{\downarrow x} = (\phi \otimes e_y)^{\downarrow x} = \phi \otimes e_y^{\downarrow x} = \phi \otimes e_x = \phi$$

by the combination and stability axioms. So, vacuous extension indeed does not add or change otherwise information. Now, more generally, for $d(\phi) = x$ and $y \in D$ arbitrary, we define the operation

$$\phi^{\to y} = (\phi^{\uparrow x \vee y})^{\downarrow y}.$$

This is called the *transport operation*; it permits to transport a piece of information from its original domain to any other domain. Note that projection and vacuous extension are just special cases of this transport operation, namely for $y \leq d(\phi)$ or $y \geq d(\phi)$ respectively. Note further that $\phi^{\uparrow x \vee y} = \phi \otimes e_{x \vee y} = \phi \otimes e_y \otimes e_{x \vee y} = \phi \otimes e_y$, hence

$$\phi^{\to y} = (\phi \otimes e_y)^{\downarrow y} = \phi^{\downarrow x \wedge y} \otimes e_y = (\phi^{\downarrow x \wedge y})^{\uparrow y}.$$

In the following lemma we collect some properties of the transport operation.

Lemma 2.   *1.* $(\phi^{\to y})^{\to z} = (\phi^{\to y \wedge z})^{\to z}$.
   2. If $d(\phi) = x$, *then* $(\phi \otimes \psi)^{\to x} = \phi \otimes \psi^{\to x}$.
   3. If $d(\phi) = x$, *then* $\phi^{\to x} = \phi$.
   4. If $d(\phi) = x$, *then* $\phi \otimes \phi^{\to y} = \phi^{\uparrow x \vee y}$.

Proof. (1) If $y \leq z$, then we claim that $\phi^{\to y} = (\phi^{\to z})^{\to y}$. In fact, assume $d(\phi) = x$, then

$$(\phi^{\to z})^{\to y} = ((\phi^{\uparrow x \vee z})^{\downarrow z})^{\downarrow y} = (\phi^{\uparrow x \vee z})^{\downarrow y}$$
$$= (((\phi^{\uparrow x \vee y})^{\uparrow x \vee z})^{\downarrow x \vee y})^{\downarrow y} = (\phi^{\uparrow x \vee y})^{\downarrow y} = \phi^{\to y}.$$

In order to prove (1) we apply this result and obtain

$$(\phi^{\rightarrow y \wedge z})^{\rightarrow z} = ((\phi^{\rightarrow y})^{\rightarrow y \wedge z})^{\rightarrow z} = ((\phi^{\rightarrow y})^{\downarrow y \wedge z})^{\uparrow z} = (\phi^{\rightarrow y})^{\rightarrow z}.$$

(2) This follows from the combination axiom

$$(\phi \otimes \psi)^{\rightarrow x} = (\phi \otimes \psi)^{\downarrow x} = \phi \otimes \psi^{\downarrow x \wedge y} = \phi \otimes e_x \otimes \psi^{\downarrow x \wedge y}$$
$$= \phi \otimes (\psi^{\downarrow x \wedge y})^{\uparrow x} = \phi \otimes \psi^{\rightarrow x}.$$

(3) This follows since $\phi^{\rightarrow x} = \phi^{\downarrow x} = \phi$.

(4) Here we have, using the idempotency axiom,

$$\phi \otimes \phi^{\rightarrow y} = \phi \otimes (\phi^{\downarrow x \wedge y})^{\uparrow y} = \phi \otimes \phi^{\downarrow x \wedge y} \otimes e_y = \phi \otimes e_y = \phi^{\uparrow x \vee y}. \qquad \square$$

These properties of transport are similar to the transitivity, combination, projection and idempotency axioms of the information algebra. In fact, they could replace them.

2.5 Domain-Free Information Algebras

Assume that, in an information algebra (Φ, D), for two elements $\phi, \psi \in \Phi$ with domains $d(\phi) = x$ and $d(\psi) = y$ it holds that

$$\phi^{\rightarrow y} = \psi, \quad \psi^{\rightarrow x} = \phi. \tag{3}$$

Then ϕ and ψ represent in some sense the *same information*, in particular $\phi^{\downarrow x \wedge y} = \psi^{\downarrow x \wedge y}$ and $\phi^{\uparrow x \vee y} = \psi^{\uparrow x \vee y}$. We write $\phi \equiv \psi$ if (3) holds. This is clearly an *equivalence relation*. Moreover it is a *congruence* in the information algebra (Φ, D) in the following sense [9]: First $\phi_1 \equiv \phi_2$ and $\psi_1 \equiv \psi_2$ imply

$$\phi_1 \otimes \psi_1 \equiv \phi_2 \otimes \psi_2,$$

and secondly, also for any $z \in D$,

$$\phi_1^{\rightarrow z} \equiv \phi_2^{\rightarrow z}.$$

In fact in the last relation equality holds.

Let then $\Phi/\equiv$ denote the *equivalence classes* $[\phi]$ of this congruence in Φ. Then, in this quotient algebra the following two operations are well defined:

1. *Combination:* $[\phi] \otimes [\psi] = [\phi \otimes \psi]$.
2. *Focussing:* $[\phi]^{\Rightarrow x} = [\phi^{\rightarrow x}]$.

In the two-sorted algebra $(\Phi/\equiv, D)$ with the two operations just defined, the following properties hold:

Theorem 1. *Let $\Psi = \Phi/\equiv$ and denote generic elements of Ψ by $\psi, \eta, \ldots$ etc. Then*

1. *Semigroup: Ψ is associative and commutative under combination.*
2. *Support: If $\psi \in \Psi$, then there is a $x \in D$ such that $\psi = \psi^{\Rightarrow x}$.*

3. *Neutrality: There is a neutral element e such that $\psi \otimes e = \psi$ for all $\psi \in \Psi$ and $e^{\Rightarrow x} = e$.*
4. *Nullity: There is a null element z such that $\psi \otimes z = z$ for all $\psi \in \Psi$ and $z^{\Rightarrow x} = z$.*
5. *Transitivity: If $\psi \in \Psi$ and $x, y \in D$, then $(\psi^{\Rightarrow x})^{\Rightarrow y} = \psi^{\Rightarrow x \wedge y}$.*
6. *Combination: If $\psi, \eta \in \Psi$ and $x \in D$, then $(\psi^{\Rightarrow x} \otimes \eta)^{\Rightarrow x} = \psi^{\Rightarrow x} \otimes \eta^{\Rightarrow x}$.*
7. *Idempotency: If $\psi \in \Psi$ and $x \in D$, then $\psi \otimes \psi^{\Rightarrow x} = \psi$.*

Proof. (1) Associativity and commutativity of combination in Ψ is inherited from Φ.

(2) By Lemma 2 (3) we have $[\phi] = [\phi^{\rightarrow x}] = [\phi]^{\Rightarrow x}$ if $d(\phi) = x$.

(3) The equivalence class $[e_y]$ is the neutral element and $[e_y]^{\Rightarrow x} = [e_y^{\rightarrow x}] = [e_x]$ proves that the neutral element is stable under focussing.

(4) The equivalence class $[z_y]$ is the null element, and $[z_y]^{\Rightarrow x} = [z_y^{\rightarrow x}] = [z_x]$ proves the stability of the null element under focussing.

(5) By Lemma 2 (1) we have $(\phi^{\rightarrow x})^{\rightarrow y} = (\phi^{\rightarrow x \wedge y})^{\rightarrow y}$. Since $(\phi^{\rightarrow x \wedge y})^{\rightarrow y} \equiv \phi^{\rightarrow x \wedge y}$ we obtain $([\phi]^{\Rightarrow x})^{\Rightarrow y} = [(\phi^{\rightarrow x})^{\rightarrow y}] = [\phi^{\rightarrow x \wedge y}] = [\phi]^{\Rightarrow x \wedge y}$.

(6) Since $d(\phi^{\rightarrow x}) = x$, we obtain, using Lemma 2 (2)

$$([\phi]^{\Rightarrow x} \otimes [\psi])^{\Rightarrow x} = [(\phi^{\rightarrow x} \otimes \psi)^{\rightarrow x}] = [\phi^{\rightarrow x} \otimes \psi^{\rightarrow x}]$$
$$= [\phi]^{\Rightarrow x} \otimes [\psi]^{\Rightarrow x}.$$

(7) This follows from Lemma 2 (4). In fact, if $d(\phi) = y$, then $[\phi] \otimes [\phi]^{\Rightarrow x} = [\phi \otimes \phi^{\rightarrow x}] = [\phi^{\uparrow x \vee y}] = [\phi]$.   □

A two-sorted algebra (Ψ, D) with the operations of combination and focussing, satisfying the properties of Theorem 1, is called a *domain-free information algebra*. Theorem 1 says that any information algebra induces a domain-free information algebra. In order to distinguish the original algebra form the domain-free one, we call it a *labeled information algebra*.

In a domain-free information algebra (Ψ, D) a domain $x \in D$ is called a support of $\psi \in \Psi$, if $\psi = \psi^{\Rightarrow x}$. This means that no information is lost, when ϕ is focussed on domain x or, in other words, the whole information in ϕ is *carried* by domain x. According to the support property (2) in Theorem 1 any element of Ψ has a support. Here are a few properties of supports:

Lemma 3. *1. x is a support of $\psi^{\Rightarrow x}$.*
2. If x and y are supports of ψ, then $x \wedge y$ is a support of ψ.
3. If x is a support of ψ and $x \leq y$, then y is a support of ψ.
4. If x is a support of ψ, y a support of η, then $x \vee y$ is a support of $\psi \otimes \eta$.

Proof. (1) By transitivity (Theorem 1 (5)) we have $(\psi^{\Rightarrow x})^{\Rightarrow x} = \psi^{\Rightarrow x \wedge x} = \psi^{\Rightarrow x}$.

(2) Again, by (5) of Theorem 1, we obtain $\psi^{\Rightarrow x \wedge y} = (\psi^{\Rightarrow x})^{\Rightarrow y} = \psi^{\Rightarrow y} = \psi$.

(3) If $x \leq y$, then $x = x \wedge y$. So, once more by Theorem 1 (5), we conclude that $\psi^{\Rightarrow y} = (\psi^{\Rightarrow x})^{\Rightarrow y} = \psi^{\Rightarrow x \wedge y} = \psi^{\Rightarrow x} = \psi$.

(4) By (6) of Theorem 1, and (3) just proved, we see that $(\psi \otimes \eta)^{\Rightarrow x \vee y} = \psi^{\Rightarrow x \vee y} \otimes \eta^{\Rightarrow x \vee y} = \psi \otimes \eta$.   □

If (Ψ, D) is a domain-free information, then define Ψ^* to be the set of all pairs (ψ, x), where $\psi \in \Psi$ and x is a support of ψ. We define then the following operations:

1. *Labeling:* $d(\psi, x) = x$.
2. *Combination:* $(\psi, x) \otimes (\eta, y) = (\psi \otimes \eta, x \vee y)$.
3. *Projection:* $(\psi, x)^{\downarrow y} = (\psi^{\Rightarrow y}, y)$ for $y \leq x$.

It is easy to verify that the two-sorted algebra (Ψ^*, D) with these operations forms a labeled information algebra. It has been shown elsewhere [9] that its domain-free version $(\Psi^*/ \equiv, D)$ is then essentially identical to the original algebra (Ψ, D). Conversely, if $(\Psi, D) = (\Phi/ \equiv, D)$ for a labeled algebra (Φ, D), then (Ψ^*, D) is essentially identical to (Φ, D) (in fact, *isomorph,* [9]). Thus labeled and domain-free algebras are different versions of the same structure. We may switch at our convenience between the two forms.

3   Some Examples

3.1   Propositional Logic

At the beginning we have shown that relational algebra is an example of a (labeled) information algebra. In this section we want to discuss further examples, especially systems related to *logic*. In the view proposed here, logic offers a *language* to describe information which refers to models or structures. We illustrate this first with *propositional logic* as a prototype case.

The vocabulary of propositional logic is formed by a countable set of variables $P = \{p_1, p_2, \ldots\}$, the constants $\bot, \top$ and the the connectors $\neg, \wedge$. Formulae of the language are:

1. Each element of P, $\bot$ and $\top$ are formulae (*atomic formulae*).
2. If f and g are formulae, then so are $\neg f$, $f \wedge g$.
3. All formulae are generated from atomic formulae by finitely often applying rule 2.

A valuation is a mapping $v : P \rightarrow \{\mathbf{f}, \mathbf{t}\}$ which assigns each propositional variable a truth value $\mathbf{f}$ (false) or $\mathbf{t}$ (true). A valuation assigns a truth value $\hat{v}(f)$ to any formula f by the following inductively defined process:

1. If f is a propositional variable, then $\hat{v}(f) = v(f)$.
2. $\hat{v}(\bot) = \mathbf{f}$ and $\hat{v}(\top) = \mathbf{t}$.
3. $\hat{v}(\neg f) = \begin{cases} \mathbf{f} \text{ if } \hat{v}(f) = \mathbf{t}, \\ \mathbf{t} \text{ if } \hat{v}(f) = \mathbf{f}. \end{cases}$
4. $\hat{v}(f \wedge g) = \begin{cases} \mathbf{t} \text{ if } \hat{v}(f) = \hat{v}(g) = \mathbf{t}, \\ \mathbf{f} \text{ otherwise.} \end{cases}$

A valuation v, under which a formula f evaluates to true, i.e for which $\hat{v}(f) = \mathbf{t}$, is said to *satisfy* the formula, or to be a *model* of the formula, which is denoted as $v \models f$. Let $M(f)$ be the set of all models of a propositional formula f. Since a valuation can also be seen as a sequence $v_1, v_2, \ldots$ of elements of $\{\mathbf{f}, \mathbf{t}\}$, the set of models $M(f)$ can be considered to be a subset of $\{\mathbf{f}, \mathbf{t}\}^\infty$.

In a given problem or context one may assume that there is some true but unknown truth assignment in the real world. The elements of $\{\mathbf{f}, \mathbf{t}\}^\infty$ are then *possible worlds*. A formula f of propositional logic can then be seen as an information about the unknown real world in that it postulates that the real world must be among its models $M(f)$. We are now going to associate an information algebra of models to propositional formulae.

Let D be the lattice of *finite subsets* of $\omega = \{1, 2, \ldots\}$. For any valuation $v \in \{\mathbf{f}, \mathbf{t}\}^\infty$ and any finite subset $x \in D$, we define $v^{\downarrow x}$ to be the x-tuple $v(i), i \in x$. We define an x-equivalence between two valuations v and w by $v \equiv_x w$ if $v^{\downarrow x} = w^{\downarrow x}$. The equivalence classes of this x-equivalence are denoted by $[v]_x$. For any subset A of $\{\mathbf{f}, \mathbf{t}\}^\infty$ let

$$A^{\Rightarrow x} = \bigcup_{v \in A} [v]_x.$$

A subset $\phi \subseteq \{\mathbf{f}, \mathbf{t}\}^\infty$ is called *cylindric* over x, if $\phi = \phi^{\Rightarrow x}$. Let then Φ_x be the family of x-cylindric subsets of $\{\mathbf{f}, \mathbf{t}\}^\infty$ and

$$\Phi = \bigcup_{x \in D} \Phi_x.$$

We claim then that (Φ, D) with intersection as combination $\otimes$ and the focussing operation $\Rightarrow$ defined above is a (domain-free) information algebra.

Let f be a propositional formula and $var(f)$ the set of propositional variables occurring in f. Then its set of models $M(f)$ belongs to $\Phi_{var(f)}$. So, any propositional formula f determines an element $\phi = M(f)$ of the information algebra Φ, its set of models $M(f)$ is the information it describes. Note that $M(f \wedge g) = M(f) \cap M(g)$, conjunction corresponds to combination. Focussing is more complicated. If g is a formula such that $M(g) = M(f)^{\Rightarrow x}$, then g is obtained form f by variable forgetting or existential quantification, we refer to [11] for more details on this algebra. Two formulae f and g are logically equivalent, if $M(f) = M(g)$. Equivalent formulae describe the same information. Below, in Subsection 3.4, it will also be shown to be an instance of a more general logic system related to information algebras.

3.2   Quantifier Algebras

If Φ is a *Boolean algebra* with minimal element $\bot$, then an *existential quantifier* is a mapping $\exists : \Phi \to \Phi$ subject to the following conditions:

1. $\exists \bot = \bot$,
2. $\phi \wedge \exists \phi = \phi$,
3. $\exists(\phi \wedge \exists \psi) = \exists \phi \wedge \exists \psi$.

More generally, let D be a lattice of subsets of some set I. Assume that there is an existential quantifier $\exists(J)$ for every subset $J \in D$ on the Boolean algebra Φ, and that

1. $\exists(\emptyset)\phi = \phi,$
2. if $J, K \in D$, then $\exists(J \cup K)\phi = \exists(J)(\exists(K)\phi).$

Then (Φ, D) is termed a *quantifier algebra* over D. If we take the meet operation of the Boolean algebra for combination and define $\phi^{\Rightarrow I - J} = \exists(J)\phi$, then (Φ, D^c) is a *domain-free information algebra*. Here D^c is the lattice of subsets $I - J$ for all $J \in D$.

If, for all $i \in I$, $\exists(i)$ is a an existential quantifier, and $\exists(i)\exists(j) = \exists(j)\exists(i)$, if $i \neq j$, one can define $\exists(J) = \exists(i_1) \cdots \exists(i_m)$ if $J = \{1, \ldots, m\}$ and $\exists(\emptyset)(\phi) = \phi$. Then (Φ, D) is a quantifier algebra.

For instance, let Φ be the powerset of some cartesian product of a family of sets U_i for $i \in I$ and D a lattice of subsets of I. The mapping $\exists(J)$ is defined by

$$\exists(J)A = \{b \in \prod_{i \in I} U_i : \exists a \in A \text{ such that } b_i = a_i, \forall i \notin J\}.$$

It can be shown that this is an existential quantifier and (Φ, D) forms a quantifier algebra [17]. It is clear that the operation $\exists(\{i\})$ is similar to *variable elimination*. More generally, existential quantification is related to focussing, as we be seen in the next example (Section 3.3). Note also that it is sufficient for Φ to be a *semilattice* in order to define existential quantification and then a quantifier algebra (Φ, D).

3.3 Predicate Logic

Another information algebra is associated with predicate logic. The vocabulary of predicate logic consists of a countable set of variables $X_1, X_2, \ldots$ and a countable set of predicate symbols $P_1, P_2, \ldots$, the logical constants $\bot, \top$ and $\wedge, \neg, \exists$. Each predicate symbol has a definite rank $\rho = 0, 1, 2, \ldots$. We refer to a predicate with rank ρ as a ρ-place predicate. Formulae of predicate logic are built using the following rules:

1. $P_i X_{i_1} \ldots X_{i_\rho}$, where ρ is the rank of P_i, $\bot$ and $\top$ are (atomic) formulae.
2. If f is a formula, then $\neg f$ and $\exists X_i f$ are formulae.
3. If f and g are formulae, then $f \wedge g$ is a formula.

The predicate language $\mathcal{L}$ consists of all formulae which are obtained by applying a finite number of times these rules.

In order to define an interpretation of formulae of predicate logic, we choose a *relational structure* $\mathcal{R} = (U, R_1, R_2, \ldots)$ where U is a non-empty set, the *universe*, and R_i are relations among elements of U with the arity equal to the rank ρ of P_i, i.e. subsets of U^ρ. A *valuation* is a mapping $v : \omega \to U$, which assigns each variable X_i a value $v(i) \in U$ for $i \in \omega = \{1, 2, \ldots\}$. The set of valuations is U^ω, i.e. the set of sequences $v(1), v(2), \ldots$. We define for a valuation v and an index $i \in \omega$

$$v^{\Rightarrow i} = \{u \in U^\omega : u(j) = v(j) \text{ for } j \neq i\}.$$

Valuations are used to assign a *truth value* $\hat{v}(f)$ to each formula $f \in \mathcal{L}$. This truth assignment is defined inductively as follows:

1. $\hat{v}(\bot) = \mathbf{f}$, $\hat{v}(\top) = \mathbf{t}$
2. $\hat{v}(P_i X_{i_1} \ldots X_{i_\rho}) = \mathbf{t}$, if $(v(i_1), \ldots, v(i_\rho)) \in R_i$, and $\hat{v}(P_i X_{i_1} \ldots X_{i_\rho}) = \mathbf{f}$ otherwise.
3. $\hat{v}(\neg f) = \mathbf{f}$, if $\hat{v}(f) = \mathbf{t}$, and $\hat{v}(\neg f) = \mathbf{t}$, if $\hat{v}(f) = \mathbf{f}$.
4. $\hat{v}(\exists X_i f) = \mathbf{t}$, if there is a valuation $u \in v^{\Rightarrow i}$ such that $\hat{u}(f) = \mathbf{t}$, and $\hat{v}(\exists X_i f) = \mathbf{f}$ otherwise.
5. $\hat{v}(f \wedge g) = \mathbf{t}$, if $\hat{v}(f) = \hat{v}(g) = \mathbf{t}$, and $\hat{v}(f \wedge g) = \mathbf{f}$ otherwise.

A valuation v is called a *model* of a formula f in the structure $\mathcal{R}$, if $\hat{v}(f) = \mathbf{t}$. We write then $v \models_{\mathcal{R}} f$. Given a structure $\mathcal{R}$, we assign finally to each formula $f \in \mathcal{L}$ the set of its models,

$$\hat{r}_{\mathcal{R}}(f) = \{v \in U^\omega : v \models_{\mathcal{R}} f\}.$$

We consider this set as the information relative to the unknown values of the variables $X_1, X_2, \ldots$ expressed by the formula f. Let Φ be the family of all sets $\hat{r}_{\mathcal{R}}(f)$ for all $f \in \mathcal{L}$. If we define as usual $f \vee g = \neg(\neg f \wedge \neg g)$, then it is easy to see that

$$\hat{r}_{\mathcal{R}}(f \wedge g) = \hat{r}_{\mathcal{R}}(f) \cap \hat{r}_{\mathcal{R}}(g),$$
$$\hat{r}_{\mathcal{R}}(f \vee g) = \hat{r}_{\mathcal{R}}(f) \cup \hat{r}_{\mathcal{R}}(g),$$
$$\hat{r}_{\mathcal{R}}(\neg f) = (\hat{r}_{\mathcal{R}}(f))^c.$$

The family Φ is thus a *Boolean algebra*. Further we see that

$$\hat{r}_{\mathcal{R}}(\exists X_i f) = \bigcup_{v \in \hat{r}_{\mathcal{R}}(f)} v^{\Rightarrow i}.$$

We may denote the right hand side as $\exists(i)\hat{r}_{\mathcal{R}}(f)$. Clearly, for all $i \in \omega$ this is a quantifier on the Boolean algebra Φ in the sense of the previous example. Hence we may derive an existential quantifier $\exists(J)$ for any finite subset of ω. If D is the lattice of finite subsets of ω, then (Φ, D) is a *quantifier algebra* and so a domain-free information algebra. Combination is intersection, focussing is related to existential quantification, as explained in the previous example.

Two formulae f and g of predicate logic are said to be equivalent relative to the structure $\mathcal{R}$, written $f \equiv_{\mathcal{R}} g$, if $\hat{r}_{\mathcal{R}}(f) = \hat{r}_{\mathcal{R}}(g)$. So, equivalent formulae describe the same information. This induces an equivalence relation on $\mathcal{L}$. We may then introduce combination and existential quantification in $\mathcal{L}/\equiv_{\mathcal{R}}$ as follows: if $[f]_{\mathcal{R}}$ denotes the equivalence classes,

$$[f]_{\mathcal{R}} \otimes [g]_{\mathcal{R}} = [f \wedge g]_{\mathcal{R}},$$
$$\exists(J)[f]_{\mathcal{R}} = [\exists(J)f]_{\mathcal{R}},$$

where $\exists(J)f = \exists X_{i_1}(\ldots \exists X_{i_k})\ldots)$ if $J = \{i_1, \ldots, i_k\}$. Then $(\mathcal{L}/\equiv_{\mathcal{R}}, D)$ inherits the properties of an information algebra from (Φ, D). So, the information algebra of structures is reflected in a corresponding information algebra of formulae. These algebras are reducts of *cylindric algebras* [8] or *polyadic* or also *Halmos algebras* [6, 17] introduced for the algebraic study of predicate logic.

3.4   Contexts

Here we consider a general system, which captures the two previous logic examples as well as many other logic and related systems. It is also closely related to the work of [1] on information flow. A *context* is a triple $(\mathcal{L}, \mathcal{M}, \models)$, where $\mathcal{L}$ can be thought of as a set of sentences, a language, $\mathcal{M}$ a set of structures or models and $\models \subseteq \mathcal{L} \times \mathcal{M}$ is a binary relation between sentences and models. This corresponds to *classifications* in [1], where the terms *types* and *tokens* are used instead of sentences and models. Finally, in formal concept analysis the elements are considered as *attributes* and *objects* [3]. We write $m \models s$ instead of $(s, m) \in \models$. The idea is of course that models m *satisfy* sentences s, and thus give some semantics to the language $\mathcal{L}$. An example is provided by *propositional logic*, where $\mathcal{L}$ is a propositional language, the elements of $\mathcal{M}$ are valuations, and $m \models s$ means that m satisfies s or m is a model of s. Similarly, *predicate logic*, together with a structure to interpret the formulae, provides another example of a context.

In a context a set of sentences $X \subseteq \mathcal{L}$ determines a set of possible models, namely the set of models satisfying all sentences of X,

$$\hat{r}(X) = \{m \in \mathcal{M} : \forall s \in X, m \models s\}.$$

If we define also similarly for a subset A of $\mathcal{M}$,

$$\check{r}(A) = \{s \in \mathcal{L} : \forall m \in A, m \models s\},$$

then $\check{r}(A)$ is the set of all sentences whose models contain A.

The following dual pairs of properties of these operators are well known [3]:

$$X \subseteq \check{r}(\hat{r}(X)), \ \ A \subseteq \hat{r}(\check{r}(A)),$$
$$X \subseteq Y \Rightarrow \hat{r}(X) \supseteq \hat{r}(Y), \ \ A \subseteq B \Rightarrow \check{r}(A) \supseteq \check{r}(B),$$
$$\hat{r}(X) = \hat{r}(\check{r}(\hat{r}(X))), \ \ \check{r}(A) = \check{r}(\hat{r}(\check{r}(A))),$$
$$\hat{r}(\bigcup_{j \in J} X_j) = \bigcap_{j \in J} \hat{r}(X_j), \ \ \check{r}(\bigcup_{j \in J} A_j) = \bigcap_{j \in J} \check{r}(A_j).$$

We define further for $X \subseteq \mathcal{L}$ and $A \subseteq \mathcal{M}$,

$$C_{\models}(X) = \check{r}(\hat{r}(X)), \quad C^{\models}(A) = \hat{r}(\check{r}(A)).$$

It follows from the properties above that $C_{\models}$ and $C^{\models}$ are *closure* or *consequence operators*, i.e.

1. $X \subseteq C_{\models}(X)$,
2. $C_{\models}(C_{\models}(X)) = C_{\models}(X)$,
3. If $X \subseteq Y$, then $C_{\models}(X) \subseteq C_{\models}(Y)$,

and similarly for $C^{\models}$. Sets $X \subseteq \mathcal{L}$ and $A \subseteq \mathcal{M}$ are called $\models$-*closed* if $X = C_{\models}(X)$ or $A = C^{\models}(A)$ respectively. We obtain then

$$\hat{r}(X) = C^{\models}(\hat{r}(X)), \quad \check{r}(A) = C_{\models}(\check{r}(A)).$$

So, any set of sentences X determines a $\models$-closed set $\hat{r}(X)$ of models as information. In the same way, any set of models A determines a $\models$-closed set $\check{r}(A)$ of sentences, which could be called the *theory* of A. In particular, $\models$-closed sets of models and theories are in a one-to-one relation, i.e. if $A = \hat{r}(X)$ and $X = \check{r}(A)$, then both A and X must be $\models$-closed.

In the case of *propositional logic*, $C_\models(X)$ is the set of all *logical consequences* of X or the *theory* of X. In this case, as in predicate logic and in many other cases, all subsets of models are closed. This is not the case in the following example: Let $X_i, i \in \omega = \{1, 2, \ldots\}$, be a countable family of variables, $\mathcal{F}$ a field and let $\mathcal{L}$ be the family of *linear equations* of the form

$$\sum_{i \in I} a_i X_i = a_0, \ I \text{ a finite subset of } \omega \text{ and } a_0, a_i \in \mathcal{F}.$$

Further, let $\mathcal{M} = \mathcal{F}^\omega$. Define $m \models s$, for $m \in \mathcal{M}$ and $s \in \mathcal{L}$, if m satisfies the linear equations s, i.e. if

$$\sum_{i \in I} a_i m_i = a_0.$$

Then, for a subset X of $\mathcal{L}$ the closed set $\hat{r}(X)$ is the linear solution manifold of the system of equations X in $\mathcal{M}$. So here, $\models$-closed sets are linear manifolds, and $C^\models(A)$ is the linear manifold spanned by $A \subseteq \mathcal{M}$. If *linear inequalities* in an ordered field, instead of linear equations are considered, then, in the same way, the $\models$-closed sets are *convex polyhedra*.

Consider the set of all $\models$-closed subsets of $\mathcal{M}$. For two elements $\phi = \hat{r}(X)$ and $\psi = \hat{r}(Y)$ we define then a combination operation

$$\phi \otimes \psi = \hat{r}(X \cup Y) = \hat{r}(X) \cap \hat{r}(Y) = \phi \cap \psi. \tag{4}$$

In fact, this operation could be defined for arbitrary families of sets $X_i \in \mathcal{L}$. So, information is combined either by the union of the sentences which define the information or by intersection of their model sets.

If we want to extend this semigroup to an information algebra, we must add a *domain structure* and a corresponding *focussing operation*. Let D be a *lattice* and, for any $x \in D$, let $\equiv_x$ be an *equivalence relation* in $\mathcal{M}$ such that

$$x \leq y \Rightarrow \equiv_x \supseteq \equiv_y . \tag{5}$$

A triple $(\mathcal{M}, D, \equiv_{x \in D})$, where D is a lattice and $\equiv_x$ are equivalence relations in $\mathcal{M}$ satisfying the condition above, is called a *similarity model structure* in [23]. For any model $m \in \mathcal{M}$ and $x \in D$, define

$$m^{\Rightarrow x} = \{n \in \mathcal{M} : n \equiv_x m\}.$$

Further, for a subset A of $\mathcal{M}$ let

$$A^{\Rightarrow x} = \bigcup_{m \in A} m^{\Rightarrow x}. \tag{6}$$

A set of models A such that $A = A^{\Rightarrow x}$ is called *cylindric* over x or x-closed. We require now two additional conditions:

1. *Closure:* If A is $\models$-closed and $x \in D$, then $A^{\Rightarrow x}$ is $\models$-closed.
2. *Independence:* If for two models m, n it holds that $m \equiv_{x \wedge y} n$, then there is a model l such that $l \equiv_x m$ and $l \equiv_y n$.

The closure property implies that the family of $\models$-closed sets is closed under the *focussing* operation $\Rightarrow$. The independence property guarantees that indeed a (domain-free) *information algebra* can be associated with contexts, as we shall show below.

In propositional and predicate logic, as in many other cases, the language $\mathcal{L}$ is defined over a set of *variables* $X_1, X_2, \ldots$ with domains $U_1, U_2, \ldots$. Models are valuations $v(i) \in U_i$. A similarity model structure is then defined for instance for finite subsets x of variables by $v \equiv_x u$ if $v(i) = u(i)$ for all $i \in x$. It can be verified that this structure satisfies the closure and independence properties above. This corresponds essentially to a *multivariate domain*.

Let Φ be the set of all cylindric sets which are $\models$-closed. The following lemma collects two important properties of cylindric, $\models$-closed sets.

Lemma 4. *For $x, y \in D$ and $\phi, \psi \in \Phi$, the following holds:*

1. *If ϕ is x-closed and ψ is y-closed, then $\phi \otimes \psi$ is $x \vee y$-closed.*
2. *If ϕ is x-closed, then $\phi^{\Rightarrow y}$ is $x \wedge y$-closed.*

Proof. (1) We claim that if $x \leq y$, then ϕ is x-closed implies ϕ is y-closed. In fact, suppose $\phi = \phi^{\Rightarrow x} = \cup_{n \in \phi} n^{\Rightarrow x}$. Then

$$\phi^{\Rightarrow y} = \bigcup_{m \in \phi} m^{\Rightarrow y} = \bigcup_{n \in \phi} \bigcup_{m \in n^{\Rightarrow x}} m^{\Rightarrow y}$$
$$= \bigcup_{n \in \phi} n^{\Rightarrow x} = \phi^{\Rightarrow x} = \phi.$$

Therefore, if ϕ and ψ are x- and y-closed respectively, both are $x \vee y$-closed and so is $\phi \otimes \psi = \phi \cap \psi$.

(2) We claim that $(m^{\Rightarrow x})^{\Rightarrow y} = m^{\Rightarrow x \wedge y}$, which then implies property 2 immediately. In fact, if $n \equiv_{x \wedge y} m$, then, by the independence property above, there is an l such that $l \in m^{\Rightarrow x}$ and $n \in l^{\Rightarrow y}$. But this means that $n \in (m^{\Rightarrow x})^{\Rightarrow y}$. Conversely, if $n \in (m^{\Rightarrow x})^{\Rightarrow y}$, then there is a l such that $n \equiv_y l \equiv_x m$. By the monotonicity property (5) it follows that $n \equiv_{x \wedge y} l \equiv_{x \wedge y} m$, hence $n \in m^{\Rightarrow x \wedge y}$. $\square$

After this preparation it can be shown that (Φ, D) forms an information algebra.

Theorem 2. *The two-sorted algebra (Φ, D) with combination $\otimes$ and focussing $\Rightarrow$ defined above by (4) and (6) respectively, is a domain-free information algebra if the closure and independence properties are satisfied.*

Proof. We verify properties (1) to (7) of Theorem 1 above. The semigroup properties holds for intersection, hence for combination and $\mathcal{M}$ is the neutral element of combination, whereas the empty set is the null element. Transitivity follows from property 2 of Lemma 4 above, because $\phi^{\Rightarrow x}$ is x-closed. Combination is verified as follows:

$$(\phi^{\Rightarrow x} \otimes \psi)^{\Rightarrow x} = \bigcup_{m \in (\phi^{\Rightarrow x} \cap \psi)} m^{\Rightarrow x}$$

$$= \left(\bigcup_{m \in \phi^{\Rightarrow x}} m^{\Rightarrow x} \right) \cap \left(\bigcup_{m \in \psi} m^{\Rightarrow x} \right)$$

$$= \phi^{\Rightarrow x} \otimes \psi^{\Rightarrow x}.$$

The support axiom holds since the elements of Φ are cylindric, and the idempotency axioms is evident.   □

We have represented the information algebra associated with a context in terms of models. But we could also represent it in terms of theories. If $\phi = \hat{r}(X)$ and $\psi = \hat{r}(Y)$, then we could consider the associated theories $C_{\models}(X)$ and $C_{\models}(Y)$ and define combination by

$$C_{\models}(X) \otimes C_{\models}(Y) = \check{r}(\phi \otimes \psi)$$
$$= C_{\models}(X \cup Y)$$
$$= C_{\models}(C_{\models}(X) \cup C_{\models}(Y)).$$

Further focussing could be defined as follows:

$$C_{\models}(X)^{\Rightarrow x} = \check{r}(\hat{r}(X)^{\Rightarrow x}).$$

This gives then a domain-free information algebra of theories associated to the algebra of models. Predicate logic provides an example with the algebra of structures and the algebra of formulae.

For any $x \in D$ we define

$$\mathcal{M}_x = \{m^{\Rightarrow x} : m \in \mathcal{M}\}, \quad \mathcal{L}_x = \{s \in \mathcal{L} : \hat{r}(\{s\}) = (\hat{r}(\{s\}))^{\Rightarrow x}\}.$$

Furthermore, we define a relation $\models_x$ between $\mathcal{M}_x$ and $\mathcal{L}_x$ by $m^{\Rightarrow x} \models_x s$ if $m \models s$ for all $m \in m^{\Rightarrow x}$. Then $(\mathcal{L}_x, \mathcal{M}_x, \models_x)$ is a context. Note that cylindric sets $A^{\Rightarrow x}$ over x can, in a natural way, also be considered as a subset of $\mathcal{M}_x$, namely the set consisting of elements $m^{\Rightarrow x}$ for all $m \in A$. Further, by the closure property, if A is $\models$-closed, then $A^{\Rightarrow x}$ is $\models_x$-closed.

Consider two elements $x, y \in D$ such that $x \leq y$. Then it follows from (5) that $m^{\Rightarrow x} \supseteq m^{\Rightarrow y}$ and $\mathcal{L}_x \subseteq \mathcal{L}_y$. We define now a contravariant pair of mappings

$$g : \mathcal{M}_y \to \mathcal{M}_x$$
$$\mathcal{L}_y \leftarrow \mathcal{L}_x : f$$

by $g(m^{\Rightarrow y}) = m^{\Rightarrow x}$, and $f(s) = s$. It can be verified, that this pair of mappings satisfies the following condition

$$g(m^{\Rightarrow y}) \models_x s \Leftrightarrow m^{\Rightarrow y} \models_y f(s).$$

A contravariant pair of mappings between two contexts $(\mathcal{L}_x, \mathcal{M}_x, \models_x)$ and $(\mathcal{L}_y, \mathcal{M}_y, \models_y)$ satisfying this condition is termed a *context morphism*. It corresponds in essence to the *infomorphism* introduced in [1]. If we consider contexts $(\mathcal{L}_x, \mathcal{M}_x, \models_x)$ and $(\mathcal{L}_y, \mathcal{M}_y, \models_y)$ together with the context $(\mathcal{L}_{x \vee y}, \mathcal{M}_{x \vee y}, \models_{x \vee y})$, and add the context morphisms between the first two contexts and the third one, then we have what is called a *channel* in [1]. In fact, it allows to *transport information* (in the sense discussed in Section 2.4) from $(\mathcal{L}_x, \mathcal{M}_x, \models_x)$ to $(\mathcal{L}_y, \mathcal{M}_y, \models_y)$ and vice versa.

3.5   Lattice Induced Algebras

Let A be a *distributive, complete lattice* with supremum (join) and infimum (meet) denoted as usual by $\vee$ and $\wedge$. Let further r denote a finite set of variables $X_1, X_2, \ldots$ and U_i the domain of variable X_i. To a set $s \subseteq r$ of variables the cartesian product

$$U_s = \prod_{i \in r} U_i$$

is assigned as domain. The elements of U_s are tuples with domain s. We adopt the convention that the domain of the empty set of variable $U_\emptyset$ consists of a single tuple, denoted by $\diamond$. We use lower case, bold-face letters such as $\mathbf{x}, \mathbf{y}, \ldots$ to denote tuples. In order to emphasize the decomposition of a tuple $\mathbf{x}$ with domain s into components belonging to two disjoint subsets t and $s - t$ of s, we write $\mathbf{x} = (\mathbf{x}^{\downarrow t}, \mathbf{x}^{\downarrow s-t})$. A *valuation* ϕ with domain s is a mapping $\phi : U_s \to A$. The domain of a valuation ϕ is denoted by $d(\phi)$. The set of all valuations with domain s is denoted by Φ_s. Let then

$$\Phi = \bigcup_{s \subseteq r} \Phi_s.$$

Further let D be the lattice of subsets of r. We now use the lattice operations in A to define two operations in the pair (Φ, D):

1. *Combination:* $\otimes : \Phi \times \Phi \to \Phi$ defined for $\mathbf{x} \in U_{d(\phi) \cup d(\psi)}$ by

$$\phi \otimes \psi(\mathbf{x}) = \phi(\mathbf{x}^{\downarrow d(\phi)}) \wedge \psi(\mathbf{x}^{\downarrow d(\psi)}).$$

2. *Projection:* $\downarrow : \Phi \times D \to \Phi$ defined for all $\phi \in \Phi$ and $t \subseteq d(\phi)$ for $\mathbf{x} \in U_t$ by

$$\phi^{\downarrow t}(\mathbf{x}) = \bigvee_{\mathbf{z} \in U_{d(\phi)} : \mathbf{z}^{\downarrow t} = \mathbf{x}} \phi(\mathbf{z}).$$

It has been shown elsewhere that (Φ, D) with the two operations defined above is a (labeled) information algebra [13]. Examples for the lattice A include the Boolean lattice $\{0, 1\}$ (in which case valuations describe *constraints* or *subsets*), or the interval $[0, 1]$ with $\max, \min$ as lattice operations. This is used in fuzzy set theory. More general distributive lattices can be used to express qualitative membership of elements to *fuzzy sets*.

4   Order and Measure of Information

4.1   Partial Orders of Information Content

In an information algebra, the elements may be ordered by information content. The idea is that a piece of information is more informative than another one, if their combination yields the first. More precisely, let (Φ, D) be a *domain-free information algebra*. Then, for $\phi, \psi \in \Phi$ we define $\phi \geq \psi$ if $\phi \otimes \psi = \phi$, i.e. ϕ is *more informative* than ψ, if combining the latter with the first one gives nothing new. It can easily be verified that this relation is a *partial order* in Φ. Here are are a few elementary properties of this order which are proven in [9]:

1. $e \leq \phi$,
2. $\phi, \psi \leq \phi \otimes \psi$,
3. $\phi^{\Rightarrow x} \leq \phi$,
4. $\phi \leq \psi$ implies $\phi^{\Rightarrow x} \leq \psi^{\Rightarrow x}$,
5. $\phi \leq \psi$ implies $\phi \otimes \eta \leq \psi \otimes \eta$,
6. $\phi^{\Rightarrow x} \otimes \psi^{\Rightarrow x} \leq (\phi \otimes \psi)^{\Rightarrow x}$,
7. $x \leq y$ implies $\phi^{\Rightarrow x} \leq \phi^{\Rightarrow y}$.

In particular, it can also be verified that $\phi \otimes \psi = \sup\{\phi, \psi\}$. Therefore, Φ is also a *semilattice* and we write sometimes $\phi \otimes \psi = \phi \vee \psi$, if we want to stress order-theoretic issues.

This order reflects the absolute information content of the elements of Φ. It is also interesting to compare the information contents of the elements of Φ with respect to a determined question, i.e. a given domain $x \in D$. For this purpose we define $\phi \leq_x \psi$, if $\phi^{\Rightarrow x} \leq \psi^{\Rightarrow x}$. So, ϕ is less informative than ψ, *relative to a domain x*, if its part relating to x is less informative than the part of ψ relating to x. The relation $\leq_x$ is a *preorder* (reflexive and transitive, but not antisymmetric) on Φ. This is equivalent to a similar order defined on *labeled information algebras*, where again $\phi \geq \psi$ if $\phi \otimes \psi = \phi$. Then, for $x \in D$, we define $\phi \leq_x \psi$, if $\phi^{\rightarrow x} \leq \psi^{\rightarrow x}$.

In the case of *propositional logic*, a propositional formula f is more informative than a formula g, if $M(f) \subseteq M(g)$, since combination is intersection. This means that f is more informative than g, if, and only if, the latter is a *logical consequence* of the former, i.e. if $f \models g$. Similarly in predicate logic, a predicate formula f is more informative than g, relative to a structure $\mathcal{R}$, if $\hat{r}_{\mathcal{R}}(f) \subseteq \hat{r}_{\mathcal{R}}(g)$, i.e. again if g is a logical consequence of f, i.e. $f \models_{\mathcal{R}} g$. In a lattice-induced information algebra, a valuation v with domain x is more informative than another valuation u with the same domain, if $v(\mathbf{x}) \leq u(\mathbf{x})$ for all $\mathbf{x} \in U_x$. This is a kind of fuzzy subset relation, generalizing the ordinary subset relation.

These partial orders describe qualitative comparisons of information content between pieces of information. We may also try to measure quantitatively the content of an information. This is discussed below in Section 4.4.

4.2   Boolean Information Algebras

In the case of a relational algebra, for two relations R and S with the same domain x, R is more informative than S, if $R \subseteq S$. This makes sense in many cases: For instance if somebody is expected in Zurich on a flight from London, and information on possible flights is given by a list of flights from London to Zurich, then the smaller the list, the more information is obtained, the less uncertainty remains. If however we look for a flight we could take from London to Zurich, then obviously we feel to dispose of more information the longer the list of possible flights we obtain. So, information content of a relation seems to depend on the question we are interested in. This is related to the Boolean nature of the relational algebra. In order to elucidate this issue we introduce Boolean information algebra in this section.

Let (Φ, D) be a domain-free information algebra, such that in particular Φ is a semilattice relative to the partial order on information content. We assume now in addition that Φ is not only a semilattice but a *Boolean algebra*. This means that Φ has a bottom and a top element e and z and is a *distributive lattice*, where not only the supremum $\phi \vee \psi$ exists relative to the order, but also the infimum $\phi \wedge \psi$ and the distributive laws hold between these two operations. Further there is a complement ϕ^c for each element $\phi \in \Phi$ such that $\phi \wedge \phi^c = e$ and $\phi \vee \phi^c = z$. Then (Φ, D) is called a *Boolean information algebra*. For instance the information algebras associated with propositional and predicate logic are Boolean.

In a Boolean algebra there exists a well known *duality* which carries over to Boolean information algebras. If (Φ, D) is a Boolean information algebra, then we define the following *dual operations* of combination and focussing:

1. *Dual Combination:* $\phi \otimes_d \psi = (\phi^c \otimes \psi^c)^c$,
2. *Dual Focussing:* $\phi^{\Rightarrow_d x} = ((\phi^c)^{\Rightarrow x})^c$.

Note that by de Morgan's law $\phi \otimes_d \psi = \phi \wedge \psi$. Similar relations hold also in the labeled version of the Boolean information algebra. It can be verified that (Φ, D) with these dual operations is still a Boolean information algebra and the mapping $\phi \to \phi^c$ is an isomorphism between dual Boolean information algebras.

Now, in the dual algebra, the partial order $\leq_d$ is defined as usual. Then, clearly, $\phi \leq_d \psi$ if, and only if, $\phi \geq \psi$.

From a domain-free Boolean algebra we may derive in the usual way (see Section 2.5) the associated labeled information algebra. This algebra as a whole is no more a Boolean algebra. Only the elements associated with a support $x \in D$ form still Boolean algebras. More precisely, a *labeled* information algebra (Φ, D) is called *Boolean*, if the following two properties hold:

1. $\forall x \in D$, the semilattice Φ_x is Boolean.
2. $\forall x, y \in D$ and $\phi, \psi \in \Phi_x$ is holds that

$$(\phi \wedge \psi) \otimes e_y = ((\phi \otimes e_y) \wedge (\psi \otimes e_y)).$$

The labeled algebra derived from a domain-free Boolean algebra certainly satisfies these properties. So does for example relational algebra, seen as a labeled information algebra.

Although Φ itself is not a Boolean algebra, it is still possible to define a *dual* algebra, using duality within the Boolean algebras Φ_x. So dual combination is defined for $\phi, \psi \in \Phi$ as

$$\phi \otimes_d \psi = (\phi^c \otimes \psi^c)^c.$$

Similarly, dual marginalization is defined for $\phi \in \Phi$ and $x \leq d(\phi)$ as

$$\phi^{\downarrow_d x} = ((\phi^c)^{\downarrow x})^c.$$

Note that the dual neutral elements are the original null elements z_x. So, dual vacuous extension is defined as follows for $x \geq d(\phi)$,

$$\phi^{\uparrow_d x} = \phi \otimes_d z_x = (\phi^c \otimes e_x)^c = ((\phi^c)^{\uparrow x})^c.$$

This allows finally the introduction of a dual transport operation, for $\phi \in \Phi$, with $d(\phi) = y$,

$$\phi^{\to_d x} = (\phi^{\uparrow_d x \vee y})^{\downarrow_d x} = ((\phi^c)^{\uparrow x \vee y})^{\downarrow x})^c = ((\phi^c)^{\to x})^c.$$

This results in a dual labeled information algebra, which is isomorph to the original one by the mapping $\phi \mapsto \phi^c$. We warn that the dual partial order $\leq_d$ induced in this dual algebra is *not* the inverse of the order $\leq$ in the original algebra. However, the dual order accounts for the issue addressed at the beginning of the section: according to the question one is interested in, one should either consider the one or the other of the dual algebras.

4.3　Atomic Algebras

In many cases there are for every domain x *most informative information pieces* representing the finest possible answers to the question posed by the domain. In relational algebra for example the one-tuple relations over a domain x represent such *atomic* information. In this section we study more generally information algebras with atomic information pieces.

For this purpose it is more convenient to work with a *labeled* information algebra (Φ, D). Remember now that the algebra has *null elements*, i.e. for all $x \in D$ there is a (necessarily unique) element z_x such that $\phi^{\to x} \otimes z_x = z_x$ for all $\phi \in \Phi$. We further have $z_x^{\to y} = z_y$. These null elements represent *contradictory information*. In fact, if $\phi \otimes \psi = z_x$, the combination of this pieces of information with further pieces yields again the contradiction. In relational algebra these null elements are represented by the empty relations, in propositional and predicate logic the logical constant $\perp$ (falsity), which has no models, represents contradiction.

Now, an *atom* in a domain x is a *maximal element* different form z_x among the elements Φ_x with domain x:

Definition 1. *An element $\alpha \in \Phi_x$ is called an atom on x if*

1. $\alpha \neq z_x$,
2. *for all $\phi \in \Phi_x$, $\alpha \leq \phi$ implies either $\alpha = \phi$ or $\phi = z_x$.*

Here are a few elementary properties of atoms which are proven in [9]:

1. If α is an atom on x, and $y \leq x$, then $\alpha^{\downarrow y}$ is an atom on y.
2. If α is an atom on x, and $d(\phi) = x$, then either $\phi \leq \alpha$ or $\alpha \otimes \phi = z_x$.
3. If α and β are atoms on x, then either $\alpha = \beta$ or $\alpha \otimes \beta = z_x$.

Denote the set of all atoms in Φ_x by $At_x(\Phi)$ and the set of all atoms in Φ by $At(\Phi)$. Furthermore let for any $\phi \in \Phi$

$$At(\phi) = \{\alpha \in At(\Phi) : d(\alpha) = d(\phi), \phi \leq \alpha\}.$$

If $\alpha \in At(\phi)$ we say also that α is an atom of ϕ or contained in ϕ. This terminology will be justified below.

We are now especially interested in information algebras, where each element is *composed* by all the atoms it contains. The following definition gives a more precise meaning to this idea:

Definition 2. *A labeled information algebra* (Φ, D) *is called atomic, if for all* $\phi \in \Phi$, $\phi \neq z_{d(\phi)}$,

$$\phi = \wedge At(\phi),$$

i.e. each information is the infimum of the atoms it contains.

The labeled versions of the information algebras associated with propositional logic and predicate logic are atomic: In the case of propositional logic, the elements of Φ_x can be considered as subsets of the Boolean cube $\{\mathbf{t}, \mathbf{f}\}^{|x|}$ and the atoms are tuples $t : x \to \{\mathbf{t}, \mathbf{f}\}$. Therefore each element of Φ_x is simply the set of the tuples it contains. Similarly, in the case of predicate logic, the elements of Φ_x can be considered as subsets of the cartesian product $U^{|x|}$ and the atoms are tuples $t : x \to U$. In the case of information algebras related to contexts, atoms exist, if $m^{\Rightarrow x}$ is $\models$-closed for all $m \in \mathcal{M}$ and $x \in D$. Then, if a cylindric set A is $\models$-closed,

$$A = A^{\Rightarrow x} = \bigcup_{m \in A} m^{\Rightarrow x}.$$

Hence, again, each element of Φ_x is simply the set of the atoms it contains. The example of linear manifolds shows however that not every set of atoms forms necessarily an element of Φ.

These examples reflect in fact a more general situation: We claim that the set $At(\Phi)$ of all atoms of an atomic information algebra (Φ, D) forms itself an information algebra, very similar to a relational algebra. We note first, that atoms behave with respect to projection like ordinary tuples in relational algebra. In fact, the following lemma summarizes the basic properties of atoms:

Lemma 5. *If a labeled information algebra* (Φ, D) *is atomic, then its atoms* α, β *in* $At(\Phi)$ *satisfy the following properties:*

1. *If $x \leq d(\alpha)$, then $\alpha^{\downarrow x} \in At(\Phi)$ and $d(\alpha^{\downarrow x}) = x$.*
2. *If $x \leq y \leq d(\alpha)$, then $(\alpha^{\downarrow y})^{\downarrow x} = \alpha^{\downarrow x}$.*
3. *If $d(\alpha) = x$, then $\alpha^{\downarrow x} = \alpha$.*
4. *If $d(\alpha) = x$, $d(\beta) = y$ and $\alpha^{\downarrow x \wedge y} = \beta^{\downarrow x \wedge y}$, then there exists a $\gamma \in At(\Phi)$ with $d(\gamma) = x \vee y$ and $\gamma^{\downarrow x} = \alpha$, $\gamma^{\downarrow y} = \beta$.*
5. *If $d(\alpha) = x$ and $x \leq y$, then there exists a $\beta \in At(\Phi)$ such that $d(\beta) = y$ and $\beta^{\downarrow x} = \alpha$.*

Proof. Properties (1) to (3) follow from the axioms of an information algebra, since atoms are elements of the algebra.

Let $\gamma = \alpha \otimes \beta$. Then $\gamma^{\downarrow x} = \alpha$ by the combination and idempotency axioms, considering that $\alpha^{\downarrow x \wedge y} = \beta^{\downarrow x \wedge y}$. Similarly, $\gamma^{\downarrow y} = \beta$. Assume that $\gamma = z_{x \vee y}$. But then $\alpha = z_x$, which is excluded, since α is an atom. Hence we conclude that $\gamma \neq z_{x \vee y}$. Therefore, since (Φ, D) is atomic, $At(\gamma)$ is not empty. Let $\eta \in At(\gamma)$. Then it follows from $\gamma \leq \eta$, that $\alpha = \gamma^{\downarrow x} \leq \eta^{\downarrow x}$. But since α is an atom, either $\alpha = \eta^{\downarrow x}$ or $\eta^{\downarrow x} = z_x$. The latter case is excluded, since η is an atom. Similarly $\beta = \eta^{\downarrow y}$. So property (4) is satisfied by η.

Further, $At(\alpha^{\uparrow y})$ is not empty either. Thus, let $\beta \in At(\alpha^{\uparrow y})$. Then $d(\beta) = y$ and $\alpha^{\uparrow y} \leq \beta$. This implies $\alpha = (\alpha^{\uparrow y})^{\downarrow x} \leq \beta^{\downarrow x}$. Since α is an atom, it holds that either $\alpha = \beta^{\downarrow x}$ or $\beta^{\downarrow x} = z_x$. But the latter case is excluded because β is an atom. So property (5) is satisfied by β. $\square$

Of course, ordinary tuples in relational algebra satisfy these properties too. That is why we may consider atoms as *generalized tuples*. As with relational algebra, we define *generalized relations* over x to be subsets R of $At(\Phi)$ such that $d(\alpha) = x$ for all $\alpha \in R$. The *domain* of α is supposed to be attached to R. It is denoted by $d(R)$. For a generalized relation R and $x \leq d(R)$, the *projection* of R onto x is defined as

$$\pi_x(R) = \{\alpha^{\downarrow x} : \alpha \in R\}.$$

The *join* of a generalized relation R over x and a generalized relation S over y is defined as follows:

$$R \bowtie S = \{\alpha \in At(\Phi) : d(\alpha) = x \vee y, \alpha^{\downarrow x} \in R, \alpha^{\downarrow y} \in S\}.$$

It is easily possible that the set on the right hand side is empty. We attach the empty set with the domain $x \vee y$ and call it $Z_{x \vee y}$, the *empty* relation on $x \vee y$. We assign it the domain $d(Z_{x \vee y}) = x \vee y$. Finally, for $x \in D$, the *full* relation over x is

$$E_x = \{\alpha \in At(\Phi) : d(\phi) = x\} = At_x(\Phi).$$

This is the neutral element for the join operation between generalized relations on x. Note that $R \bowtie S = R \cap S$ if R and S are relations over the same domain.

Let $\mathcal{R}_\Phi$ be the set of all generalized relations of atoms of the information algebra (Φ, D). Then these generalized relations form a labeled information algebra.

Theorem 3. *The two-sorted algebra* $(\mathcal{R}_\Phi, D)$ *with the operations of projection and join defined above forms a labeled information algebra.*

This can easily be verified. In fact, it satisfies the same properties as an ordinary relational algebra summarized in Section 2.1. Furthermore, just as ordinary relational algebra, it forms the labeled version of a *Boolean information algebra*. It turns out that the atomic algebra (Φ, D) is part of its associated generalized relational algebra.

Assume the labeled information algebra (Φ, D) to be atomic.

Theorem 4. *The mapping* $At : \Phi \to \mathcal{R}_\Phi$ *defined by* $\phi \mapsto At(\phi)$ *is an embedding of* (Φ, D) *into* $(\mathcal{R}_\Phi, D)$.

Proof. We have first to show the following:

1. $At(\phi \otimes \psi) = At(\phi) \bowtie At(\psi)$,
2. $At(\phi^{\downarrow x}) = \pi_x(At(\phi))$.
3. $At(e_x) = E_x$.
4. $At(z_x) = Z_x$.

(1) Let $d(\phi) = x$ and $d(\psi) = y$. Consider an atom $\alpha \in At(\phi \otimes \psi)$. Then it follows that $\phi \leq \phi \otimes \psi \leq \alpha$, hence $\phi = \phi^{\downarrow x} \leq \alpha^{\downarrow x}$. Thus $\alpha^{\downarrow x} \in At(\phi)$. Similarly $\alpha^{\downarrow y} \in At(\psi)$, hence $\alpha \in At(\phi) \bowtie At(\psi)$. Conversely, assume $\alpha \in At(\phi) \bowtie At(\psi)$. Then $d(\alpha) = x \vee y$ and $\phi \leq \alpha^{\downarrow x} \leq \alpha$ and $\psi \leq \alpha^{\downarrow y} \leq \alpha$, hence $\phi \otimes \psi \leq \alpha$, which means that $\alpha \in At(\phi \otimes \psi)$. This proves (1).

(2) Let $\alpha \in At(\phi^{\downarrow x})$, such that $d(\alpha) = x$ and $\phi^{\downarrow x} \leq \alpha$. We have $\phi \leq \phi \otimes \alpha$. Suppose that $\phi \otimes \alpha = z_y$, if $d(\phi) = y \geq x$. Then

$$\alpha = \phi^{\downarrow x} \otimes \alpha = (\phi \otimes \alpha)^{\downarrow x} = z_x.$$

But this is excluded, because α is an atom. Therefore $\phi \otimes \alpha \neq z_y$. Since Φ is atomic there is a $\beta \in At(\phi \otimes \alpha)$ with $d(\beta) = y$ and $\phi \leq \beta$, hence $\beta \in At(\phi)$. But we have also $\alpha = (\phi \otimes \alpha)^{\downarrow x} \leq \beta^{\downarrow x}$. Since $\beta^{\downarrow x}$ is also an atom, we must have $\alpha = \beta^{\downarrow x}$ and therefore $\alpha \in \pi_x(At(\phi))$. Conversely, if $\beta \in \pi_x(At(\phi))$, then $\beta = \gamma^{\downarrow x}$ for some atom $\gamma \in At(\phi)$. But $\phi \leq \gamma$, hence $\phi^{\downarrow x} \leq \beta$ and therefore $\beta \in At(\phi^{\downarrow x})$. So (2) holds.

(3), (4) follow directly from the definition of At.

It remains to show that the mapping At is one-to-one. Assume $At(\phi) = At(\psi)$. Then $\phi = \wedge At(\phi) = \wedge At(\psi) = \psi$. □

The information algebras associated with propositional logic, for instance, coincide with their relational version. But this is not the case in general. The information algebras associated with predicate logic are proper subalgebras of the relational information algebra of relations over U.

In the case of an *atomic Boolean information algebra* (Φ, D) there is also a dual notion of the concept of an atom. A dual atom on x is a maximal element on x with respect to the dual order $\leq_d$. Let $At_d(\Phi)$ denote the set of dual atoms. If $\alpha \in At_d(\Phi)$ and $d(\alpha) = x$, then $\alpha \neq z_x^c = e_x$, hence $\alpha^c \neq z_x$. Further, assume $\alpha^c \leq \phi$ for a $\phi \in \Phi_x$. Then $\alpha \leq_d \phi^c$, hence either $\phi^c = \alpha$, i.e. $\phi = \alpha^c$, or $\phi^c = z_x^c$, i.e. $\phi = z_x$. Thus if α is a dual atom, then α^c is an atom.

Lemma 6. *Let (Φ, D) be an atomic Boolean information algebra. Then the following holds:*

1. $At(\phi) \cap At(\phi^c) = \emptyset$.
2. *If* $d(\phi) = x$, *then* $At(\phi) \cup At(\phi^c) = At_x(\Phi)$.

Proof. (1) Suppose there is an atom α on x such that $\phi \leq \alpha$ and $\phi^c \leq \alpha$. Taking the join of both sides we obtain $z_x \leq \alpha$, which is impossible.

(2) If $\alpha \in At_x(\Phi)$, and $d(\phi) = x$, then either $\phi \leq \alpha$ and $\alpha \in At(\phi)$ or $\alpha \vee \phi = z_x$. But in the latter case $\phi^c \leq \alpha$ and $\alpha \in At(\phi^c)$.   □

Note further that $\alpha \in At(\phi)$, i.e. $\phi \leq \alpha$, implies $\phi^c \leq_d \alpha^c$, hence $\alpha^c \in At_d(\phi^c)$ and vice versa. Hence the cardinality of the two sets $At(\phi)$ and $At_d(\phi^c)$ are the same. This implies also that the sets of all atoms on x, $At_x(\Phi) = At(e_x)$ and $At_{d,x}(\Phi) = At_d(z_x)$ have the same cardinality.

4.4 Measure of Information Content

Shannon, in his information theory, introduces a *quantitative measure* of information. He measures the information about a transmitted symbol by the reduction of uncertainty, when the transmitted symbol becomes known. So, in our context, we may say that Shannon considers a fixed *question*, namely what symbol out of a (finite) alphabet is selected for transmission. The uncertainty is measured by the *entropy* of the alphabet [19]. Once the symbol to be transmitted is known, the uncertainty is reduced to zero. Therefore the entropy measures the information gained by knowing the symbol.

This basic idea can be applied in our context too, if the labeled information algebra (Φ, D) is *atomic*. The first point to stress is that the information content of an element $\phi \in \Phi$ is measured relative to its domain $d(\phi)$, i.e. relative to the question it refers to. We assume further that for all domains $x \in D$ the total number $At(e_x)$ of atoms of the domain is *finite*. An atom of a domain x is the finest, i.e. the maximal information one may obtain about the domain. Assuming the number of atoms finite means that this information can be coded by a number of bits bounded by $\lceil \log_2 |At(e_x)| \rceil$, whereas an infinite number of atoms would mean that the information in an atom cannot be coded into a finite memory. Then the total uncertainty associated with a domain x can be measured by $\log |At(e_x)|$, the *Hartley measure*. This corresponds to the entropy of $At(e_x)$ under an assumed uniform probability distributions over the atoms. However, we shall avoid here probabilistic considerations, since there is no random experiment involved in our discussion. Usually the logarithm is taken to base 2, but any other base serves our purpose too, since it involves only a shift of scale in the measurement of uncertainty and information. Once information ϕ with $d(\phi) = x$ is given, the uncertainty concerning the possible atoms is reduced to $\log |At(\phi)| \leq \log |At(e_x)|$. So, the information content of ϕ relative to the domain (question) x can be defined as the reduction of uncertainty obtained by ϕ with respect to knowing nothing (i.e. knowing only the vacuous information e_x),

$$i(\phi) = \log |At(e_x)| - \log |At(\phi)| = -\log \frac{|At(\phi)|}{|At(e_x)|}. \tag{7}$$

We may consider

$$p(\phi) = \frac{|At(\phi)|}{|At(e_x)|}$$

as the probability of ϕ, or, more precisely, the probability that an atom in $At(\phi)$ is selected out of the atoms of $At(e_x)$, when all atoms have the same chance to be selected. Then we obtain $i(\phi) = -\log p(\phi)$, which corresponds to an often proposed definition of the information content of an "event" observed in a random experiment. But, once more, we prefer at this place to not refer to probabilistic considerations, since, in our view, information, in the first place at least, has nothing to do with probability, although in applications probability may play an important role (as for example in communication theory). We note that $i(e_x) = 0$, the vacuous information carries no information. Further we obtain $i(z_x) = \infty$ by (7) since $At(z_x) = \emptyset$. Note that z_x is in fact not really an information about a possible atom, since it contains no atom at all. We could as well convene that the information content of z_x is not defined.

More generally, any element $\phi \in \Phi$ contains possibly information about any other domain $y \neq d(\phi)$. In fact, it is natural to define the information content of ϕ relative to domain y by

$$i(\phi; y) = i(\phi^{\rightarrow y}).$$

Clearly and consistently we see that $i(\phi; x) = i(\phi)$, if $d(\phi) = x$. Further, if $[\phi]$ is the class of equivalent information elements (see Section 2.5), then all elements of the class have the same information content $i(\phi; y)$ with respect to any domain y. This means that we may assign a measure of information content by defining $i([\phi]; y) = i(\phi; y)$ also to the elements of the domain-free version of an information algebra.

The next theorem shows that our quantitative measure of information respects the qualitative orders of information introduced above (Section 4.1):

Theorem 5. *Let (Φ, D) be an atomic information algebra, with finite sets of atoms $At_x(\Phi)$. Then, for all $x \in D$ and $\phi, \psi \in \Phi$, the inequalities $\phi \leq \psi$, $[\phi] \leq [\psi]$ and $\phi \leq_x \psi$ imply $i(\phi; x) \leq i(\psi; x)$.*

Proof. Both $\phi \leq \psi$ and $[\phi] \leq [\psi]$ imply $\phi \leq_x \psi$. The latter implies $At(\phi^{\rightarrow x}) \supseteq At(\psi^{\rightarrow x})$, hence $|At(\phi^{\rightarrow x})| \geq |At(\psi^{\rightarrow x})|$, and therefore $i(\phi; x) = i(\phi^{\rightarrow x}) \leq i(\psi^{\rightarrow x}) = i(\psi; x)$. $\square$

From this theorem a number of simple results may be derived, which follow from the properties of the partial order: For all $x, y, z \in D$ and $\phi, \psi \in \Phi$:

1. $i(\phi; x), i(\psi; x) \leq i(\phi \otimes \psi; x)$,
2. $i(\phi^{\rightarrow y}; x) \leq i(\phi; x)$,
3. $\phi \leq \psi$ implies $i(\phi^{\rightarrow y}; x) \leq i(\psi^{\rightarrow y}; x)$,
4. $\phi_1 \leq \phi_2$ and $\psi_1 \leq \psi_2$ imply $i(\phi_1 \otimes \psi_1 : x) \leq i(\phi_2 \otimes \psi_2 : x)$,
5. $i(\phi^{\rightarrow y} \otimes \psi^{\rightarrow y}; x) \leq i((\phi \otimes \psi)^{\rightarrow y}; x)$,
6. $x \leq y$ implies $i(\phi^{\rightarrow x}; z) \leq i(\phi^{\rightarrow y}; z)$.

When the partial order in the domain-free algebra is considered, similar results are obtained. So far we measure the information content of an element ϕ on a domain x by the reduction of the uncertainty with respect to initial ignorance, i.e. vacuous information. More generally, we may also measure the relative information content of a piece of information relative to another, previous piece of information ψ. If information ψ is already given, the remaining uncertainty is $\log |At(\psi)|$. When a new information ϕ arrives, the total information is $\psi \otimes \phi$, and the remaining uncertainty $\log |At(\psi \otimes \phi)|$. If ϕ and ψ have the same domain $d(\phi) = d(\psi) = x$, then the relative information content of ϕ relative to ψ relating to the domain x can then be measured by

$$i(\phi|\psi) = \log |At(\psi)| - \log |At(\psi \otimes \phi)| = -\log \frac{|At(\psi \otimes \phi)|}{|At(\psi)|}.$$

Since by Theorem 4 in this case $At(\psi \otimes \phi) = At(\phi) \cap At(\psi)$, we may also consider $i(\phi|\psi) = -\log p(\phi|\psi)$, i.e. as the negative logarithm of the *conditional probability* of ϕ given ψ, with the usual assumption of uniform probability distribution over the atoms of $At(e_x)$. As before we may extend this definition of relative information measure to any domain y and information elements ψ and ϕ on any domains

$$i(\phi|\psi; x) = \log |At(\psi^{\to x})| - \log |At((\psi \otimes \phi)^{\to x})| = -\log \frac{|At((\psi \otimes \phi)^{\to x})|}{|At(\psi^{\to x})|}.$$

Note however that in general $i(\phi|\psi; x) \neq i(\phi^{\to x}|\psi^{\to x})$. Further, $\psi \leq \psi \otimes \phi$ implies also $\psi^{\to x} \leq (\psi \otimes \phi)^{\to x}$; therefore we conclude that $i(\phi|\psi; x) \geq 0$. It may be that $\psi \otimes \phi = z_y$, which means that ϕ and ψ are *incompatible* or *contradictory* pieces of information. Correspondingly we obtain in this case $i(\phi|\psi; x) = \infty$ for all domains x. This is simply the mathematical expression for the fact that such two pieces of information can not hold at the same time. Note further that $i(\phi; x) = i(\phi|e_x; x)$.

The following result shows that the measure of a combined information can be obtained as the sum of the measure of the first information and the relative information of the second relative to the first one. This is called the *chaining theorem*.

Theorem 6. *For all $x \in D$ and $\phi, \psi \in \Phi$ it holds that*

$$i(\phi \otimes \psi; x) = i(\phi; x) + i(\psi|\phi; x).$$

Proof. We have

$$\begin{aligned}
i(\phi \otimes \psi; x) &= \log |At(e_x)| - \log |At(\phi \otimes \psi)^{\to x}| \\
&= (\log |At(e_x)| - \log |At(\phi^{\to x})|) \\
&\quad + (\log |At(\phi^{\to x})| - \log |At(\phi \otimes \psi)^{\to x}|) \\
&= i(\phi; x) + i(\psi|\phi; x).
\end{aligned}$$

$\square$

This result can easily be generalized to the combination of $n \geq 2$ information elements.

Corollary 1. *For all $x \in D$ and $\phi_1, \ldots, \phi_n \in \Phi$,*

$$i(\phi_1 \otimes \cdots \otimes \phi_n; x) = i(\phi_1; x) + i(\phi_2|\phi_1; x) + \cdots + i(\phi_n|\phi_1 \otimes \cdots \otimes \phi_{n-1}; x).$$

Here are a few simple results about relative information.

Lemma 7. *For all $x \in D$ and $\phi, \psi \in \Phi$*

1. *If $\phi \geq \psi$, then $i(\phi|\psi; x) \leq i(\phi; x)$.*
2. *$\phi_1 \leq \phi_2$ implies $i(\phi_1|\psi; x) \leq i(\phi_2|\psi; x)$.*
3. *$\phi \leq \psi$ implies $i(\phi|\psi; x) = 0$.*

Proof. (1) We note that $At(e_x) \supseteq At(\psi^{\rightarrow x})$ and $At((\phi \otimes \psi)^{\rightarrow x}) = At(\phi^{\rightarrow x})$. Then

$$\begin{aligned}
i(\phi|\psi; x) &= \log|At(\psi^{\rightarrow x})| - \log|At((\phi \otimes \psi)^{\rightarrow x})| \\
&\leq \log|At(e_x)| - \log|At(\phi^{\rightarrow x})| \\
&= i(\phi; x).
\end{aligned}$$

(2) follows since $(\phi_1 \otimes \psi)^{\rightarrow x} \leq (\phi_2 \otimes \psi)^{\rightarrow x}$, hence $At((\phi_1 \otimes \psi)^{\rightarrow x}) \supseteq At((\phi_2 \otimes \psi)^{\rightarrow x})$.

(3) follows because in this case $\phi \otimes \psi = \psi$.   □

Suppose $i(\phi|\psi; x) = i(\phi; x)$ and $i(\psi|\phi; x) = i(\psi; x)$. In this case, knowing ψ contributes nothing to the information represented by ϕ and, similarly, knowing ϕ contributes nothing to the information represented by ψ. Therefore we say that ϕ and ψ are *independent* pieces of information relative to x and we write $\phi||\psi; x$. In this case, by the Chaining Theorem 6, the following *additivity property* holds,

$$i(\phi \otimes \psi; x) = i(\phi; x) + i(\psi; x).$$

Independent information simply adds up.

An important special case are *atomic Boolean information algebras*. We may define there a *dual information measure* for an element $\phi \in \Phi$ with $d(\phi) = x$,

$$i_d(\phi) = \log|At_d(z_x)| - \log|At_d(\phi)|,$$

since z_x is the dual neutral element, hence the dual vacuous information. This dual measure makes sense: In relational databases for instance, if a relation indicates all the flights by which a person can arrive, then the first measure applies, the smaller the relation the more information is available. When however the relations represents all the flights which a person may select for her trip, then the dual measure applies, the larger the relation, the more information is given. We have seen that $|At_d(\phi)| = |At(\phi^c)|$ (Section 4.3), hence

$$i_d(\phi) = \log|At(e_x)| - \log|At(\phi^c)| = i(\phi^c).$$

This duality relation between the dual information measures holds also for the general information measure relative to a domain

$$i_d(\phi; x) = i_d(\phi^{\to_d x}) = i((\phi^{\to_d x})^c) = i((\phi^c)^{\to x}) = i(\phi^c; x).$$

It holds also for relative information. By the dual chaining theorem,

$$\begin{aligned} i_d(\phi|\psi; x) &= i_d(\phi \otimes_d \psi; x) - i_d(\psi; x) \\ &= i((\phi \otimes_d \psi)^c; x) - i(\psi^c; x) \\ &= i(\phi^c \otimes \psi^c; x) - i(\psi^c; x) \\ &= i(\phi^c|\psi^c; x). \end{aligned}$$

We illustrate these concepts in the important case of information algebras with *multivariate domains*, where the results can be considerably sharpened. Assume thus that (Φ, D) is an atomic information algebra, where D is a lattice of subsets of some set r such that $x, y \in D$ implies $x - y \in D$. This is the case for instance for the lattice of finite subsets of an arbitrary set r. We introduce two further assumptions:

– Every atom $\alpha \in At_{x \cup y}(\Phi)$ on the domain $x \cup y$ has a decomposition of the form

$$\alpha = \alpha^{\downarrow x} \otimes \alpha^{\downarrow y}. \tag{8}$$

– For every $\eta \in \Phi_t$ we have

$$\eta^{\downarrow \emptyset} = \begin{cases} e_\emptyset, & \text{if } \eta \neq z_t, \\ z_\emptyset, & \text{else.} \end{cases} \tag{9}$$

Note that the combination of atoms is, in the general case, not necessarily an atom. This condition is satisfied, whenever Φ contains subsets of cartesian products, i.e. in the case of propositional and predicate logic, relational algebra and linear manifolds. As before we assume that the atom sets $At_x(\Phi)$ are finite for all $x \in D$. This is the case for propositional logic, predicate logic and relational algebra with finite domains and linear manifolds over product spaces of finite (or Galois) fields. In this case the following basic result holds:

Lemma 8. *Let (Φ, D) is an atomic information algebra, where D is a lattice of subsets of some set r, and such that conditions (8) and (9) hold. Then, if for $x, y \in D$ with $x \cap y = \emptyset$, and*

$$\phi = \phi^{\downarrow x} \otimes \phi^{\downarrow y}, \tag{10}$$

it holds that

$$i(\phi) = i(\phi^{\downarrow x}) + i(\phi^{\downarrow y}).$$

Proof. From (10) and Theorem 4 it follows that

$$At(\phi) = At(\phi^{\downarrow x}) \bowtie At(\phi^{\downarrow y})$$
$$= \{\alpha \in At_{x \cup y}(\Phi) : \alpha = \alpha^{\downarrow x} \otimes \alpha^{\downarrow y}, \alpha^{\downarrow x} \in At(\phi^{\downarrow x}), \alpha^{\downarrow y} \in At(\phi^{\downarrow y})\}.$$

From this we conclude that $|At(\phi)| = |At(\phi^{\downarrow x})| \cdot |At(\phi^{\downarrow y})|$. Similarly, it follows that $|At(e_{x \cup y})| = |At(e_x)| \cdot |At(e_y)|$. Therefore we obtain

$$i(\phi) = \log|At(e_{x \cup y})| - \log|At(\phi)|$$
$$= (\log|At(e_x)| + \log|At(e_y)|) - (\log|At(\phi^{\downarrow x})| + \log|At(\phi^{\downarrow y})|)$$
$$= (\log|At(e_x)| - \log|At(\phi^{\downarrow x})|) + (\log|At(e_y)| - \log|At(\phi^{\downarrow y})|)$$
$$= i(\phi^{\downarrow x}) + i(\phi^{\downarrow y}). \qquad \square$$

This result allows to introduce an absolute information measure into the domain-free version of the information algebra. In fact, if $\phi \equiv \psi$, then $\phi^{\uparrow x \cup y} = \psi^{\uparrow x \cup y}$, if $d(\phi) = x$ and $d(\psi) = y$. Let $z = x \cup y - x$. Then, since $\phi^{\uparrow x \cup y} = \phi \otimes e_z = (\phi^{\uparrow x \cup y})^{\downarrow x} \otimes (\phi^{\uparrow x \cup y})^{\downarrow z}$, by the previous Lemma 8 $i(\phi^{\uparrow x \cup y}) = i(\phi)$. Similarly we obtain $i(\psi^{\uparrow x \cup y}) = i(\psi)$, hence $i(\phi) \stackrel{'}{=} i(\psi)$. Define then the *absolute information measure* $i([\phi]) = i(\phi)$. The absolute information measure respects the partial information order in the domain-free information algebra $\Phi/\equiv$. Indeed, if $[\phi] \leq [\psi]$, then $[\phi] \otimes [\psi] = [\phi \otimes \psi] = [\psi]$. We have then $i([\phi]) = i(\phi^{\uparrow x \cup y}) \leq i(\phi \otimes \psi) = i(\psi^{\uparrow x \cup y}) = i([\psi])$, if $d(\phi) = x$ and $d(\psi) = y$.

In a similar way we define the relative information measure

$$i([\phi]|[\psi]) = \log|At(\psi^{\uparrow x \cup y})| - \log|At(\phi \otimes \psi)| = i([\phi] \otimes [\psi]) - i([\psi]) \geq 0.$$

Thus the absolute chaining theorem holds too,

$$i([\phi] \otimes [\psi]) = i([\phi]) + i([\psi]|[\phi]).$$

As before, we may call $[\phi]$ and $[\psi]$ *independent*, if the addition property

$$i([\phi] \otimes [\psi]) = i([\phi]) + i([\psi])$$

or, equivalently, $i([\phi]) = i([\phi]|[\psi])$ and $i([\psi]) = i([\psi]|[\phi])$ hold. This is the case if there are supports x and y of $[\phi]$ and $[\psi]$ respectively, such that $x \cap y = \emptyset$.

5   Conclusion

Information algebras represent a structure which captures essential features of any concept of "information". The presentation here focuses on its basic theory. There are many more aspects: One is computation: How are pieces of information combined and focussed on the domains of interest? This is the problem of *query processing* where *local computation methods* can be applied. A lot of work has been done with respect to this problem. In particular domains like query processing in relational algebra, solving linear equations, for instance in coding theory, consequence finding in logic, etc have been studied extensively. Whereas each of these domains has its particularities which can and must be explored, information algebra offers a common background on which *generic* methods can be developed [9, 12] . In this respect we refer to [18] which describes a generic software for local computation, permitting instantiations with any information (or rather valuation) algebra.

Another issue is *approximation* of 'infinite" information by "finite" information. This is modeled by *compact* information algebras [9]. More precisely, the question arises when an information algebra is *effective*, i.e. when the operations of combination and focussing can effectively be computed on a computer. Similar questions arise in domain theory, a theory which has close links to compact information algebras, see for instance [22]. In this same context, one can ask how information and its algebra is related to deduction. One is used that information may allow the inference of further information; in fact projection is a deduction procedure. Particularly in logic, i.e. in contexts, this became clear. In fact, it turns out that, similar to domain theory, information algebra may be equivalently be replaced by a system based on entailment, similar to *information systems* in the sense of Scott [22]. This means that *logic* in a wider sense is a general way to express and treat information. In a similar way, it can also be shown that any information algebra, in some precise sense, is part of *generalized relational algebra*. Thus, there are two general and complementary ways to see information in general, as (generalized) *relations* or as *logic*. These relations are discussed in [9].

Finally comes up the idea of "uncertain information", a term often used, but rarely, if ever, precisely defined. In the framework of information algebras, uncertain information can be represented by random variables taking values in an information algebra [9, 10]. This is closely related to *probabilistic argumentation systems* [5, 9]. Here, probability theory is combined with logic, in the way that the latter serves to prove hypotheses under certain assumptions and the former permits to compute the probability that those assumptions are valid. This brings the theory more into the realm of Shannon's entropy based information theory. Also it generalizes the concept of *random sets* [16], which are usually considered as random variables taking values in an algebra of closed sets of some topological space.

References

1. Barwise, J., Seligman, J.: Information Flow: The Logic of Distributed Systems. Cambridge Tracts in Theoretical Computer Science, vol. 44. Cambridge University Press, Cambridge (1997)
2. Beeri, C., Fagin, R., Maier, D., Yannakakis, M.: On the desirability of acyclic database schemes. Journal of the ACM 30(3), 479–513 (1983)
3. Davey, B.A., Priestley, H.A.: Introduction to Lattices and Order. Cambridge University Press, Cambridge (1990)
4. Grätzer, G.: General Lattice Theory. Academic Press, London (1978)

5. Haenni, R., Kohlas, J., Lehmann, N.: Probabilistic argumentation systems. In: Kohlas, J., Moral, S. (eds.) Handbook of Defeasible Reasoning and Uncertainty Management Systems. Algorithms for Uncertainty and Defeasible Reasoning, vol. 5, pp. 221–287. Kluwer, Dordrecht (2000)
6. Halmos, P.R.: Algebraic Logic. Chelsea, New York (1962)
7. Paul, R.: Halmos and Steven Givant. Logic as Algebra. Dolciani Mathematical Expositions No. 21, Mathematical Association of America (1998)
8. Henkin, L., Monk, J.D., Tarski, A.: Cylindric Algebras. North-Holland, Amsterdam (1971)
9. Kohlas, J.: Information Algebras: Generic Structures for Inference. Springer, Heidelberg (2003)
10. Kohlas, J.: Uncertain information: random variables in graded semilattices. Int. J. Approx. Reason. (2007) doi:10.1016/j.ijar.2006.12.005
11. Kohlas, J., Haenni, R., Moral, S.: Propositional information systems. Journal of Logic and Computation 9(5), 651–681 (1999)
12. Kohlas, J., Shenoy, P.P.: Computation in valuation algebras. In: Kohlas, J., Moral, S. (eds.) Handbook of Defeasible Reasoning and Uncertainty Management Systems. Algorithms for Uncertainty and Defeasible Reasoning, vol. 5, pp. 5–39. Kluwer, Dordrecht (2000)
13. Kohlas, J., Wilson, N.: Exact and approximate local computation in semiring induced valuation algebras. Technical Report 06-06, Department of Informatics, University of Fribourg (2006)
14. Lauritzen, S.L., Spiegelhalter, D.J.: Local computations with probabilities on graphical structures and their application to expert systems. J. of Royal Stat. Soc. 50(2), 157–224 (1988)
15. Maier, D.: The Theory of Relational Databases. Pitman, London (1983)
16. Molchanov, I.: Theory of Random Sets. Springer, London (2005)
17. Plotkin, B.I.: Universal Algebra, Algebraic Logic, and Databases. Mathematics and its applications, vol. 272. Kluwer Academic Publishers, Dordrecht (1994)
18. Pouly, M.: Nenok 1.1 user guide. Technical Report 06-02, Department of Informatics, University of Fribourg (2006)
19. Shannon, C.E.: A mathematical theory of communications. The Bell System Technical Journal 27, 379–432 (1948)
20. Shenoy, P.P., Shafer, G.: Axioms for probability and belief-function proagation. In: Shachter, R.D., Levitt, T.S., Kanal, L.N., Lemmer, J.F. (eds.) Uncertainty in Artificial Intelligence 4. Machine intelligence and pattern recognition, vol. 9, pp. 169–198. Elsevier, Amsterdam (1990)
21. Shenoy, P.P., Shafer, G.: Axioms for probability and belief function propagation. In: Lemmer, J.F., Shachter, R.D., Levitt, T.S., Kanal, L.N. (eds.) Uncertainty in Artif. Intell. 4, pp. 169–198. North-Holland, Amsterdam (1990)
22. Stoltenberg-Hansen, V., Lindstroem, I., Griftor, E.: Mathematical Theory of Domains. Cambridge University Press, Cambridge (1994)
23. Wilson, N., Mengin, J.: Logical deduction using the local computation framework. In: Hunter, A., Parsons, S. (eds.) ECSQARU 1999. LNCS, vol. 1638, pp. 386–396. Springer, Heidelberg (1999)

Uncertain Information*

Jürg Kohlas and Christian Eichenberger

University of Fribourg
Department of Informatics DIUF
Bd de Pérolles 90
1700 Fribourg, Switzerland
juerg.kohlas@unifr.ch, christianmarkus.eichenberger@unifr.ch
http://diuf.unifr.ch/tcs/

1 Introduction

In the previous chapter Information Algebra, an algebraic structure capturing
the idea that pieces of information refer to precise questions and that they can
be combined and focussed on other questions is presented and discussed. A prototype of such *information algebras* is relational algebra. But also various kind
of logic systems induce information algebras. In this chapter, this framework
will be used to study *uncertain information*. It is often the case that a piece
of information is known to be valid under certain assumptions, but it is not
altogether sure that these assumptions really hold. Varying the assumptions
leads to different information. Given such an *uncertain* body of information,
assumption-based reasoning permits to deduce certain conclusions or to prove
certain hypotheses under some assumptions. This kind of *assumption-based inference* can be carried further if the varying likelihood of different assumptions
is described by a probability measure on the assumptions. Then, it is possible
to measure the degree of support of a hypothesis by the probability that the
assumptions supporting the hypothesis hold. A prototype system of such a *probabilistic argumentation system* based on propositional logic is described in [3].
Another example will be described in Section 2 of this chapter. It is shown that
this way to model uncertain information leads to a theory which generalizes the
well-known Dempster-Shafer theory [6, 19].

Such probabilistic assumption-based information can be represented in a general way by random variables taking values in an information algebra, i.e. mappings from a probability space, whose elements represent uncertain assumptions,
into an information algebra. In order to avoid measure-theoretic complications,
we shall assume *finite* sample spaces, that is finite sets of possible assumptions.
The random variables are then called simple. More general random variables
are discussed in [6, 8]. In Section 3, simple random variables are defined. It is
shown that they form themselves an information algebra. This emphasizes the
fact that assumption-based information is information, although uncertain information. Next, in Section 4, the probability distributions associated with random

* Research supported by grant No. 200020–109510 of the Swiss National Foundation
for Research.

G. Sommaruga (Ed.): Formal Theories of Information, LNCS 5363, pp. 128–160, 2009.

variables are introduced. They arise from the probabilities of the assumptions supporting hypotheses. This defines *support functions* mapping the elements of the underlying information algebra to degrees of support, that is numbers in the interval $[0, 1]$. Related to support functions are possibility functions, which measure the probability that a hypothesis cannot be rejected, thus remains possible, although not asserted. These functions representing distribution functions of the random variables can, in the case of simple random variables, also be defined in terms of *basic probability assignments* (bpa). This relates the present theory of random variables in information algebras to the Dempster-Shafer theory [19]. It turns out that an algebra of bpas can be defined, which satisfies all axioms of an information algebra, except the idempotency axiom. Finally, it is discussed how the algebraic properties of uncertain information is linked to *Bayesian theorems*.

Uncertain assumption-based information being information, that is part of an information algebra of random variables, implies that an *order* between its elements can be defined. This order is induced by the algebra and reflects a comparison of random variables with respect to information content. This is a general issue in information algebras (see the previous chapter Information Algebra). But, in the case of assumption-based information, the information is not only related to the elements of the underlying information algebra, the range of the mapping, but also to the assumptions. This is explained in Section 5. Based on this observation, the algebra of random variables can be extended or generalized admitting also varying spaces of assumptions, although only in the case where the underlying information algebra is *Boolean*. This then also permits to introduce a *measure of information content* already proposed along general lines in the chapter Information Algebra, first neglecting the probability measure on assumptions. However, the particular semantics of assumption-based information needs to be taken into account. It turns out that this measure respects the order of information among random variables. The probability measure on the assumptions is additional information and can also be incorporated into the information measure. This quantitative measure of information content is defined as the *reduction of uncertainty* by the random variable with respect to full ignorance. Uncertainty is measured by Shannon's concept of *entropy*, adapted to random variables with values in an information algebra. This finally establishes the link of the theory of information algebra with Shannon's information theory. In particular, lossy channels and decoding could be described in terms of uncertain information. But this is not developed in detail in this chapter.

2   Probabilistic Assumption-Based Inference

2.1   Functional Models

In this section, we examine, by means of an introductory example, a particular way of how uncertain information can arise. This should serve as a motivation and illustration for the general, abstract model presented in the subsequent sections. The example is drawn from statistics. Based on a functional model, which describes how data is generated in some experiment or process, an assumption-based inference approach is presented. This represents the logical part of the inference. On top of this inference, a probability analysis allows to measure the likelihood of the possible deductions. The inference can be captured in an object called a *hint*, which represents the information drawn from the experiment.

Functional models describe the process by which a data x is generated from a parameter θ and some random element ω. The set of possible values, i.e. the domain of the data x, is denoted by X, whereas the domain of the parameter θ is denoted by Θ, and the domain of the random element ω is denoted by Ω. Unless otherwise stated, we assume that the sets X, Θ and Ω are finite. This is in order to simplify the mathematics, but is not a necessary assumption. The data generation process is specified by a function

$$f: \; \Theta \times \Omega \to X, \tag{1}$$

which relates the data with the perturbation and the parameter. In other words, if $\theta \in \Theta$ is the correct value of the parameter and the random element $\omega \in \Omega$ occurs, then the data x is uniquely determined by the function f according to the equation

$$x = f(\theta, \omega). \tag{2}$$

We assume that the function f is known, and we also assume that the probability distribution of the random element ω is known. This probability distribution is denoted by $p(\omega)$ and the corresponding probability measure on Ω is denoted by P. Note that the probability measure P does not depend on θ. The function f together with the probability measure P constitute a *functional model* for a statistical experiment.

In a functional model, if we assume a parameter θ, then, from the probabilities $p(\omega)$, we can compute the probabilities for the data x, namely

$$p_\theta(x) = \sum_{\omega : x = f(\theta, \omega)} p(\omega).$$

This shows that a functional model induces a parametric family of probability distributions on the sample space X, an object which is usually assumed a priori in modelling statistical experiments. These probability measures are the *statistical specification* associated with the functional model. As shown in [17], we emphasize that *different* functional models may induce the *same* statistical specification $p_\theta(x)$. This means that functional models contain *more* information than the family of distributions $p_\theta(x)$.

We illustrate the idea of functional models by two examples.

Example 1 (Sensor). Consider a sensor which supervises a dangerous event e. If this event occurs, it sends an alarm signal a. However, there is the possibility that the sensor fails, a possibility denoted by ω. To complete the notation, we denote the non-occurrence of the event e by $\neg e$, the non-occurrence of the alarm signal by $\neg a$ and, finally, the case of an intact sensor by $\neg \omega$. Now we may assume that, if the event e occurs *and* the sensor is intact, then the alarm sounds. On the other hand, if the event e has not occurred and the sensor is intact, then no alarm occurs. Finally, if the sensor fails, no alarm is possible. If we consider the observation space $X = \{a, \neg a\}$, the parameter space $\Theta = \{e, \neg e\}$ and the disturbance space $\Omega = \{\omega, \neg\omega\}$, then these assumptions can be captured in the following functional model:

$$f(e, \neg\omega) = a, \quad f(\neg e, \neg\omega) = \neg a, \quad f(e, \omega) = f(\neg e, \omega) = \neg a.$$

If we assume a probability p for the failure of the sensor, $p(\omega) = p$, then the functional model is complete.

Example 2 (Poisson Process). Suppose that traffic is observed at a given point over time. Cars are passing at an average number of $\lambda \cdot \Delta t$ during a time interval Δt. This corresponds to a *Poisson process* where the time interval between two subsequent cars is distributed exponentially with parameter λ. In order to get information about the unknown parameter λ, we may fix a time interval of some given length, say Δt, and then observe the number of cars passing in this time interval. Here, we have already two ingredients of a functional model describing this process, namely the parameter and the observation. The random elements ω are responsible for the random time intervals between passing cars. In fact, if ω is a random variable distributed according to the *exponential distribution* with unit parameter, given by the density function

$$e^{-t}, \tag{3}$$

then $\mu = \omega/\lambda$ is a random variable with exponential distribution with parameter λ, which means a mean time of $1/\lambda$ between two subsequent cars. So the count of cars passing during the time interval Δt is given by the largest value i such that

$$\sum_{k=1}^{i} \mu_i \leq \Delta t. \tag{4}$$

Here, the μ_i are stochastically independent random variables, each with exponential distribution with parameter λ. Thus, we may define the following *functional model* for the counting x of passing cars in a Poisson process:

$$x = i, \text{ if } \frac{1}{\lambda} \sum_{k=1}^{i} \omega_k \leq \Delta t < \frac{1}{\lambda} \sum_{k=1}^{i+1} \omega_k, \text{ for } i = 0, 1, \dots. \tag{5}$$

Here, the ω_k are assumed to be stochastically independent and exponentially distributed with parameter 1. This model explains how the counts in a time interval of length Δt are generated. In fact, in this model, the random element is given by an infinite sequence of random variables $\omega_1, \omega_2, \dots$.

2.2   Assumption-Based Reasoning

Consider an experiment represented by a functional model $x = f(\theta, \omega)$ with given probabilities $p(\omega)$ for the random elements. Suppose that the outcome of the experiment is observed to be x. Given this data x and the experiment, what can be inferred about the value of the unknown parameter θ? To answer this question, we use the principles of assumption-based reasoning. The basic idea of assumption-based reasoning is to assume that a random element ω generated the data and then determine the consequences of this assumption on the parameter. The consequences about θ are then evaluated according to the probabilities $p(\omega)$ of the assumptions ω in Ω.

Some random elements ω in Ω may become impossible after x has been observed. In fact, if, for an $\omega \in \Omega$, there is no $\theta \in \Theta$ such that $x = f(\theta, \omega)$, then this ω is clearly impossible: It cannot have generated the actual observation x. Therefore, the observation x induces an event in Ω, namely the event

$$v_x = \{\omega \in \Omega : \text{there is a } \theta \in \Theta \text{ such that } x = f(\theta, \omega)\}. \tag{6}$$

Since it is known that v_x has occurred, we need to condition the probability measure P on the event v_x, which leads to the revised probabilities

$$p'(\omega) = \frac{p(\omega)}{P(v_x)}$$

for all $\omega \in v_x$. These probabilities define a probability measure P' on v_x given by the equation

$$P'(A) = \sum_{\omega \in A} p'(\omega)$$

for all subsets $A \subseteq v_x$. It is still unknown which random element ω in v_x generated the observation x, but $p'(\omega)$ is the probability that this element is ω. Nevertheless, let us assume for the time being that ω caused the observation x. Then, according to the function f relating the parameter and the random element with the data, the possible values for the parameter θ can be logically restricted to the set

$$T_x(\omega) = \{\theta \in \Theta : x = f(\theta, \omega)\}. \tag{7}$$

Note that, in general, $T_x(\omega)$ may contain several elements, but it could also be a one-element subset of Θ. It is also possible that $T_x(\omega) = \Theta$, in which case the observation x does not carry any information about θ, assuming that ω caused the observations x. Therefore, in general, even if the chance element generating the observation x were known, it would still not be possible to identify the value of the parameter unambiguously. This analysis shows that an observation x in a functional model generates the structure

$$\mathcal{H}_x = (v_x, P', T_x, \Theta), \tag{8}$$

which we call a *hint*.

This approach to inference can be illustrated by the two examples introduced at the end of Section 2.1.

Example 3 (Sensor). Assume that, in the sensor described in Example 1, an alarm a sounds. Then, we may use the functional model specified there to judge how likely it is that the event e occurred. First, it is clear that the sensor must be intact since, otherwise, there would be no alarm. Hence, we have $v_a = \{\neg\omega\}$. But then, $T_a(\neg\omega) = \{e\}$, hence the event e has occurred for sure. So this is a very reliable inference.

If, on the other hand, no alarm sounds, how sure can we be that the event e has not occurred? We remark that the observation $\neg a$ is compatible both with ω (sensor failed) as well as with $\neg\omega$ (sensor intact). Hence, we have $v_{\neg a} = \Omega = \{\omega, \neg\omega\}$. If we assume that the system is intact, then we conclude that e has not occurred, i.e. $T_{\neg a}(\neg\omega) = \{\neg e\}$. If we assume, on the other hand, that the sensor failed, then both possibilities, e and $\neg e$, remain possible, whence $T_{\neg a}(\omega) = \{e, \neg e\}$. So, in this case, we do not arrive at a unique and clear conclusion. We remark, however, that we may assume that a reliable sensor was with high probability functioning correctly, so we have some guarantee that the event e did not occur. This informal consideration will be made more precise below.

Example 4 (Inference for a Poisson Process). We refer to the functional model (5) defined in Example 2 for the counting of passing cars. Although in this case neither the parameter space nor the space of disturbances ω is finite, the inference can nevertheless be carried out in the way described in this section. If we observe a given value x, then, according to (5), we must have

$$\sum_{k=1}^{x} \omega_k \le \Delta t \cdot \lambda < \sum_{k=1}^{x+1} \omega_k.$$

First, we note that the observation x excludes no sequence $\omega_1, \omega_2, \ldots$, hence no conditioning of the underlying probabilities is needed. If we assume a given sequence $\omega = \omega_1, \omega_2, \ldots$ we then conclude that the unknown parameter must be in the interval

$$T_x(\omega) = [\sum_{k=1}^{x} \omega_k, \sum_{k=1}^{x+1} \omega_k). \tag{9}$$

Below, we shall see that these findings can be used to quantify the support for certain hypotheses about the unknown parameter.

2.3   Hints

The hint defined in (8) is an instance of a more general concept of a hint. In general, if Θ denotes the set of possible answers to a question of interest, then a hint on Θ is a quadruple of the form

$$\mathcal{H} = (\Omega, P, \Gamma, \Theta),$$

where Ω is a set of assumptions, P is a probability measure on Ω reflecting the probability of the different assumptions and Γ is a mapping from the assumptions to the power set of Θ,

$$\Gamma : \; \Omega \longrightarrow 2^{\Theta}.$$

For an assumption $\omega \in \Omega$, the subset $\Gamma(\omega)$ is the smallest subset of Θ that is known for sure to contain the correct answer to the question of interest. In other words, if the assumption ω is correct, then the answer is certainly within $\Gamma(\omega)$. The theory of hints is presented in detail in [12]. Intuitively, a hint represents a piece of information regarding the correct value in Θ. It can be used to evaluate the validity of certain hypotheses regarding Θ. A hypothesis is a subset H of Θ and it is true if, and only if, it contains the correct answer to the question of interest.

The most important concept for the evaluation of the hypothesis H is the *degree of support* of H, which is defined by

$$sp(H) = P(\{\omega \in \Omega : \Gamma(\omega) \subseteq H\}). \tag{10}$$

If the assumption ω is correct and $\Gamma(\omega) \subseteq H$, then H is necessarily true because the correct answer is in $\Gamma(\omega)$ by definition of the mapping Γ. The degree of support of H is the probability of the assumptions that are capable of proving H. Such assumptions are called *arguments* for the validity of H and $sp(H)$ measures the strength of these arguments. The arguments represent the logical evaluation of H, whereas $sp(H)$ represents the quantitative evaluation of H.

Similarly, the *degree of plausibility* of H, which is defined as

$$pl(H) = P(\{\omega \in \Omega : \Gamma(\omega) \cap H \neq \emptyset\}), \tag{11}$$

measures the level of compatibility between the hypothesis H and the assumptions. Obviously, degrees of support and plausibility for hypotheses lead to the functions

$$sp : \; 2^{\Theta} \longrightarrow [0,1], \quad pl : \; 2^{\Theta} \longrightarrow [0,1],$$

which are called the support and plausibility functions associated with the hint. It can easily be shown that

$$pl(H) = 1 - sp(H^c) \tag{12}$$

and sp is a belief function in the sense of the Dempster-Shafer theory of evidence [19].

We are going to reconsider the examples above for an illustration.

Example 5 (Sensor). Let p be the probability that the sensor fails, that is the probability of the event ω. If there is an alarm, then $\neg\omega$ must hold, as we have seen above (see Example 1), hence $p'(\neg\omega) = 1$. The event e has occurred for sure since $T(\neg\omega) = \{e\}$. Hence, we conclude that $sp(\{e\}) = pl(\{e\}) = 1$. On the other hand, if there is no alarm, then the assumption that the sensor is intact supports the hypothesis that the event e has not occurred. Hence, we have $sp(\{\neg e\}) = 1 - p$. There is no support for the hypothesis that event e occurred, hence $sp(\{e\}) = 0$. Both assumptions, that the sensor is intact, as well as that it failed, are compatible with the hypothesis that the event e has not occurred. This implies that $pl(\{\neg e\}) = 1$. However, only the assumption that the sensor failed is compatible with the hypothesis of an event e; therefore, $pl(\{e\}) = p$. So, if the sensor is reliable, i.e. if the probability p of failure is small, then, by (12), we have a lot of support for the hypothesis $\neg e$ that no dangerous event e occurred, whereas the alternative e has only small plausibility.

Example 6 (Inference for a Poisson Process). In Example 4, we determined that, for a given observation of x cars passing and a sequence ω_k of disturbances, the unknown parameter must be in the interval $T_x(\omega)$ as given in (9). This now allows to determine the degree of support for intervals by

$$sp_x(u \leq \lambda \leq v) = P\left(u \leq \sum_{k=1}^{x} \omega_k < \sum_{k=1}^{x+1} \omega_k \leq v\right).$$

It is convenient to introduce the distribution function

$$Q_x(u, v) = P\left(\sum_{k=1}^{x} \omega_k \leq u \leq v \leq \sum_{k=1}^{x+1} \omega_k\right)$$

$$= \frac{1}{x!} u^x e^{-v} \text{ for } 0 \leq u \leq v.$$

This distribution function has the density function

$$q_x(u, v) = \frac{\partial^2}{\partial u \partial v} Q_x(u, v) = -Q_{x-1}(u, v) \text{ for } x \geq 1.$$

For $x = 0$, we have $Q_0(u, v) = e^{-v}$ with a density $q_0(u, v) = e^{-v}$. Further, we define

$$F_x(u) = P\left(\sum_{k=1}^{x} \omega_k \leq u\right),$$

$$G_x(v) = P\left(v \leq \sum_{k=1}^{x} \omega_k\right).$$

Both these functions have densities too, namely

$$f_x(u) = \frac{\partial}{\partial u} Q_x(u, v)|_{v=u} = q_x(u, u),$$

$$g_x(v) = \frac{\partial}{\partial v} Q_x(u, v)|_{u=v} = q_{x+1}(v, v).$$

This permits to compute $sp_x(u \leq \lambda \leq v)$, since

$$sp_x(u \leq \lambda \leq v) = 1 - (F_x(u) + G_{x+1}(v) - Q_x(u, v)).$$

Similarly, we compute the degree of plausibility for the unknown parameter as

$$pl_x(u \leq \lambda \leq v) = 1 - \left(P\left(v < \sum_{k=1}^{x}\omega_k\right) + P\left(\sum_{k=1}^{x+1}\omega_k < u\right)\right)$$
$$= 1 - ((1 - F_x(v)) + (1 - G_{x+1}(u)))$$
$$= F_x(v) + G_{x+1}(u) - 1.$$

In particular, we may determine the plausibility of singletons,

$$pl_x(u \leq \lambda \leq u) = P\left(\sum_{k=1}^{x}\omega_k \leq u \leq \sum_{k=1}^{x+1}\omega_k\right) = \frac{1}{x!}u^x e^{-u}.$$

This could be used to determine the most plausible estimate of the unknown parameter λ, i.e. for maximum likelihood estimation.

When there are several hints relative to the same domain Θ, we may ask how we should combine these hints in order to obtain a single hint reflecting the pooled information contained in all the hints? For the sake of simplicity, we only consider the combination of two hints; the generalization to more than two hints is straightforward. So let

$$\mathcal{H}_1 = (\Omega_1, P_1, \Gamma_1, \Theta), \quad \mathcal{H}_2 = (\Omega_2, P_2, \Gamma_2, \Theta) \tag{13}$$

be two hints on Θ. If the assumptions of the two hints are independent, then the prior probability of a pair of assumptions (ω_1, ω_2) in $\Omega_1 \times \Omega_2$ is $P_1(\omega_1) \cdot P_2(\omega_2)$. If the intersection $\Gamma_1(\omega_1) \cap \Gamma_2(\omega_2)$ is empty, then this pair of assumptions is called contradictory because, by construction of the mappings Γ_1 and Γ_2, it is impossible that both ω_1 and ω_2 are correct. Therefore, if

$$C = \{(\omega_1, \omega_2) \in \Omega_1 \times \Omega_2 : \Gamma_1(\omega_1) \cap \Gamma_2(\omega_2) = \emptyset\}$$

denotes the set of contradictory pairs of assumptions, then the correct pair must be in the set

$$\Omega = (\Omega_1 \times \Omega_2) - C. \tag{14}$$

Since the correct pair of assumptions is in Ω, we must condition the product probability measure $P_1 P_2$ on the set Ω, which we assume non-empty; if $\Omega = \emptyset$, then the two hints cannot be combined. Specifically, if we define

$$K = \sum_{(\omega_1, \omega_2) \in C} P_1(\omega_1)P_2(\omega_2),$$

then the conditioning operation results in a new probability space (Ω, P) where

$$P(\omega_1, \omega_2) = \frac{P_1(\omega_1)P_2(\omega_2)}{1 - K} \tag{15}$$

for all $(\omega_1, \omega_2) \in \Omega$. Furthermore, if (ω_1, ω_2) is the correct pair of assumptions, then the correct value in Θ must necessarily be in the set

$$\Gamma(\omega_1, \omega_2) = \Gamma_1(\omega_1) \cap \Gamma_2(\omega_2). \tag{16}$$

This is the smallest subset of Θ that is known for sure to contain the correct answer if both ω_1 and ω_2 are assumed to be the respective correct assumptions. Therefore, we define the combination of the two hints $\mathcal{H}_1$ and $\mathcal{H}_2$ as the hint

$$\mathcal{H}_1 \oplus \mathcal{H}_2 = (\Omega, P, \Gamma, \Theta),$$

where Ω is defined in (14), P is defined in (15) and Γ is defined in (16).

There is another notion regarding hints that we need to introduce, namely the operation of marginalization of a hint. Consider a hint on a domain that is the Cartesian product of two domains, namely a hint of the form

$$\mathcal{H} = (\Omega, P, \Gamma, \Theta_1 \times \Theta_2).$$

In order to evaluate hypotheses regarding the domain Θ_1, we define the *marginalization* of $\mathcal{H}$ to Θ_1 as the hint

$$\mathcal{H}^{\downarrow \Theta_1} = (\Omega, P, \Gamma_1, \Theta_1)$$

where

$$\Gamma_1(\omega) = \{\theta_1 \in \Theta_1 : \exists \, \theta_2 \in \Theta_2 \text{ such that } (\theta_1, \theta_2) \in \Gamma(\omega)\}.$$

The set $\Gamma_1(\omega)$ is the projection of $\Gamma(\omega)$ to the domain Θ_1. The definition of the mapping Γ_1 is justified by the fact that the projection $\Gamma_1(\omega)$ is the smallest subset of Θ_1 that is known for sure to contain the correct value in Θ_1 when we know for sure that the correct joint value is in $\Gamma(\omega)$. The marginalized hint $\mathcal{H}^{\downarrow \Theta_1}$ can then be used as usual to evaluate hypotheses H that are subsets of Θ_1.

We refer to [13] for a systematic presentation of the theory of hints. The approach of assumption-based reasoning to statistical inference as indicated here is developed in detail in [10]. Here, we hint at two further applications of probabilistic assumption-based reasoning in order to complete the picture.

In statistics, often, linear equations with stochastic disturbance terms are considered, systems like

$$\sum_{j=1}^{m} a_{ij} x_j + \omega_i = z_i, \text{ for } i = 1, \ldots, n.$$

This can be considered as a particular functional model, where real-valued parameters $x = (x_1, \ldots, x_m)$ together with stochastic disturbance terms $\omega = (\omega_1, \ldots, \omega_n)$ determine outcomes $z = (z_1, \ldots, z_n)$. Without going into details, we remark that, given observed outcomes z, the resulting hint about the parameters x contains a mapping $T_z(\omega)$, which maps disturbances ω to linear affine subspaces of the m-dimensional real vector space. These subspaces form an information algebra [6]. Here, we have an example of a random variable with values in an information algebra, a subject which we take up in the following sections as a generalization of the concept of hints presented above. For the analysis of this kind of linear systems with assumption-based reasoning, we refer to [10].

A further popular example is related to *propositional logic*. Information is described by formulae of a propositional language, where some propositional symbols are declared as *assumptions*. This permits to express uncertainty about facts or relations and leads to hints which map into subsets of Boolean cubes $\{0,1\}^n$, representing models of propositional formulae, which, again, form an information algebra. We refer to [3] for a detailed presentation of propositional argumentation systems.

3   Random Variables in Information Algebras

3.1   Simple Random Variables

In this section, we are proposing a generalization of the concept of a hint in the framework of information algebras. Note that a hint can be seen as a random set, i.e. a random variable with values in the power set of some set Θ. In this spirit, we may, more generally, consider random variables with values in an information algebra. We refer to the chapter Information Algebra for the theory of information algebras.

We consider a domain-free information algebra (Φ, D). Uncertain information can be represented by random variables taking values in such an information algebra. For example, probabilistic, propositional assumption-based systems as discussed in the previous Section 2 lead to random sets in the information algebra of subsets of Boolean cubes. Therefore, we develop in this section the elements of a theory of such random variables. For a more complete presentation, we refer to [6, 8].

Let $(\Omega, \mathcal{A}, P)$ be a probability space. It is well known that a mapping from such a space to any algebraic structure inherits this structure [4]. This will be exploited here.

Let $\mathcal{B} = \{B_1, \ldots, B_n\}$ be any finite partition of Ω such that all B_i belong to $\mathcal{A}$. A mapping $\Delta : \Omega \to \Phi$ such that

$$\Delta(\omega) = \phi_i \ \forall \omega \in B_i, \ i = 1, \ldots, n,$$

is called a *simple random variable* in (Φ, D). Among simple random variables in (Φ, D), we can define the operations of combination and focussing:

1. *Combination:* Let Δ_1 and Δ_2 be simple random variables in (Φ, D). Then, let $\Delta_1 \otimes \Delta_2$ be defined by

$$(\Delta_1 \otimes \Delta_2)(\omega) = \Delta_1(\omega) \otimes \Delta_2(\omega).$$

2. *Focussing:* Let Δ be a simple random variable in (Φ, D) and $x \in D$, then let $\Delta^{\Rightarrow x}$ be defined by

$$(\Delta^{\Rightarrow x})(\omega) = (\Delta(\omega))^{\Rightarrow x}.$$

Clearly, both $\Delta_1 \otimes \Delta_2$ and $\Delta^{\Rightarrow x}$ are simple random variables in (Φ, D). Let $\mathcal{R}_s$ denote the set of simple random variables in (Φ, D). With these operations of combination and focussing, $(\mathcal{R}_s, D)$ is an information algebra. The axioms are simply inherited from (Φ, D) and easily verified. The neutral element of this algebra is the random variable $E(\omega) = e$ for all $\omega \in \Omega$, the *vacuous random variable*. Further, since for any $\phi \in \Phi$, the mapping $\Delta_\phi(\omega) = \phi$ for all $\omega \in \Omega$ is a simple random variable, the algebra (Φ, D) is embedded in the algebra $(\mathcal{R}_s, D)$. A random variable like $\Delta_\phi(\omega) = \phi$, which takes a fixed value ϕ with probability one, is called *deterministic* or *degenerate*. We remark that $\Delta_1 \leq \Delta_2$ in $\mathcal{R}_s$ if, and only if, for all $\omega \in \Omega$ it holds that $\Delta_1(\omega) \leq \Delta_2(\omega)$.

Since simple random variables form themselves an information algebra containing the original one, it follows that simple random variables are pieces of information in this same sense. We remark that more general random variables can be defined [6, 8]. However, for our our discussion of uncertain information, simple random variables suffice.

There are two important special cases of simple random variables: If, for a random variable Δ defined relative to a partition $\mathcal{B} = \{B_1, \ldots, B_n\}$, it holds that $\phi_i \neq \phi_j$ for $i \neq j$, then the variable is called *canonical*. It is a simple matter to transform any random variable Δ into an associated canonical one: Take the union of all elements $B_i \in \mathcal{B}$ with the same value ϕ_j, which yields a new partition $\mathcal{B}'$ of Ω. Define $\Delta'(\omega) = \Delta(\omega)$. Then, Δ' is the *canonical version* of Δ, and we write $\Delta^\rightarrow = \Delta'$. We may consider the set of *canonical random variables*, $\mathcal{R}_{s,c}$, and define combination and focussing in this set as follows:

$$\Delta_1 \otimes_c \Delta_2 = (\Delta_1 \otimes \Delta_2)^\rightarrow,$$
$$\Delta^{\Rightarrow_c x} = (\Delta^{\Rightarrow x})^\rightarrow.$$

Then, $(\mathcal{R}_{s,c}, D)$ is still an information algebra under these modified operations. We remark also that $(\Delta_1 \otimes \Delta_2)^\rightarrow = (\Delta_1^\rightarrow \otimes \Delta_2^\rightarrow)^\rightarrow$ and $((\Delta^\rightarrow)^{\Rightarrow x})^\rightarrow = (\Delta^{\Rightarrow x})^\rightarrow$.

Secondly, if $\Delta(\omega) \neq z$ for all $\omega \in \Omega$, then Δ is called *normalized*. We can associate a normalized random variable $\Delta^\downarrow$ with any simple random variable Δ provided that $P(\{\omega \in \Omega : \Delta(\omega) = z\}) > 0$. In fact, let $\Omega^\downarrow = \{\omega \in \Omega : \Delta(\omega) \neq z\}$. This is a set with probability $P(\Omega^\downarrow) = 1 - P(\{\omega \in \Omega : \Delta(\omega) = z\}) > 0$. We then consider the new probability space $(\Omega, \mathcal{A}, P')$, where P' is the probability measure on $\mathcal{A}$ defined by

$$P'(A) = \frac{P(A \cap \Omega^\downarrow)}{P(\Omega^\downarrow)}. \tag{17}$$

On this new probability space, define $\Delta^\downarrow(\omega) = \Delta(\omega)$. Clearly, it holds that $(\Delta^\rightarrow)^\downarrow = (\Delta^\downarrow)^\rightarrow$.

The idea behind normalization becomes clear when we consider the *combination* of (normalized) random variables Δ_1 and Δ_2, which represents some (uncertain) information with the following interpretation: One of the $\omega \in \Omega$ must be the correct, but unknown assumption of the information; if ω happens to be the correct assumption, then, under the first random variable, the information

$\Delta_1(\omega)$ can be asserted, and, under the second variable, the information $\Delta_2(\omega)$. Thus, together, still under the assumption ω, the information $\Delta_1(\omega) \otimes \Delta_2(\omega)$ can be assumed. However, it is possible that $\Delta_1(\omega) \otimes \Delta_2(\omega) = z$ even if both Δ_1 and Δ_2 are *normalized*. Since the null information z represents a *contradiction*, in view of the information given by the variables Δ_1 and Δ_2, the assumption ω cannot hold, can (and must) be excluded. This amounts to normalizing the random variable $\Delta_1 \otimes \Delta_2$, by excluding all $\omega \in \Omega$ for which the combination results in a contradiction, and then to condition (i.e. normalize) the probability on the non-contradictory assumptions. This is a generalization of the combination rule introduced in the previous Section 2.3.

A simple random variable Δ takes a *finite* number of different values, say $\{\phi_1, \ldots, \phi_n\}$, and thus induces a finite partition $\mathcal{P}_\Delta$ with blocks $B_i = \{\omega : \Delta(\omega) = \phi_i\} \in \mathcal{A}$ of Ω. We call two partitions $\mathcal{P}_1$ and $\mathcal{P}_2$ *stochastically independent*, and write $\mathcal{P}_1 \perp \mathcal{P}_2$, if every block $B_{1,i}$ of $\mathcal{P}_1$ has a non-empty intersection with every block $B_{2,j}$ of $\mathcal{P}_2$ and if, furthermore, for any such pair of blocks,

$$P(B_{1,i} \cap B_{2,j}) = P(B_{1,i}) \cdot P(B_{2,j}).$$

Similarly, two random variables Δ_1 and Δ_2 are called *stochastically independent*, written $\Delta_1 \perp \Delta_2$, if their partitions are stochastically independent, $\mathcal{P}_{\Delta_1} \perp \mathcal{P}_{\Delta_2}$. This is an important property of random variables.

3.2   Labeled Random Variables

To the domain-free information algebra $(\mathcal{R}_s, D)$, we may associate a *labeled information algebra* (see the chapter Information Algebra). Its elements are pairs (Δ, x), where x is a support of Δ, i.e. it holds that $\Delta^{\Rightarrow x}(\omega) = \Delta(\omega)$ for all $\omega \in \Omega$. This means that x is support for all possible values ϕ_i of Δ. Therefore, a labeled element (Δ, x) can as well be considered as a mapping Δ_x between Ω and the labeled version of the information algebra (Φ, D), where $\Delta_x(\omega) = (\phi_i, x)$ for $\omega \in B_i$. The operations of combination and projection in the labeled algebra are then defined as follows, in correspondence with the usual approach:

$$(\Delta_1, x) \otimes (\Delta_2, y)(\omega) = (\Delta_1(\omega) \otimes \Delta_2(\omega), x \vee y),$$
$$(\Delta, x)^{\downarrow y}(\omega) = (\Delta^{\Rightarrow x}(\omega), y), \text{ for } y \leq x.$$

For some purposes, we need to work with the labeled version of simple random variables.

If (Φ, D) is a *domain-free* information algebra and (Φ^*, D) the associated *labeled* algebra, i.e. the algebra of pairs (ϕ, x) with x a support of ϕ, then the labeled simple random variables (Δ, x) can also be considered as mappings $\Delta^* : \Omega \to \Phi_x^*$, where Φ_x^* is the set of all $(\phi, x) \in \Phi^*$. The mapping is defined by $\Delta^*(\omega) = (\Delta(\omega), x)$. This is well defined since x being a support of Δ means that x is a support of $\Delta(\omega)$ for all $\omega \in \Omega$.

This observation indicates that we may define labeled random variables also directly: Let (Φ, D) be a *labeled* information algebra and Φ_x the members of Φ with label $d(\phi) = x$. A simple random variable with label x is then a mapping

$\Delta : \Omega \to \Phi_x$ such that there is a finite decomposition of $\mathcal{B} = \{B_1, \ldots, B_n\}$ such that all B_i belong to $\mathcal{A}$ and

$$\Delta(\omega) = \phi_i \in \Phi_x \ \forall \omega \in B_i, i = 1, \ldots, n.$$

The label of such a random variable is $d(\Delta) = x$. As before, it is easily verified that such *labeled* random variables form themselves a labeled information algebra. In fact, define the operations of combination and projection in the natural way as follows:

1. *Combination:* Let Δ_1 and Δ_2 be simple random variables in (Φ, D). Then, let $\Delta_1 \otimes \Delta_2$ be defined by

$$(\Delta_1 \otimes \Delta_2)(\omega) = \Delta_1(\omega) \otimes \Delta_2(\omega).$$

2. *Projection:* Let Δ be a simple random variable in (Φ, D) and $x \in D$ such that $x \leq d(\Delta)$, then let $\Delta^{\downarrow x}$ be defined by

$$(\Delta^{\downarrow x})(\omega) = (\Delta(\omega))^{\downarrow x}.$$

In the same way as the domain-free version of the algebra (Φ, D) is obtained, the domain-free version of the algebra of labeled simple random variables can be obtained. We abstain from presenting the details. Important is only that we may switch between the two equivalent versions of simple random variables, the domain-free and the labeled version, as conveniently as with any other information algebra. Some issues are more conveniently discussed in the domain-free version; for others the labeled version is better adapted.

3.3 Variables in Atomic Information Algebras

In Section 4.3 of the chapter Information Algebra in this volume, we have introduced the concept of an *atom* in a domain x of a labeled information algebra: It is a finest non-zero piece of information in domain x. An atom α in a domain x is contained in some piece of information $\phi \in \Phi$ of the same domain if $\phi \leq \alpha$. Let $At(\phi)$ denote the set of atoms α contained in ϕ; we also write $\alpha \in \phi$ for $\alpha \in At(\phi)$. The labeled information algebra (Φ, D) is called *atomic* if, for all $\phi \in \Phi$,

$$\phi = \wedge At(\phi).$$

In the chapter Information Algebra in this volume, we have seen that the atoms can be considered as generalized tuples and that the set of atoms $\mathcal{R}_\Phi$ of Φ forms a generalized relational algebra, that is essentially a subset algebra. By replacing ϕ by its set of atoms $At(\phi)$, the algebra (Φ, D) is embedded into the relational algebra $(\mathcal{R}_\Phi, D)$. Note that the algebra $(\mathcal{R}_\Phi, D)$ is in fact an *atomic Boolean information algebra*. In particular, any element of $\mathcal{R}_\Phi$ has a complement. We refer to the chapter Information Algebra for more details on Boolean information algebras.

If Δ is now a labeled simple random variable with values in a domain x of an atomic labeled information algebra, then $\Delta(\omega)$ may be replaced by $At(\Delta(\omega))$. This means that the mapping $\Delta : \Omega \to \Phi_x$ can be changed to the mapping $\Delta' : \Omega \to At_x(\Phi)$, where $At_x(\Phi)$ is the set of atoms in domain x, defined by

$$\Delta'(\omega) = At(\Delta(\omega)).$$

Thus, any simple random variable in an atomic information algebra (Φ, D) can as well be considered as a random variable in the associated atomic Boolean information algebra $(\mathcal{R}_\Phi, D)$. This is essentially a *random set* in the set $At_x(\Phi) = At(e_x)$, the set of all atoms in domain x.

Of particular interest are random variables Δ whose values are either z_x or *atoms*, $\Delta(\omega) \in At_x(\Phi)$ for all $\omega \in \Omega$. Such variables are called *precise* since, for each possible assumption ω, a most precise piece of information in domain x is given. As we shall see later, these variables have a number of interesting properties. For instance, let Δ be a precise simple random variable defined on a partition $\mathcal{B} = \{B_1, \ldots, B_n\}$ such that its value on B_i is $\alpha_i \in At_x(\Phi)$, and let Γ any other random variable on the domain x. Then, for the combined variable $\Delta \otimes \Gamma$, we obtain, for $\omega \in B_i$, $i = 1, \ldots, n$,

$$(\Delta \otimes \Gamma)(\omega) = \begin{cases} \alpha_i \text{ if } \alpha_i \in At(\Gamma(\omega)), \\ z_x \text{ otherwise.} \end{cases}$$

Thus, the combined variable is still *precise*: Within a fixed domain x, the precise random variables are absorbing, i.e. their combination with any other variable on the same domain results in a precise variable. Furthermore, the combined variable has the same values as the precise one. This remarkable result has some far reaching consequences, as we shall see later in Section 4.4.

4   Probability Distributions in Information Algebras

4.1   Support and Possibility

This section is devoted to the study of the *probability distribution* of simple random variables. The starting point is the following question: Given a random variable Δ on a domain-free information algebra (Φ, D) and an element $\phi \in \Phi$, under which assumptions can the information ϕ be asserted? And how likely is it that these assumptions hold?

If $\omega \in \Omega$ is an assumption such that $\Delta(\omega) \geq \phi$, then $\Delta(\omega)$ implies ϕ. In this case we say that ω is an assumption which *supports* ϕ given Δ. Supporting assumptions are arguments in favor of ϕ. Therefore, we define for every $\phi \in \Phi$ the set

$$qs_\Delta(\phi) = \{\omega \in \Omega : \phi \leq \Delta(\omega)\}$$

of assumptions supporting ϕ. However, if $\Delta(\omega) = z$, then ω supports every $\phi \in \Phi$ since $\phi \leq z$. The null element z represents the *contradiction*, which implies everything. In a consistent theory, contradictions must be excluded. Thus, assuming that our information is consistent, we may conclude that assumptions such that $\Delta(\omega) = z$ are not really possible assumptions and must be eliminated. This corresponds, of course, to the normalization of the random variable Δ. Once more, we refer to Section 2 for an illustration and justification. Let

$$qs_\Delta(z) = \{\omega \in \Omega : \Delta(\omega) = z\}.$$

We assume that $qs_\Delta(z)$ is not equal to Ω; otherwise Δ is called the *null* variable representing a contradiction. In other words, we assume that proper information is never fully contradictory. If we eliminate the contradictory assumptions from $qs_\Delta(\phi)$, we obtain the *support set*

$$s_\Delta(\phi) = \{\omega \in \Omega : \phi \leq \Delta(\omega) \neq z\} = qs_\Delta(\phi) - qs_\Delta(z)$$

of ϕ, which is the set of assumptions properly supporting ϕ, and the mapping $s_\Delta : \Phi \to 2^\Omega$ is called the *allocation of support* induced by Δ. The set $qs(\phi)$ will be called *quasi-support set* to underline that it may contain contradictory assumptions. This set has little interest from a semantic point of view, but it has some importance for technical and especially for computational purposes. These concepts capture the essence of probabilistic argumentation systems, as exemplified in the context of propositional logic in the previous Section 2, where further references can be found.

Here are the basic properties of allocations of support:

Theorem 1. *If Δ is a simple random variable in an information algebra (Φ, D), then the following holds for the associated allocation of support s_Δ:*

1. $qs_\Delta(e) = \Omega$, $s_\Delta(z) = \emptyset$;
2. *if Δ is normalized, then $qs_\Delta(z) = \emptyset$;*
3. *for any pair $\phi, \psi \in \Phi$,*

$$qs_\Delta(\phi \otimes \psi) = qs_\Delta(\phi) \cap qs_\Delta(\psi),$$
$$s_\Delta(\phi \otimes \psi) = s_\Delta(\phi) \cap s_\Delta(\psi).$$

Proof. (1) and (2) follow immediately from the definition of the allocation of support. (3) follows since $\phi \otimes \psi \leq \Delta(\omega)$ if, and only if, $\phi \leq \Delta(\omega)$ and $\psi \leq \Delta(\omega)$. $\square$

Knowing the assumptions supporting a hypothesis ϕ is already interesting and important. It is the part logic can provide. On top of this, it is important to know how *likely* it is that a supporting assumption holds. This is the part probability adds. If we know or may assume that the information is consistent, then we should condition the original probability measure P in Ω on the event $qs_\Delta^c(z)$. This then leads to the probability space $(qs_\Delta^c(z), \mathcal{A} \cap qs_\Delta^c(z), P')$, where $P'(A) = P(A)/P(qs_\Delta^c(z))$. The likelihood of the assumptions supporting $\phi \in \Phi$ can then be measured by

$$sp_\Delta(\phi) = P'(s_\Delta(\phi)).$$

The value $sp_\Delta(\omega)$ is called *degree of support* of ϕ associated with the random variable Δ. The function $sp : \Phi \to [0,1]$ is called the *support function* of Δ. It corresponds to the *distribution function* of ordinary random variables.

It is, for technical reasons, convenient to define also the *degree of quasi-support* by

$$qsp_\Delta(\phi) = P(qs_\Delta(\phi)).$$

Then, the degree of support can also be expressed in terms of degrees of quasi-support by

$$sp_\Delta(\phi) = \frac{qsp_\Delta(\phi) - qsp(z)}{1 - qsp_\Delta(z)}.$$

This is the form which is usually used in applications [9].

Alternatively, we can also ask under which assumptions $\omega \in \Omega$ a hypothesis ϕ is possible given Δ, i.e. not excluded, although not necessarily supported. If $\Delta(\omega)$ is such that, combined with ϕ, it leads to a contradiction, i.e. if $\Delta(\omega) \otimes \phi = z$, then, under ω, the information ϕ is excluded by a consistency consideration as above. So we define the set

$$p_\Delta(\phi) = \{\omega \in \Omega : \Delta(\omega) \otimes \phi \neq z\}.$$

This is the set of assumptions under which ϕ is not excluded. Therefore, we call it the *possibility set* of ϕ. Again, $p_\Delta(\phi)$ is a measurable set, and we can define the *degree of possibility*, also sometimes called *degree of plausibility* (e.g. in [19]),

$$pl_\Delta(\phi) = P'(p_\Delta(\phi)).$$

If $\omega \in p_\Delta^c(\phi)$, then, under this assumption, ϕ is impossible, is excluded. So the set $p_\Delta^c(\phi)$ contains arguments *against* ϕ and

$$do_\Delta(\phi) = P'(p_\Delta^c(\phi)) = 1 - pl_\Delta(\phi).$$

can be called the *degree of doubt* about ϕ. Note that $s_\Delta(\phi) \subseteq p_\Delta(\phi)$ since $\phi \leq \Delta(\omega) \neq z$ implies $\phi \otimes \Delta(\omega) = \Delta(\omega) \neq z$. Hence, we see that, for all $\phi \in \Phi$, we have that $sp_\Delta(\phi) \leq pl_\Delta(\phi)$.

In the particular case of a *Boolean information algebra*, we have $\phi \leq \psi$ if, and only if, $\phi^c \otimes \psi = z$. This implies that $qs_\Delta(\phi) = p_\Delta^c(\phi^c)$, hence,

$$pl_\Delta(\phi) = 1 - sp_\Delta(\phi^c). \tag{18}$$

A piece of information is the more plausible, the less its negation or complement is supported.

In particular, we are going to consider the case of *precise* random variables (see Section 3.3): These are variables Δ with values in an atomic, labeled information algebra (Φ, D) such that for all $\omega \in \Omega$ the value $\Delta(\omega)$ is either an *atom* of Φ_x (i.e. $\Delta(\omega) \in At_x(\Phi)$) or $\Delta(\omega) = z_x$. In such a case, we remark that $s_\Delta(\phi) =$

$\{\omega \in \Omega : \Delta(\omega) \in At(\phi)\}$ for any $\phi \in \Phi_x$. On the other hand, for any atom α of domain x and $\phi \in \Phi_x$, it holds that either $\alpha \in At(\phi)$ or $\alpha \otimes \phi = z_x$. This implies that $\Delta(\omega) \otimes \phi \neq z_x$ holds if, and only if, $\Delta(\omega) \in At(\phi)$. Therefore, it follows that for precise random variables $s_\Delta(\phi) = p_\Delta(\phi)$ and hence $sp_\Delta(\phi) = pl_\Delta(\phi)$ for all $\phi \in \Phi_x$. Further, we note that

$$pl_\Delta(\phi) = P'(p_\Delta(\phi)) = \sum_{\alpha \in At(\phi)} pl_\Delta(\alpha).$$

So the support or plausibiliy of atoms is sufficient to determine the support and plausibility of any other element $\phi \in \Phi_x$. Since, in the case of a normalized variable Δ,

$$\sum_{\alpha \in At_x(\Phi)} pl_\Delta(\alpha) = pl_\Delta(e_x) = 1,$$

it follows that the plausibilities of atoms determine in this case a probability measure on the set of atoms in domain x.

If the information algebra (Φ, D) is in addition *Boolean*, then (18) shows that $sp_\Delta(\phi) = 1 - sp_\Delta(\phi^c)$ and this implies that $sp_\Delta = pl_\Delta$ is a *probability measure* on the Boolean information algebra.

4.2 Basic Probability Assignments

Consider the canonical version $\Delta^\rightarrow$ of a random variable Δ. Let $\mathcal{B} = \{B_1, \ldots, B_n\}$ be its associated partition of Ω. Then, we define the probabilities

$$m_\Delta(\phi_i) = P(B_i), \text{ for } i = 1, \ldots, n.$$

The $m_\Delta(\phi_i)$ are called *basic probability assignments* (bpa). This term comes from the Dempster-Shafer theory of evidence (DS theory) [19], which corresponds to our theory of random variables when the information algebra of subsets of a finite set is considered. It then immediately follows that

$$qsp_\Delta(\phi) = \sum_{i:\phi \leq \phi_i} m_\Delta(\phi_i).$$

So it is sufficient to know the bpas of a random variable to determine its support function. In many applications, people only work with the bpas and do not care about the underlying random variables [19]. However, implicitly, it is then generally assumed that the underlying variables are *stochastically independent*, although this is often not rigorously and explicitly stated.

We may also consider *normalized* bpas, corresponding to the normalized variable $\Delta^\downarrow$,

$$m_{\Delta^\downarrow}(\phi) = \frac{m_\Delta(\phi)}{1 - m_\Delta(z)}.$$

In terms of this normalized bpa, the degrees of support become

$$sp_\Delta(\phi) = \sum_{i:\phi \le \psi_i} m_{\Delta^\downarrow}(\psi_i).$$

Similarly, the degree of plausibility may also be obtained from the normalized bpa,

$$pl_\Delta(\phi) = \sum_{i:\phi \otimes \psi_i \ne z} m_{\Delta^\downarrow}(\psi_i).$$

The bpas, as the degrees of quasi-support, are not important semantically, but convenient concepts especially for computational purposes [3].

Assume that the two random variables Δ_1 and Δ_2 are stochastically independent. Then, it is evident that $\Delta_1^\rightarrow$ and $\Delta_2^\rightarrow$ are also stochastically independent. The following theorem shows how the bpa of the combined random variable $\Delta_1 \otimes \Delta_2$ can be computed from the bpas of the individual variables.

Theorem 2.   *Let $m_{\Delta_1}(\phi_{1,i})$ for $i = 1, \ldots, m$ and $m_{\Delta_2}(\phi_{2,j})$ for $j = 1, \ldots, n$ be the bpas of the two stochastically independent random variables Δ_1 and Δ_2. Then, the bpa of $\Delta_1 \otimes \Delta_2$ is given by*

$$m_{\Delta_1 \otimes \Delta_2}(\phi) = \sum_{i,j:\phi_{1,i} \otimes \phi_{2,j} = \phi} m_{\Delta_1}(\phi_{1,i}) \cdot m_{\Delta_2}(\phi_{2,j}). \tag{19}$$

Proof. The canonical version of the combined random variables $(\Delta_1 \otimes \Delta_2)^\rightarrow = (\Delta_1^\rightarrow \otimes \Delta_2^\rightarrow)^\rightarrow$ has as possible values the combinations $\phi_{1,i} \otimes \phi_{2,j}$. Each such combination has assigned the probability $m_{\Delta_1}(\phi_{1,i}) \cdot m_{\Delta_2}(\phi_{2,j})$ of the underlying intersection $B_{1,i} \cap B_{2,j}$ of the blocks of the two orthogonal partitions associated with $\Delta_1^\rightarrow$ and $\Delta_1^\rightarrow$. So, when the canonical version of $\Delta_1^\rightarrow \otimes \Delta_2^\rightarrow$ is taken, these probabilities sum up over all pairs i, j such that $\phi_{1,i} \otimes \phi_{2,j}$ takes the same value ϕ.   □

Of course, in (19), only a finite number of ϕs have a non-empty sum on the right. The method of combining the bpas proposed in Theorem 2 is called the (non-normalized) *Dempster rule* since it has first been proposed in the context of multivalued mappings in [2]. More precisely, Dempster proposed a *normalized* version of this rule, where the combined variable is normalized, $(\Delta_1 \otimes \Delta_2)^\downarrow$. Then, the rule is

$$m_{(\Delta_1 \otimes \Delta_2)^\downarrow}(\phi) = \frac{\sum_{i,j:\phi_{1,i} \otimes \phi_{2,j} = \phi} m_{\Delta_1}(\phi_{1,i}) \cdot m_{\Delta_2}(\phi_{2,j})}{\sum_{i,j:\phi_{1,i} \otimes \phi_{2,j} \ne z} m_{\Delta_1}(\phi_{1,i}) \cdot m_{\Delta_2}(\phi_{2,j})}. \tag{20}$$

Here, the combinations which result in a contradiction are eliminated and the probability is renormalized.

A particular case is the combination of a random variable with a *degenerate variable* Δ_ϕ associated with an element $\phi \in \Phi$. Let Δ be a random variable on Φ with bpa $m_\Delta(\phi_i)$ for $i = 1, \ldots, m$, and $\phi \in \Phi$. To ϕ corresponds a degenerate random variable Δ_ϕ with bpa $m(\phi) = 1$ and $m(\psi) = 0$ for all $\psi \neq \phi$. Then, by (19), we get

$$m_{\Delta \otimes \Delta_\phi}(\psi) = \sum_{i:\phi_i \otimes \phi = \psi} m_\Delta(\phi_i). \tag{21}$$

This is Dempster's (non-normalized) *rule of conditioning*. In the *normalized* case, the rule becomes

$$m_{\Delta \otimes \Delta_\phi}(\psi) = \frac{\sum_{i:\phi_i \otimes \phi = \psi} m_\Delta(\phi_i)}{\sum_{i:\phi_i \otimes \phi \neq z} m_\Delta(\phi_i)}. \tag{22}$$

We can also compute the bpas of a focussed random variable from the bpas of the original variable. This is shown in the next theorem.

Theorem 3. *Let $m_\Delta(\phi_i)$ for $i = 1, \ldots, m$, be the bpas of a random variable Δ. Then, the bpas of $\Delta^{\Rightarrow x}$ are given by*

$$m_{\Delta^{\Rightarrow x}}(\phi) = \sum_{i:\phi_i^{\Rightarrow x} = \phi} m_\Delta(\phi_i). \tag{23}$$

Proof. This follows since the partition of $(\Delta^{\Rightarrow x})^\rightarrow$ contains the unions of the blocks B_i of the partition of $\Delta^\rightarrow$ for which $\phi_i^{\Rightarrow x} = \phi$.   □

Again, in (23), only a finite number of ϕs have a non-empty sum on the right. So we see that the operations with simple random variables are reflected by their associated bpas. In fact, in the next section, we shall consider an algebra of bpas, similar to an information algebra, based on these operations.

4.3   Algebra of Bpas

On a domain-free information algebra (Φ, D), we may define bpas as functions $m : \Phi \to [0, 1]$ such that $m(\phi) > 0$ only for a finite number $\phi \in \Phi$, and such that

$$\sum_{\phi \in \Phi} m(\phi) = 1.$$

The elements $\phi \in \Phi$ for which $m(\phi) > 0$ are called *focal elements* of m. If further $m(z) = 0$, we call the bpa *normalized*. Clearly, we may always associate with a bpa a simple random variable: If $m(\phi) > 0$ for $\phi = \phi_1, \ldots, \phi_m$, then define $\Omega = \{1, \ldots, m\}$ and by $p(i) = m(\phi_i)$ a probability measure P is determined on Ω. If $\Delta(i) = \phi_i$, then Δ becomes a simple random variable whose bpa is exactly m.

Let $\mathcal{M}$ be the family of bpas on Φ. In $\mathcal{M}$, we define the operations of combination and focussing following the pattern of Dempster's rule (Theorem 2) and of (23):

1. *Combination:* for $m_1, m_2 \in \mathcal{M}$, let $m_1 \otimes m_2$ be defined by

$$m_1 \otimes m_2(\phi) = \sum_{\phi_1 \otimes \phi_2 = \phi} m_1(\phi_1) \cdot m_2(\phi_2);$$

2. *Focussing:* for $m \in \mathcal{M}$ and $x \in D$, let $m^{\Rightarrow x}$ be defined by

$$m^{\Rightarrow x}(\phi) = \sum_{\psi^{\Rightarrow x} = \phi} m(\psi).$$

Clearly, both $m_1 \otimes m_2$ and $m^{\Rightarrow x}$ are bpas on Φ. Combination is modelled according to the combination of *stochastically independent* random variables. Often, in applications, people work with bpas according to these, or equivalent, operations. It is then implicitly assumed that combined bpas are "independent", a concept whose rigorous meaning is defined by stochastic independence of underlying random variables, as shown in the previous Section 4.2. In fact, using these operations, we operate in an algebra which satisfies all axioms of an information algebra, except the *idempotency axiom*. Such an algebra is called a *valuation algebra* in [6, 14].

Theorem 4. *The two-sorted algebra $(\mathcal{M}, D)$, with the operations of combination and focussing defined above, satisfies all axioms of a domain-free information algebra, except the idempotency axiom.*

For a proof of this theorem, we refer to [6].

The properties of a valuation algebra, as those of information algebras, permit efficient computation methods for solving query problems. In fact, the axioms of valuation algebras were proposed in [21] as sufficient for local computation algorithms similar to those known for probability networks [15]. See [6] for a discussion of several architectures of local computation based on valuation algebras. Usually however, the labeled version of the algebra is used for computational purposes. We remark that the algebra has also a null element, given by the bpa $m(z) = 1$.

So far, we have defined combination by the non-normalized Dempster rule. We could also have used the normalized version and still obtained a valuation algebra [6].

4.4   Bayesian Theorems

Dempster's rule of combination becomes especially simple in the setting of precise random variables in a labeled information algebra. Let therefore (Φ, D) be a labeled atomic information algebra, Δ a precise random variable with focal elements $\alpha_1, \ldots, \alpha_n \in At_x(\Phi)$ and Γ an arbitrary random variable with values $\phi_1, \ldots, \phi_m \in \Phi_x$. In Section 3.3, we have shown that the combination $\Delta \otimes \Gamma$ is still a precise variable with $\alpha_1, \ldots, \alpha_n$ as possible values. Further, we have seen in Section 4.1 that the plausibility or support of any element $\phi \in \Phi_x$ is determined by the plausibilities of atoms in domain x. The following theorem shows how the plausibilities of atoms for the combination $\Delta \otimes \Gamma$ can be obtained.

Theorem 5. *Let Δ a precise random variable with values $\alpha_1, \ldots, \alpha_n \in At_x(\Phi)$ and Γ an arbitrary random variable. If these two random variables are stochastically independent, then*

$$pl_{\Delta \otimes \Gamma}(\alpha_i) = pl_\Delta(\alpha_i) \cdot pl_\Gamma(\alpha_i)$$

for the unnormalized *combination* $\otimes$.

Proof. We first note that in the case of a precise random variable the bpa corresponds to the plausibilities of atoms, that is $m_\Delta(\alpha_i) = pl_\Delta(\alpha_i)$. Further, for an arbitrary random variable Γ with focal elements $\phi_1, \ldots, \phi_m$ in Φ_x, we have

$$m_\Gamma(\alpha_i) = \sum_{j:\alpha_i \in At(\phi_j)} m_\Gamma(\phi_j).$$

Note that $m_\Gamma(\alpha_i) \neq 0$ if, and only if, there is a j such that $\alpha_i \in At(\phi_j)$. Then, by Dempster's rule (19),

$$pl_{\Delta \otimes \Gamma}(\alpha_i) = m_{\Delta \otimes \Gamma}(\alpha_i)$$
$$= \sum_{j:\alpha_i \in At(\phi_j)} m_\Delta(\alpha_i) \cdot m_\Gamma(\phi_j)$$
$$= m_\Delta(\alpha_i) \cdot \sum_{j:\alpha_i \in At(\phi_j)} m_\Gamma(\phi_j)$$
$$= pl_\Delta(\alpha_i) \cdot pl_\Gamma(\alpha_i).$$

This concludes the proof. □

It is remarkable that in this particular form of Dempster's rule of combination only the plausibilities of atoms of the random variable Γ are required, and nothing more. Of course, we can also consider the normalized version of Dempster's rule for combination. This then gives the plausibilities of the normalized version of the combination, and, by Theorem 5 and (22), we obtain

$$pl_{(\Delta \otimes \Gamma)^\downarrow}(\alpha_i) = K^{-1} \cdot pl_\Delta(\alpha_i) \cdot pl_\Gamma(\alpha_i),$$

with

$$K = \sum_{i=1}^n pl_\Delta(\alpha_i) \cdot pl_\Gamma(\alpha_i).$$

Of course, this is valid only if $K > 0$. Otherwise, the two random variables are incompatible, the information they represent is contradictory. This result 5 reproduces in some important cases the famous *Bayes theorem* of probability theory.

5   Information Order and Measure

5.1   Order between Random Variables

Random variables represent uncertain information, as has been argued in Section 3. They represent *information* since they form an information algebra; the information is *uncertain* since, depending on uncertain assumptions, different information elements in an underlying information algebra are selected. Since random variables form an information algebra (domain-free or labeled, according to convenience), there exists a *partial order* between random variables. However, we have to be careful if we want to interpret this order as a comparison of information content. In fact, if the random variable is *not normalized*, then there are assumptions which imply z, the contradiction. It has been argued in the chapter Information Algebra that z is not to be considered as information. Therefore, we propose here a more subtle point of view: Assumptions which map to z are to be considered as not possible; i.e. a random variable Δ contains information about possible assumptions, namely those in $\Omega^\downarrow = \{\omega : \Delta(\omega) \neq z\}$. This point of view is supported by the discussion and the examples in Section 2. This means that a non-normalized random variable carries not only information about Φ, but also about the space of assumptions Ω. On the other hand, this interpretation destroys the idea that the partial order induced by the information algebra of random variables represents a comparison of information content on the side of the information algebra only. This has important consequences and will be elaborated in this section.

To simplify the discussion and focus on the important points, we assume that Ω is a *finite* set. The probability measure P is then simply determined by the probabilities $p(\omega)$ of the assumptions $\omega \in \Omega$. Consider two simple random variables Δ_1 and Δ_2 in the information algebra $(\mathcal{R}_s, D)$ and assume that $\Delta_1 \leq \Delta_2$. This means, by definition (see the chapter Information Algebra), that $\Delta_1 \otimes \Delta_2 = \Delta_2$, hence, for all $\omega \in \Omega$, it holds that $\Delta_1(\omega) \otimes \Delta_2(\omega) = \Delta_2(\omega)$. Finally, this implies that, for all $\omega \in \Omega$, we have $\Delta_1(\omega) \leq \Delta_2(\omega)$. If we now compare the normalized variables $\Delta_1^\downarrow$ and $\Delta_2^\downarrow$, then we see that the following holds:

1. $\Omega_1^\downarrow \supseteq \Omega_2^\downarrow$;
2. for all $\omega \in \Omega_2^\downarrow$, it holds that $p_1^\downarrow(\omega) \leq p_2^\downarrow(\omega)$;
3. for all $\omega \in \Omega_2^\downarrow$, it holds that $\Delta_1(\omega) \leq \Delta_2(\omega)$.

The first point follows since $\Delta_2(\omega) \neq z$ implies $\Delta_1(\omega) \neq z$. The second point holds because $\Omega_1^\downarrow \supseteq \Omega_2^\downarrow$ implies $P(\Omega_1^\downarrow) \leq P(\Omega_2^\downarrow)$ and, therefore,

$$p_1^\downarrow(\omega) = \frac{p(\omega)}{P(\Omega_1^\downarrow)} \leq \frac{p(\omega)}{P(\Omega_2^\downarrow)} = p_2^\downarrow(\omega).$$

Thus, more information means in this context of uncertain information, first, that there are less possible assumptions, second, that these assumptions become more probable and, third, imply more information in Φ. This makes the idea more precise that random variables contain information both with respect to assumptions as well as with respect to the domains D of the underlying information algebra (Φ, D).

However, this information order is not necessarily reflected in a corresponding order of degrees of support and plausibility. In fact, only the relations given in the following theorem hold:

Theorem 6. *Assume $\Delta_1 \leq \Delta_2$, and let $K_1 = P(\Omega_1^\downarrow)$, $K_2 = P(\Omega_2^\downarrow)$. Then, for all $\phi \in \Phi$,*

$$sp_{\Delta_1}(\phi) \leq \frac{K_2}{K_1} \cdot sp_{\Delta_2}(\phi) + \frac{1 - K_2}{K_1},$$

$$pl_{\Delta_1}(\phi) \geq \frac{K_2}{K_1} \cdot pl_{\Delta_2}(\phi).$$

If $\Delta_1 \leq \Delta_2$, but $\Omega_1^\downarrow = \Omega_2^\downarrow = \Omega$, then $K_1 = K_2 = 1$ and

$$[sp_{\Delta_2}(\phi), pl_{\Delta_2}(\phi)] \quad \subseteq \quad [sp_{\Delta_1}(\phi), pl_{\Delta_1}(\phi)].$$

Proof. Note first that

$$s_{\Delta_1}(\phi) = \{\omega : \phi \leq \Delta_1(\omega) \neq z\} \subseteq \{\omega : \phi \leq \Delta_2(\omega) \neq z\} \cup \{\omega : \Delta_2(\omega) = z\}$$
$$= s_{\Delta_2}(\phi) \cup (\Omega_2^\downarrow)^c.$$

From this, we deduce that

$$sp_{\Delta_1}(\phi) = \frac{P(s_{\Delta_1}(\phi))}{K_1} \leq \frac{P(s_{\Delta_2}(\phi)) + (1 - K_2)}{K_1}$$
$$= \frac{K_2}{K_1} \cdot sp_{\Delta_2}(\phi) + \frac{1 - K_2}{K_1}.$$

Further, we have

$$p_{\Delta_1}(\phi) = \{\omega : \Delta_1(\omega) \otimes \phi \neq z\} \supseteq \{\omega : \Delta_2(\omega) \otimes \phi \neq z\} = p_{\Delta_2}(\phi).$$

From this, it follows that

$$pl_{\Delta_1}(\phi) = \frac{P(p_{\Delta_1}(\phi))}{K_1} \geq \frac{P(p_{\Delta_2}(\phi))}{K_1} = \frac{K_2}{K_1} \cdot pl_{\Delta_2}(\phi).$$

The last part of the theorem is an immediate consequence of these inequalities.
□

This theorem reflects a certain *non-monotonicity* of uncertain reasoning: Although $\Delta_1 \otimes \Delta_2$ has more information content than Δ_1, it is well possible that the support of a certain hypothesis ϕ decreases. New information may shed new doubt on something which was strongly supported before. This issue will be also disussed in the next section with relation to information measure.

We remark that these considerations carry over to the derived information order relative to a domain $x \in D$. We remind that, according to the general theory of information algebras (see the chapter Information Algebra), $\Delta_1 \leq_x \Delta_2$ means that $\Delta_1^{\Rightarrow x} \leq \Delta_2^{\Rightarrow x}$. The same ideas also apply to the *labeled* version of random variables. In fact, we shall here propose a slight generalization of an algebra of random variables, which underlines the fact that a random variable carries information not only about the pieces of information in the information algebra (Φ, D), but also on the assumptions in Ω.

We consider a *labeled Boolean information algebra* (Φ, D). This means in particular that, for every domain $x \in D$, the algebra Φ_x of information pieces with domain x is a Boolean algebra. Besides combination or join, there is also a meet or infimum $\phi \wedge \psi$ defined between two elements $\phi, \psi \in \Phi_x$. Furthermore, let $r = \{1, \ldots, n\}$ be an index set and Ω_i a finite set associated with index $i \in r$, the frame of the assumption i. For any subset $s \subseteq r$, define

$$\Omega_s = \prod_{i \in s} \Omega_i.$$

The elements of Ω_s will be denoted by ω, which can also be considered as a mapping $\omega_s \to \cup_{i \in s} \Omega_i$ such that $\omega(i) \in \Omega_i$. Thus, ω is called an s-tuple and $\omega(i)$ is called the i-component of ω. For an s-tuple ω and $t \subseteq s$, we define the projection $\omega^{\downarrow t}$ to be the restriction of the mapping ω to t, that is, $\omega^{\downarrow t}(i) = \omega_i$ for all $i \in t$.

Let then P_r be a probability measure on Ω_r, defined by the element probabilities $p_r(\omega)$ for all $\omega \in \Omega_r$. Further, let P_s be the associated marginal distribution on the set Ω_s, defined by

$$p_s(\omega) = \sum_{\nu \in \Omega_r : \nu^{\downarrow s} = \omega} p_r(\nu).$$

Often, the probabilities on the individual assumption frames Ω_i for $i \in r$ are assumed to be stochastically independent. Then,

$$p_r(\omega) = \prod_{i \in r} p_i(\omega(i)),$$

and, for a subset s of r, the probabilities

$$p_s(\omega) = \prod_{i \in s} p_i(\omega(i))$$

define the marginal probability measure P_s. This holds in particular for $s = r$.

An assumption-based information in this context is a labeled random variable $\Delta : \Omega_s \to \Phi_x$. As long as the domain s of the assumptions remains fixed, this is just as discussed in Section 3. But now we not only consider varying domains $x \subseteq s$, but also varying domains s of assumptions. So we consider the product lattice $\mathcal{P} \times D$ of the subset lattice of r with the lattice D. The label of Δ is then defined to be $d(\Delta) = (s, x)$.

Let the set of all such labeled random variables be denoted by $\mathcal{R}$. We are now going to introduce the operations of combination of random variables as well as projection and shall then verify that they form an information algebra. This then gives more substance to the claim above: Random variables represent information not only with respect to Φ, but also with respect to the assumptions in Ω_s. Here are the definitions:

1. *Combination:* Let Δ_1 and Δ_2 be two random variables with domains $d(\Delta_1) = (s, x)$ and $d(\Delta_2) = (t, y)$. Then, $\Delta_1 \otimes \Delta_2$ is, for $\omega \in \Omega_{s \cup t}$, defined by

$$\Delta_1 \otimes \Delta_2(\omega) = \Delta_1(\omega^{\downarrow s}) \otimes \Delta_2(\omega^{\downarrow t}).$$

Note that the combination on the right operator is the one of the algebra Φ.

2. *Projection:* Let Δ be a random variable with domain $d(\Delta) = (s, x)$ and let $t \subseteq s$, $y \leq x$. Then, $\Delta^{\downarrow t, y}$ is defined for $\omega \in \Omega_t$ by

$$\Delta^{\downarrow t, y}(\omega) = \bigwedge_{\nu \in \Omega_s : \nu^{\downarrow t} = \omega} \Delta^{\downarrow y}(\nu).$$

The idea behind this definition of the projection is that, for a given $\omega \in \Omega_t$, we must consider the possible information pieces $\Delta(\nu)$ for all ν which are compatible with ω, this is, projected to ω. Since in ω we lose information about the possible assumption, we associate with ω the infimum of all compatible pieces of information in Φ_x. This piece of information is then projected to the domain y. Note that $(\phi \wedge \psi)^{\downarrow y} = \phi^{\downarrow y} \wedge \psi^{\downarrow y}$ (see the chapter Information Algebra). This then justifies the definition above.

With these definitions, the axioms of a labeled information algebra (see the chapter Information Algebra) can be easily verified. Commutativity and associativity of combination follow from the corresponding properties in the algebra Φ and from the property $(\omega^{\downarrow t})^{\downarrow u}$ if $u \subseteq t$. The neutral element of the combination in the domain (s, x) is the mapping $E_{s,x}(\omega) = e_x$ for all $\omega \in \Omega_s$, where e_x is the neutral element in Φ_x; and the null element is $Z_{s,x}(\omega) = z_x \in \Phi_x$, where z_x is the null element in Φ_x. The neutrality and nullity axioms are easily verified. The labeling axiom holds by definition. A bit more involved is the verification of the projection, the combination and the idempotency axioms. For the projection axiom, assume that $d(\Delta) = (s, x)$ and $u \subseteq t \subseteq s$ and $z \leq y \leq x$. Then, by repeated application of the definition of projection, by the distributivity of projection over meet in the algebra Φ_x, and, by the projection axiom of Φ_x,

$$(\Delta^{\downarrow t, y})^{\downarrow u, z}(\omega) = \bigwedge_{\mu \in \Omega_t : \mu^{\downarrow u} = \omega} \left(\bigwedge_{\nu \in \Omega_s : \nu^{\downarrow t} = \mu} (\Delta(\nu))^{\downarrow y} \right)^{\downarrow z}$$

$$= \bigwedge_{\mu \in \Omega_t : \mu^{\downarrow u} = \omega} \left(\bigwedge_{\nu \in \Omega_s : \nu^{\downarrow t} = \mu} (\Delta(\nu))^{\downarrow y})^{\downarrow z} \right)$$

$$= \bigwedge_{\mu \in \Omega_t : \mu^{\downarrow u} = \omega} \left(\bigwedge_{\nu \in \Omega_s : \nu^{\downarrow t} = \mu} (\Delta(\nu))^{\downarrow z} \right).$$

At this point, we invoke the associativity of the meet operation in Φ_x and obtain

$$(\Delta^{\downarrow t, y})^{\downarrow u, z}(\omega) = \bigwedge_{\nu : (\nu^{\downarrow t})^{\downarrow u} = \omega} (\Delta(\nu))^{\downarrow z}$$

$$= \bigwedge_{\nu : \nu^{\downarrow u} = \omega} (\Delta(\nu))^{\downarrow z}$$

$$= \Delta^{\downarrow u, z}(\omega).$$

This proves the projection axiom.

In order to prove the combination axiom, assume $d(\Delta_1) = (s, x)$ and $d(\Delta_2) = (t, y)$. Then, by definition of combination and projection, it follows that, for $\omega \in \Omega_s$, using the combination axiom in Φ,

$$(\Delta_1 \otimes \Delta_2)^{\downarrow s,x}(\omega) = \bigwedge_{\nu:\nu^{\downarrow s}=\omega} (\Delta_1(\nu^{\downarrow s}) \otimes \Delta_2(\nu^{\downarrow t}))^{\downarrow x}$$

$$= \bigwedge_{\nu:\nu^{\downarrow s}=\omega} \Delta_1(\omega) \otimes (\Delta_2(\nu^{\downarrow t}))^{\downarrow x \wedge y}.$$

Now we invoke the distributive law in the Boolean algebra Φ and the idempotency of $\wedge$ and conclude

$$(\Delta_1 \otimes \Delta_2)^{\downarrow s,x}(\omega) = \Delta_1(\omega) \otimes \left(\bigwedge_{\nu:\nu^{\downarrow s}=\omega} \Delta_2(\nu^{\downarrow t})^{\downarrow x \wedge y} \right)$$

$$= \Delta_1(\omega) \otimes \left(\bigwedge_{\nu:\nu^{\downarrow s}=\omega,\nu^{\downarrow t}=\mu} \Delta_2(\mu)^{\downarrow x \wedge y} \right)$$

$$= \Delta_1(\omega) \otimes \left(\bigwedge_{\mu \in \Omega_t:\exists \nu \in \Omega_{s \cup t},\nu^{\downarrow s}=\omega,\nu^{\downarrow t}=\mu} \Delta_2(\mu)^{\downarrow x \wedge y} \right)$$

$$= \Delta_1(\omega) \otimes \left(\bigwedge_{\mu \in \Omega_t:\mu^{\downarrow s \cap t}=\omega^{\downarrow s \cap t}} \Delta_2(\mu)^{\downarrow x \wedge y} \right).$$

Then, it follows from the definition of projection,

$$(\Delta_1 \otimes \Delta_2)^{\downarrow s,x}(\omega) = \Delta_1(\omega) \otimes \Delta_2^{\downarrow s \cap t, x \wedge y}(\omega^{\downarrow s \cap t})$$

$$= \Delta_1 \otimes \Delta_2^{\downarrow s \cap t, x \wedge y}(\omega^{\downarrow s \cap t}).$$

This shows that the combination axiom holds in the algebra $(\mathcal{R}_s, \mathcal{P} \times D)$.

Finally, consider $d(\Delta) = (s, x)$ and $t \subseteq s$ and $y \leq x$. Then, using the idempotency in Φ,

$$\Delta \otimes \Delta^{\downarrow t,y}(\omega) = \Delta(\omega) \otimes \Delta^{\downarrow t,y}(\omega^{\downarrow t})$$

$$= \Delta(\omega) \otimes \left(\bigwedge_{\mu:\mu^{\downarrow t}=\omega^{\downarrow t}} (\Delta(\mu))^{\downarrow y} \right)$$

$$= \bigwedge_{\mu:\mu^{\downarrow t}=\omega^{\downarrow t}} (\Delta(\omega) \otimes (\Delta(\mu))^{\downarrow y})$$

$$= \bigwedge_{\mu:\mu^{\downarrow t}=\omega^{\downarrow t}} \Delta(\omega)$$

$$= \Delta(\omega)$$

since $\Delta(\omega) \otimes (\Delta(\mu))^{\downarrow y} \geq \Delta(\omega)$. This then proves idempotency, $\Delta \otimes \Delta^{\downarrow t,y} = \Delta$. Altogether, this shows that the algebra of assumption-based information is really an information algebra.

Theorem 7. *The algebra $(\mathcal{R}_s, \mathcal{P} \times D)$ with the operation of labeling, combination and projection as defined above is an information algebra.*

In fact, $(\mathcal{R}_s, \mathcal{P} \times D)$ is a *Boolean information algebra*. The meet between two random variables Δ_1 and Δ_2 with the same domain (s, x) is defined by $\Delta_1 \wedge \Delta_2(\omega) = \Delta_1(\omega) \wedge \Delta_2(\omega)$. If Δ is a random variable with label (s, x), then Δ^c, defined for $\omega \in \Omega_s$ by $\Delta^c(\omega) = (\Delta(\omega))^c$, is clearly the *complement* of Δ: It holds that $\Delta \otimes \Delta^c = Z_{s,x}$ and $\Delta \wedge \Delta^c = E_{s,x}$. So the random variables for a fixed domain (s, x) form a Boolean algebra. The property $(\Delta_1 \wedge \Delta_2) \otimes E_{t,y} = (\Delta_1 \otimes E_{t,y}) \wedge (\Delta_2 \otimes E_{t,y})$ is inherited from the underlying Boolean information algebra (Φ, D). The instance of assumption-based information worked out so far in depth is the case of *propositional assumption-based information systems* [3]. An application of this framework to statistical inference sketched along general lines in Section 2 is given in [10].

So the information algebra $(\mathcal{R}_s, \mathcal{P} \times D)$ underlines the fact that random variables, representing assumption-based information, contain information both with respect to the algebra Φ, as well as to the assumptions in Ω. And, as explained above, the order induced in $(\mathcal{R}_s, \mathcal{P} \times D)$ reflects the corresponding information content. With respect to Ω_s, the information is given by the set $\Omega_\Delta^\downarrow = \{\omega \in \Omega_s : \Delta(\omega) \neq z_x\}$. Neglecting the probability measure on the assumptions or, equivalently, assuming uniform probability distributions, the discussion in Section 4.4 in the chapter Information Algebra carries over to the algebra $(\mathcal{R}_s, \mathcal{P} \times D)$. This will be worked out in the following section.

Before we turn to this issue, we remark that the domain-free version of the algebra $(\mathcal{R}_s, \mathcal{P} \times D)$ can be obtained by extending a random variable Δ with domain (s, x) to a mapping $[\Delta] : \Omega_r \to \Phi/\equiv$, where $\Phi/\equiv$ is the domain-free version of the labeled algebra (Φ, D), as follows: $[\Delta](\omega) = [\Delta(\omega^{\downarrow s})]$.

5.2   Information Measures

In the chapter Information Algebra, we have seen that the information content of a piece of information can be quantitatively measured if the algebra is assumed to be finitely atomic, that is if each of its elements is represented by the finite set of atoms it contains. The algebra then essentially becomes a set or relational algebra. This can be extended to the information algebra $(\mathcal{R}_s, \mathcal{P} \times D)$ of random variables introduced above. Assume (Φ, D) finitely atomic and let $At(\phi) \subseteq At(e_x)$ denote the set of atoms in ϕ, where $At(e_x)$ is the set of all atoms in Φ_x. What are then the atoms in $(\mathcal{R}_s, \mathcal{P} \times D)$? Select an assumption $\mu \in \Omega_s$ in the domain (s, x) and an atom in α in $At(e_x)$. Define the mapping $A : \Omega_s \to \Phi_x$

$$A(\omega) = \begin{cases} \alpha & \text{if } \omega = \mu, \\ z_x & \text{otherwise.} \end{cases}$$

Clearly, the variable A is an atom in the algebra $(\mathcal{R}_s, \mathcal{P} \times D)$, that is either $A \otimes \Delta = A$, in which case we write $A \in \Delta$, or $A \otimes \Delta = Z_{s,x}$ for all random variables Δ with label (s,x). In the first case, we have $\mu \in \Omega_\Delta^\downarrow$ and $A(\mu) \in At(\Delta(\mu))$. Also, the property

$$\bigwedge_{A \in \Delta} A = \Delta$$

is inherited from the atomic algebra (Φ, D) as is easily verified. Thus, the information algebra $(\mathcal{R}_s, \mathcal{P} \times D)$ is atomic, so we may carry over the considerations in Section 4.4 of the chapter on Information algebras to the present case.

Thus, the information content of an uncertain assumption-based information, represented by a random variable $\Delta \in \mathcal{R}_s$, is measured in terms of the numbers of atoms in Δ. However, there are, at first sight, two ways of doing this: The first one, which directly carries over from ordinary information algebras, is to measure the uncertainty contained in Δ by the logarithm of the number of its atoms:

$$\log \sum_{\omega \in \Omega_\Delta^\downarrow} |At(\Delta(\omega))|.$$

However, this neglects the semantic idea behind assumption-based information: Given a possible assumption $\omega \in \Omega^\downarrow$, the measure of uncertainty remaining in the piece of information $\Delta(\omega)$ is $\log|At(\Delta(\omega))|$. Thus, the expected conditional uncertainty of $\Delta(\omega)$ is

$$\sum_{\omega \in \Omega_\Delta^\downarrow} \frac{1}{|\Omega_\Delta^\downarrow|} \log|At(\Delta(\omega))|.$$

To this is added the uncertainty about the possible assumptions, measured by $\log|\Omega_\Delta^\downarrow|$. Thus, in the spirit of the chaining theorem for entropies, the total uncertainty remaining in Δ can be measured by

$$\log|\Omega_\Delta^\downarrow| + \sum_{\omega \in \Omega_\Delta^\downarrow} \frac{1}{|\Omega_\Delta^\downarrow|} \log|At(\Delta(\omega))|. \tag{24}$$

The original uncertainty, before the assumption-based information Δ arises, is $\log\left(|\Omega| \cdot |At(e_x)|\right)$. Thus, the information contained in Δ can be defined as

$$i(\Delta) = \log\left(|\Omega| \cdot |At(e_x)|\right) - \log|\Omega_\Delta^\downarrow| - \sum_{\omega \in \Omega_\Delta^\downarrow} \frac{1}{|\Omega_\Delta^\downarrow|} \log|At(\Delta(\omega))|$$

$$= -\log\frac{|\Omega_\Delta^\downarrow|}{|\Omega|} - \sum_{\omega \in \Omega_\Delta^\downarrow} \frac{1}{|\Omega_\Delta^\downarrow|} \log\frac{|At(\Delta(\omega))|}{|At(e_x)|}.$$

The fractions $p(\Omega_\Delta^\downarrow) = \frac{|\Omega_\Delta^\downarrow|}{|\Omega|}$ and $q(\Delta(\omega)) = \frac{1}{|\Omega_\Delta^\downarrow|} \log\frac{|At(\Delta(\omega))|}{|At(e_x)|}$ can be interpreted as probabilities $p(\Omega_\Delta^\downarrow)$ and $q(\Delta(\omega))$. Then, the information in Δ can also be written as

$$i(\Delta) = -\log p(\Omega_\Delta^\downarrow) - \sum_{\omega \in \Omega_\Delta^\downarrow} \frac{1}{|\Omega_\Delta^\downarrow|} \log q(\Delta(\omega)).$$

The first term represents the part of the information on the assumptions and the second one the part on the information algebra Φ.

As in the case of deterministic information, discussed in Section 4.4, the measure of information introduced above respects the order between assumption-based information:

Theorem 8. *Let $(\mathcal{R}_s, \mathcal{P} \times D)$ be the algebra of assumption-based information. Then, for all $\Delta_1, \Delta_2 \in \mathcal{R}_s$, with the same domain $d(\Delta_1) = d(\Delta_2)$, the inequality $\Delta_1 \leq \Delta_2$ implies $i(\Delta_1) \leq i(\Delta_2)$.*

Proof. The claim follows since $\Delta_1 \leq \Delta_2$ implies $\Omega_{\Delta_1}^{\downarrow} \supseteq \Omega_{\Delta_2}^{\downarrow}$, hence $|\Omega_{\Delta_1}^{\downarrow}| \geq |\Omega_{\Delta_2}^{\downarrow}|$ and also $|At(\Delta_1(\omega))| \geq |At(\Delta_2(\omega))|$ for all $\omega \in \Omega_{\Delta_2}^{\downarrow}$. $\square$

In particular, this theorem implies that the combination of information $\Delta_1 \otimes \Delta_2$ increases information content. More generally, we may define the information in an assumption-based information Δ with respect to any domain (t, y) by $i(\Omega^{\to t, y})$, where the transport operation $\Omega^{\to t, y}$ is defined as usual (see Section 2.4 in the chapter Information Algebra). Then, similiar monotonicity properties with respect to these more general information measures hold as in the theorem above.

So far, we have not considered the probability measure on the assumptions in Ω_r. This is additional information, which reduces the uncertainty associated with a piece of assumption-based information $\Delta : \Omega_s \to \Phi_x$. In fact, the consideration above can be extended to this case of probabilistic information. As explained above (Section 5.1), the random variables Δ contain information both with respect to the domain x and the elements in Ω. Relevant for this is the normalized variable $\Delta^{\downarrow}$, where the impossible assumptions ω leading to the contradiction $\Delta(\omega) = z_x$ are eliminated. So we consider the possible assumptions in $\Omega^{\downarrow} = \{\omega \in \Omega : \Delta(\omega) \neq z\}$ and the conditional probabilities

$$p'(\omega) = \frac{p(\omega)}{K}, \text{ where } K = \sum_{\omega \in \Omega^{\downarrow}} p(\omega),$$

for all $\omega \in \Omega^{\downarrow}$. The uncertainty in the information represented by Δ can be obtained from formula (24) by simply substituting the entropy of the probability distribution $p'(\omega)$ for the uniform probability $1/|\Omega_{\Delta}^{\downarrow}|$ in the expected uncertainty on $At(e_x)$, which gives

$$H(\Delta) = -\sum_{\omega \in \Omega_{\Delta}^{\downarrow}} p'(\omega) \log p'(\omega) + \sum_{\omega \in \Omega_{\Delta}^{\downarrow}} p'(\omega) \log |At(\Delta(\omega))|. \qquad (25)$$

The information contained in Δ, including the probability information, is again measured by the difference between the uncertainty associated with full ignorance and the uncertainty left in Δ:

$$i_p(\Delta) = \log |\Omega| + \log |At(e_x)| - H(\Delta)$$
$$= \sum_{\omega \in \Omega^{\downarrow}} p'(\omega) \log (p'(\omega) \cdot |\Omega|) + \sum_{\omega \in \Omega^{\downarrow}} p'(\omega) \log \frac{|At(e_x)|}{|At(\Delta(\omega))|}.$$

The first line here shows that $i(\Delta)$ is nonnegative, since $\log\left(|\Omega|\cdot|At(e_x)|\right)$ is the maximum entropy, hence at least as large as $H(\Delta)$. We define

$$p(\omega,\alpha) = \begin{cases} \frac{p'(\omega)}{|At(\Delta(\omega))|} & \text{if } \alpha \in \Delta(\omega), \\ 0 & \text{otherwise.} \end{cases}$$

Then, we can also write

$$i_p(\Delta) = \log\left(|\Omega|\cdot|At(e_x)|\right) - \sum_{\omega,\alpha} p(\omega,\alpha)\log p(\omega,\alpha)$$

$$= \sum_{(\omega,\alpha):\alpha\in At(\Delta(\omega))} \frac{p'(\omega)}{|At(\Delta(\omega))|} \log \frac{p'(\omega)}{|At(\Delta(\omega))|}|\Omega|\cdot|At(e_x)|.$$

The last line is the Kullback-Leibler divergence or the relative entropy between the probability distribution $p(\omega,\alpha)$ and the uniform distribution on $\Omega \times At(e_x)$.

6   Conclusion

Random variables with values in an information algebra represent uncertain assumption-based information, which forms itself an information algebra. In this chapter, the situation is a little bit simplified from a mathematical point of view since only finite sets of assumptions or sample spaces are considered. This covers for example the case of propositional argumentation systems [3]. More general situations have been studied from the point of view of an application to statistical inference in [10, 17]. In this respect, especially linear systems of equations with Gaussian disturbances have been considered. In an abstract setting, the framework has been discussed in [5, 6, 7, 8, 11]. The theory resembles and is related to random set theory [16, 18], where random variables with values in certain set algebras are studied, especially in certain topological spaces.

From a practical point of view, a language is needed to describe the uncertain, assumption-based information. This can be a logical language, like propositional logic [3], or predicate logic. It can be the language of linear algebra, as in the case of linear systems of equalities with stochastic disturbances [10, 17]. In both cases, the formalism behind the language permits the derivation of supports and the computation of degrees of support for given hypotheses. In the case of logic, this will be deduction schemes. In the case of linear systems, matrix algebra can be used. In the case of stochastically independent simple random variables, the structure of the valuation algebra of bpas (Section 4.3) allows for efficient local computation schemes, just as for ordinary information algebras [6, 20].

So, whereas a theory of random variables with values in an information algebra provides a general setting to study uncertain, assumption-based information, for practical applications, a particular framework must be selected, appropriate for the given problem. Only within such a specific framework can the computations be performed which lead to practically useful results in domains like reliability and diagnostics [1]. Once more, this underlines that the theory of information algebra is a theory of computation, in contrast to Shannon's information theory, which is a theory of communication.

References

1. Anrig, B., Kohlas, J.: Model-based reliability and diagnostic: A common framework for reliability and diagnostics. Int. J. of Intell. Systems 18(10), 1001–1033 (2003)

2. Dempster, A.: Upper and lower probabilities induced by a multivalued mapping. Ann. Math. Stat. 38, 325–339 (1967)

3. Haenni, R., Kohlas, J., Lehmann, N.: Probabilistic argumentation systems. In: Kohlas, J., Moral, S. (eds.) Handbook of Defeasible Reasoning and Uncertainty Management Systems. Algorithms for Uncertainty and Defeasible Reasoning, vol. 5, pp. 221–287. Kluwer, Dordrecht (2000)

4. Kappos, D.A.: Probability Algebras and Stochastic Spaces. Academic Press, New York (1969)

5. Kohlas, J.: Computational theory for information systems. Technical Report 97–07, Institute of Informatics, University of Fribourg (1997)

6. Kohlas, J.: Information Algebras: Generic Structures for Inference. Discrete Mathematics and Theoretical Computer Science. Springer, Heidelberg (2003)

7. Kohlas, J.: Probabilistic argumentation systems: A new way to combine logic with probability. Journal of Applied Logic 1(3-4), 225–253 (2003)

8. Kohlas, J.: Uncertain information: random variables in graded semilattices. Int. J. Approx. Reason.(2007) doi:10.1016/j.ijar.2006.12.005

9. Kohlas, J., Haenni, R., Lehmann, N.: Assumption-based reasoning and probabilistic argumentation systems. In: Kohlas, J., Moral, S. (eds.) Algorithms for Uncertainty and Defeasible Reasoning (1999) (accepted for Publication)

10. Kohlas, J., Monney, P.-A.: Statistical Information. Assumption-Based Statistical Inference. Sigma Series in Stochastics, vol. 3. Heldermann, Lemgo (2008)

11. Kohlas, J., Monney, P.A.: Representation of evidence by hints. In: Yager, R.R., Kacprzyk, J., Fedrizzi, M. (eds.) Advances in the Dempster-Shafer Theory of Evidence, pp. 472–492. John Wiley & Sons, Chichester (1994)

12. Kohlas, J., Monney, P.A.: A Mathematical Theory of Hints. An Approach to the Dempster-Shafer Theory of Evidence. Lecture Notes in Economics and Mathematical Systems, vol. 425. Springer, Heidelberg (1995)

13. Kohlas, J., Monney, P.A.: A Mathematical Theory of Hints. An Approach to the Dempster-Shafer Theory of Evidence. Lecture Notes in Economics and Mathematical Systems, vol. 425. Springer, Heidelberg (1995)

14. Kohlas, J., Shenoy, P.P.: Computation in valuation algebras. In: Kohlas, J., Moral, S. (eds.) Handbook of Defeasible Reasoning and Uncertainty Management Systems, Algorithms for Uncertainty and Defeasible Reasoning, vol. 5, pp. 5–39. Kluwer, Dordrecht (2000)

15. Lauritzen, S.L., Spiegelhalter, D.J.: Local computations with probabilities on graphical structures and their application to expert systems. J. Royal Statis. Soc. B 50, 157–224 (1988)

16. Molchanov, I.: Theory of Random Sets. Springer, London (2005)

17. Monney, P.-A.: A Mathematical Theory of Arguments for Statistical Evidence. In: Contributions to Statistics. Physica-Verlag (2003)

18. Nguyen, H.: On random sets and belief functions. Journal of Mathematical Analysis and Applications 65, 531–542 (1978)

19. Shafer, G.: A Mathematical Theory of Evidence. Princeton University Press, Princeton (1976)
20. Shenoy, P.P., Shafer, G.: Axioms for probability and belief-function propagation. In: Ross, D., Shachter, T.S., Levitt, L.N. (eds.) Uncertainty in Artificial Intelligence 4. Machine intelligence and pattern recognition, vol. 9, pp. 169–198. Elsevier, Amsterdam (1990)
21. Shenoy, P.P., Shafer, G.: Axioms for probability and belief function propagation. In: Lemmer, J.F., Shachter, R.D., Levitt, T.S., Kanal, L.N. (eds.) Uncertainty in Artif. Intell., vol. 4, pp. 169–198. North-Holland, Amsterdam (1990)

Comparing Questions and Answers: A Bit of Logic, a Bit of Language, and Some Bits of Information*

Robert van Rooij

Institute for Logic, Language and Computation
University of Amsterdam
R.A.M.vanRooij@uva.nl

1 Introduction

Notions like 'entropy' and '(expected) value of observations' are widely used in science to determine which experiment to conduct to make a better informed choice between a set of scientific theories that are all consistent with the data. But these notions seem to be almost equally important for our use of language in daily life as they are for scientific inquiries.

I will make use of these notions to measure how 'good' particular questions and answers are in particular circumstances. In doing so, I will extend and/or refine the *qualitative* approach towards such measurements proposed by Groenendijk & Stokhof (1984). The refinements are due to the fact that I also take into account (i) probabilities, (ii) utilities, and (iii) the idea that we ask questions to resolve decision problems.

In this paper I will first explain Groenendijk & Stokhof's partition based analysis of questions, and then discuss their *qualitative* method of measurement. Next, I will take also *probabilities* into account, and show how a natural *quantitative* measure of informativity can be defined in terms of it. Following the lead of Communication Theory and Inductive Logic, I will then show that we can also describe a natural measure of the informative value of questions and answers in terms of *conditional entropy*, when we take into account that questions are asked to resolve *decision problems*. Finally, I will argue that to measure the value of questions and answers we should in general also take *utilities* seriously, and following standard practice in Statistical Decision Theory, I show how some intuitively appealing utility values can be specified.

---

* I appreciate it a lot that Giovanni Sommaruga invited me to submit this paper to the present volume, given that the bulk of it was written already in 2000. I would like to thank the following people for discussion and comments: Alexandru Baltag, Balder ten Cate, Paul Dekker, Roberto Festa, Jeroen Groenendijk, Emiel Krahmer, Marie Nilsenova, and Yde Vennema. Since the time that I wrote this paper, I have published two articles (van Rooij 2004a,b) that could be thought of as successors of this paper.

G. Sommaruga (Ed.): Formal Theories of Information, LNCS 5363, pp. 161–192, 2009.

2   The Semantics of Declaratives and Interrogatives

The perhaps most 'natural' conception of 'meaning', at least in its point of departure, identifies 'meaning' with *naming*. The meaning of an expression is that what the expression *refers to*, or *is about*. What meaning does is to establish a *correspondence* between expressions in a *language* and things in the (model of the) *world*. For simple expressions like proper names and simple declarative sentences, this view of meaning is natural and simple. The meaning of *John* is the object it refers to, while the meaning of a simple declarative sentence like *John is sick* could then be the *fact* that John is sick. Beyond this point of departure, things are perhaps less natural. What, for example, should be the things out in the world that a negated sentence like *John is not sick* is about, and what should *John is sick* be about if the sentence is false? Notice that to be a competent speaker of English one has to know what it means for *John is sick* to be true or false. So a minimal requirement for any theory of meaning would be that one knows the meaning of a declarative sentence if one knows under which circumstances it is, or would be, true. The proposal of formal semanticists to solve our above conceptual problems is to stick to this minimal requirement: identify the meaning of a declarative sentence with the *conditions*, or *circumstances* under which the sentence is *true*. These circumstances can, in turn, be thought of as the ways the world might have been, or *possible worlds*. Thus, the meaning of a sentence can be thought of as the set of circumstances, or possible worlds, in which it is true. This latter set is known in possible worlds semantics as the *proposition* expressed by the sentence. We will denote the meaning of any declarative sentence A by $[[A]]$, and identify it with the set of worlds in which A is true (where W is the set of all possible worlds):[1]

$$[[A]] = \{w \in W : A \text{ is true in } w\}.$$

Just as it is standard to assume that you know the meaning of a declarative sentence when you know under which circumstances this sentence is true, Hamblin (1958) argues that you know the meaning of a question when you know what counts as an appropriate answer to the question. Because we answer a question by making a statement that expresses a proposition, this means that the meaning of a question as linguistic object (interrogative sentence) can be equated with the set of propositions that would be expressed by the appropriate linguistic answers. This gives rise to the problem what an appropriate linguistic answer is to a question.

For a *yes/no*-question like *Is John sick?* it is widely agreed that it has only one appropriate true answer; *Yes* in case John is sick, and *No* when John is not sick. This means that with respect to each world a *yes/no*-question simply expresses a proposition; the proposition expressed by the true appropriate answer in that world. If we represent a *yes/no*-question simply by a formula like $?A$, where A

---

[1] Here, and elsewhere in this paper, I will assume that we analyze sentences with respect to a fixed intensional model.

is a sentence, and assume that $[[A]]^w$ denotes the truth value of A in w, the proposition expressed by question $?A$ in world w is:

$$[[?A]]^w \quad = \quad \{v \in W : \; [[A]]^v = [[A]]^w\}.$$

Given this analysis of polar interrogative sentences, the question arises what the meaning of a *wh*-question is; i.e. what counts in a world as an appropriate true answer to questions like *Who is sick?* and *Who kissed whom?*

Groenendijk & Stokhof (1984) have argued on the basis of linguistic phenomena that not only *yes/no*-questions, but also (multiple) *wh*-questions can in each world only have *one true (complete) answer*. They argue that for John to know the answer to the question *Who is sick?*, for instance, John must know of *each* (relevant) individual *whether* he or she is sick.[2]

Representing questions abstractly by $?P$, where P is an n-ary predicate, John gives in w the true and complete answer to the above question just in case he gives an answer that entails the following proposition, where $[[P]]^v$ denotes the extension of predicate P in world v:

$$[[?P]]^w \quad = \quad \{v \in W| \; [[P]]^v = [[P]]^w \}.$$

If P is a 1-ary predicate like *is sick*, $[[P]]^w$ denotes the set of individuals that are sick in w, and $[[?P]]^w$ denotes the set of worlds where P has the same extension as in world w. If P is a binary predicate like *kissed*, $[[P]]^w$ denotes the set of ordered pairs $\langle d, d' \rangle$ such that d kissed d' in w, and $[[?P]]^w$ denotes the set of worlds where the same individuals kissed each other as in world w. An interesting special case is when P is a zero-ary predicate, i.e., when P is a sentence and when the question is thus a *yes/no*-question. In that case the proposition expressed by the question in a world will be exactly the same as the proposition determined via our second equation. Thus, according to Groenendijk & Stokhof's (1982) proposal, we should not only treat single and multiple *wh*-questions in the same way, but we should analyze *yes/no*-questions in a similar way, too.

Suppose, contrary to Hamblin's suggestion, that we equate the meaning of a question with the meaning of its true answer. This would immediately allow us to define an entailment relation between questions.[3] We can just say that one question entails another, just in case the proposition expressed by the true answer to the first question entails the proposition expressed by the true answer to the second question. And given an entailment relation between questions, it seems only natural to say that one question is 'better', or 'more informative' than another exactly when the former question entails the latter.

However, the above suggested entailment-relation between questions, and the thus induced 'better than'-relation, doesn't seem to be very natural. Suppose

---

[2] This doesn't mean that everybody agrees. For a discussion of some problems, and alternative analyses of questions, see Groenendijk & Stokhof (1997).

[3] In this paper I will use the term 'question' not only for interrogative sentences, but also for the meanings they express. Something similar holds for the term 'answer'. I hope this will never lead to confusion.

that in fact both John and Mary are sick. In that case it holds that the true answer to the question *Are John and Mary sick?* entails the true answer to the question *Is John sick?*, and thus it is predicted that the first question also entails the second. But this prediction seems to be wrong; the first question does intuitively not entail the second question because when Mary were in fact not sick (although John still is), the true answer to the first question would no longer entail the true answer to the second question. What this suggests is that the entailment-relation between questions does not just depend on how the world *actually* is, but also on how the world *could have been*.

Above, we have defined the proposition expressed by a question with respect to the real world, w. The above discussion suggests that to define an entailment relation between propositions, we should *abstract* away from how the actual world looks like. We should say that one question entails another just in case knowing the true answer to the former means that you also know the true answer to the latter, *however the world looks like*. Thus, $?P_1$ entails $?P_2$, $?P_1 \models ?P_2$, just in case it holds for *every* world w that $[[?P_1]]^w$ is a subset of $[[?P_2]]^w$:

$$?P_1 \models ?P_2 \quad \text{iff} \quad \forall w : [[?P_1]]^w \subseteq [[?P_2]]^w.$$

We might also define this entailment relation between questions more directly in terms of their meanings. In order to do this, we should think of the meaning of a question itself no longer simply as a proposition, but rather as a *function* from worlds to propositions (answers):

$$[[?P]] \quad = \quad \lambda w.\{v \in W | \ [[P]]^v \ = \ [[P]]^w \ \}.$$

Notice that this function from worlds to propositions is simply equivalent to the following *set* of propositions:

$$\{\{v \in W | \ [[P]]^v \ = \ [[P]]^w \ \} | \ w \in W\}.$$

and, due to the assumption that a question has in each world only one true answer, this set of propositions *partitions* the set of worlds W. A partition of W is a set of mutually exclusive non-empty subsets of W such that their union equals W. In fact, the partition that is induced in this way by a question is exactly what Groenendijk & Stokhof (1984) have proposed to call the *semantic meaning*, or *intension*, of a question, and they distinguish it from the *extension* a question has, $[[?P]]^w = [[?P]](w)$, in the particular world w. Notice that Groenendijk & Stokhof's account is in accordance with Hamblin's proposal: the meaning of a question is represented by its set of possible appropriate answers.

We have seen that the partition semantics of questions is based on the assumption that every question can in each world have at most *one* semantic answer. Thus, if you ask me *Who of John and Mary are sick?*, I can only resolve the question according to this analysis by giving an *exhaustive* answer where I tell for both John and Mary whether they are sick or not. It might, however, be the case that I only know whether John is sick, and that I just respond by saying *(At least) John is sick*. This response will obviously not resolve the whole issue, and thus will not count as a *complete*, or semantic, answer to the question. Still, it does count as an answer to the question, although only a *partial* one. We can say that an assertion counts as a partial answer to the question iff it is a non-contradictory proposition that is incompatible with at least one cell of the partition induced by the question. In our above example, for instance, the response *(At least) John is sick* counts as a partial answer to the question, because it is incompatible with 2 of the 4 cells of the partition induced by the question. Observe that according to our characterization of partial answerhood, it holds that a complete, semantic, answer to the question also is incompatible with at least one cell of the partition, and thus also counts as a partial answer. So we see that some partial answers are more informative, and better, than others.

Suppose that Q and Q' are two partitions of the logical space that are induced by two interrogative sentences. Let us also assume for simplicity that we can equate the meaning of an interrogative sentence with the question itself. Making use of the fact that every question has according to their semantics (at most) one answer in each world, Groenendijk & Stokhof (1984) can define the entailment-relation between questions directly in terms of their meanings making use of a generalized subset-relation, '$\sqsubseteq$', between partitions. Remember that according to our above requirement, for question Q to *entail* question Q', $Q \models Q'$, it must be the case that knowing the true answer to Q means that you also know the true answer to Q', *however the world looks like*. In terms of Groenendijk & Stokhof's (1984) partition semantics this comes down to the natural requirement that for every element of Q there must be an element of Q' such that the former entails the latter, i.e. $Q \sqsubseteq Q'$:

$$Q \models Q' \quad \text{iff} \quad Q \sqsubseteq Q' \quad \text{iff} \quad \forall q \in Q : \exists q' \in Q' : q \subseteq q'.$$

According to this definition it follows, for instance, that the *wh*-question *Who of John and Mary are sick?* entails the *yes/no*-question *Is John sick?*, because every (complete) answer to the first question entails an answer to the second question. And indeed, when you know the answer to the first question, the second question can no longer be an issue. Something similar is the case for the multiple *wh*-question *Who kissed whom?* and the single *wh*-question *Who kissed Mary?*; learning the answer to the first question is *more informative* than learning the answer to the second question.

3 Comparing Questions and Answers Qualitatively

3.1 A Semantic Comparison

If somebody asks you who murdered Smith, he would not be satisfied with an answer like *The murderer of Smith*. Although this answer will obviously be true, it is unsatisfactory because the answer will *not* be *informative*. Indeed, it is generally agreed that in normal circumstances the utterance of an interrogative sentence is meant as a means to acquire information.

If the aim of the question is to get some information, it seems natural to say that Q is a *better* question than Q', if it holds that whatever the world is, knowing the true answer to question Q means that you also know the true answer to Q', i.e. $Q \sqsubseteq Q'$. As we have seen above, this would mean that the question *Who of John and Mary are sick?* is a 'better' question than *Is John sick?*, because the former question entails the latter. Notice that by adopting this approach, the *value*, or *goodness*, of a question is ultimately reduced to the pure *informativity* of the expected answer.

Not only can we compare questions to each other with respect to their 'goodness', the same can be done for *answers*. We have seen in the previous section that complete answers are special kinds of partial answers; the most informative partial answers that are true in the worlds of just one cell of a partition. This suggests, perhaps, the following proposal; say that one answer is 'better' than another, just in case the former *entails* the latter. But this would be mistaken, for it would wrongly predict that we prefer *over*informative answers to answers that are just complete. If I ask you, for instance, whether John is sick, I would be very puzzled by your answer *Yes, John is sick, and it is warm in Africa*. The second conjunct to the answer seems to be completely *irrelevant* to the issue, and thus should not be mentioned.

So it seems that we should measure the 'goodness' of an answer mostly in terms of the partition induced by the question. And indeed, this is exactly what Groenendijk & Stokhof (1984) propose. Define A_Q as the set of cells of partition Q that are compatible with answer A:[4]

$$A_Q \;=\; \{q \in Q : q \cap A \neq \emptyset\}.$$

Notice now that one partial answer can be more informative than another one because it is incompatible with more cells of the partition than the other one. Remember that the answer $A = (At\ least)\ John\ is\ sick$ counts as a partial answer to the question $Q = Who\ of\ John\ and\ Mary\ are\ sick?$, and is incompatible with 2 of the 4 cells of the partition. The answer $B = If\ Mary\ is\ not\ sick,\ then\ neither\ is\ John$ also counts as a partial answer to the question, because it is incompatible with 1 cell of the partition. But it is a weaker answer than *(At least) John is*, because it is entailed by the latter and incompatible with less cells of the partition than the former one, i.e. $A_Q \subset B_Q$. Groenendijk & Stokhof propose that when answer A is incompatible with more cells of the partition than answer B, i.e. $A_Q \subset B_Q$, the former should be counted as a better answer to the question than the latter.

But what if two answers are incompatible with the same cells of the partition, i.e. if $A_Q = B_Q$? It is possible that when two partial answers to a question are incompatible with, for example, just one cell of the partition, one of them can be more informative than the other because the former *entails* the latter. In our above example, for instance, not only *(At least) John is sick*, but also *John is sick, and it is warm in Africa* is an answer that is incompatible with just two cells of the partition induced by the question. As we have suggested above already, the former counts in that case as a better answer than the latter,

---

[4] From now on I tend to use the same notation both for a declarative sentence and the proposition it expresses. I hope this will never lead to confusion.

because it doesn't give extra *irrelevant* information. Thus, in case $A_Q = B_Q$, A is a better answer than B iff $A \supset B$.

Combining both constraints, Groenendijk & Stokhof (1984) propose that A is (quantitatively) a *better* semantic answer to question Q than B, $A >_Q B$, by defining the latter notion as follows:

$$A >_Q B \quad \text{iff either (i)} \quad A_Q \subset B_Q, \text{ or}$$
$$\text{(ii)} \quad A_Q = B_Q, \text{ and } A \supset B.$$

Lewis (1988) and Groenendijk (1999) defined a notion of *aboutness* in terms of which answers can be compared in a more direct way.[5] They say that answer A is *about* question Q just in case the following condition is satisfied:

$$A \text{ is } about \ Q \quad \text{iff} \quad \forall q \in Q : q \subseteq A \text{ or } q \cap A \neq \emptyset.$$

Thus, when A is true/false in a world w, it should be the case that A is also true/false in any world v that is an element of the same cell of the partition Q as w is. Notice that because Q is a partition, the above definition of aboutness is equivalent to the following condition:

$$A \text{ is } about \ Q \quad \text{iff} \quad \bigcup A_Q = A.$$

This notion of aboutness intuitively corresponds with the second condition in the definition of $>_Q$ that no extra *irrelevant* information should be given. Using the standard Stalnakerian (1978) assertion conditions, we might say that with respect to a certain question, an assertion is *relevant* if it is (i) consistent, (ii) informative, and (iii) is about the question. In terms of such a notion of relevance, we can re-define the above 'better than' relation, $A >_Q B$, between relevant answers A and B to question Q simply as follows:

$$A >_Q B \quad \text{iff} \quad A \subset B.$$

Notice that according to the above analysis, any *contingent* proposition satisfies the first two constraints of the above definition of *relevance*. But some contingent propositions are, of course, intuitively *irrelevant* because they are already entailed by, or inconsistent with, what is already believed by the participants of the conversation. It is only natural to expect that what is believed also influences the comparative goodness relation of answers to questions. And indeed, that turns out to be the case.

3.2   A Pragmatic Comparison

Although the above defined comparative notion of goodness of answers is quite appealing, it still can be the case that certain answers to a question can be better than others, although they are according to the above ordering relations predicted to be worse. It can even be the case that some responses to questions are predicted to be semantically *irrelevant*, because they do not even give a

---

[5] In Groenendijk (1999) the notion is called 'licencing'.

partial semantic answer to the question, but still *completely* resolve the issue. If I ask you, for instance, *What are the official languages spoken in Belgium?*, you can intuitively resolve the issue by saying *The official languages of its major neighboring countries*. The reason is, of course, that the relevance of an answer should always be determined with respect to what is believed/known by the questioner. The above answer would completely resolve my question, because I know what the major neighboring countries of Belgium are (France, Germany, and the Netherlands), and I know which official languages are spoken in those countries (French, German, and Dutch, respectively).[6]

The relevance of a question, too, depends on the relevant information state. Although the question *What is the official language of the Netherlands?* gives *semantically* rise to a non-trivial partition, I wouldn't learn much when you told me the answer. We can conclude that we should *relativize* the definitions of relevance and goodness of questions and answers to particular information states.

In comparing the 'goodness' of questions to one another, and in comparing answers, we have until now neglected what is already known or believed by the agent who asks the question. When we denote the relevant information state of the questioner by K, which is represented by a set of possible worlds, we can redefine the relevant notions. First, we can define the meaning of question $?P$ with respect to information state K, $[[?P]]_K$:

$$[[?P]]_K = \{\{v \in K|\ [[P]]^v = [[P]]^w\ \}|\ w \in K\}.$$

Then we can say that *question Q is relevant* with respect to information state K just in case Q_K is a non-singleton set. To determine whether A is a *relevant answer* to Q with respect to information state K, we first define $A_{Q,K}$, which denotes the set of cells of Q_K compatible with proposition A:

$$A_{Q,K} = \{q \in Q_K :\ q \cap A \neq \emptyset\}.$$

Now we can say that A is *about Q with respect to K*, just in case $\bigcup A_{Q,K} = (K \cap A)$. Then we call A a *relevant answer* to Q w.r.t. K iff it is *contingent* with respect to K and *about Q with respect to K*. Now we are ready to compare questions and answers with respects to information states. First questions:

Question Q is *at least as good w.r.t. K as* Q' iff $Q_K \sqsubseteq Q'_K$.

Determining the ordering relation for answers A and B that are relevant with respect to Q and K is equally straightforward:

$$A \geq_{Q,K} B \quad \text{iff} \quad (K \cap A) \subseteq (K \cap B).$$

If we want, we might also follow Groenendijk & Stokhof (1984) by also comparing answers that express the same proposition with respect to our state K. They propose that in that case one answer is better than another one, if it is *semantically* better, i.e. if it is higher on the '$>_Q$'-scale than the other one.

---

[6] Neglecting the claim of some that Frisian is an official language of the Netherlands, too.

3.3   Limitations of Qualitative Comparisons

When we would relate questions and answers with respect to the relations '$\sqsubseteq_K$' and '$\geq_{Q,K}$', respectively, both relations would give rise to *partial* orderings. This is not very surprising giving our *qualitative* method used to define them. These qualitative methods are rather *coarse grained*, and this also holds for the criterium an answer should satisfy, according to the above method, to count as a *relevant* answer. Remember that according to the proposal above, answer A can only be relevant with respect to question Q and information state K if it is inconsistent with at least one cell of the partition induced by question Q_K, i.e. if $A_{Q,K} \sqsubseteq Q_K$ and $A_{Q,K} \neq Q_K$.

Although the definition of *relevance* given in the previous subsection is quite appealing, it seems that we have more intuitions about 'relevance' than this *qualitative* notion can capture. An answer can, intuitively, sometimes be relevant, although it is consistent with *all* cells of the partition. When I would ask you *Will John come?*, and you answer by saying *Most probably not*, this response counts intuitively as a very relevant answer, although it does not rule out any of the cells induced by the question. In this case the answer changes the probability distribution of the cells of the partition, but our problem also shows up when probability doesn't play a (major) role. When I ask you the *yes/no*-question *Are John and Mary sick?*, your answer *At least John is* is compatible with both answers, but still felt to be very relevant. This suggests that the notion of *relevance* of answers should be determined with respect to a more fine-grained ordering relation than our above '$\geq_{Q,K}$'.

There is also a more direct reason why the ordering relation between answers should be defined in a more fine-grained way. It is possible that one answer that is consistent with all elements of a partition can be more relevant than another (relevant) answer that is consistent with all elements of a partition, even if the one does not entail the other: The answer *(At least) John and Mary are sick* is normally felt to be a more relevant, or informative, answer to the question *Who of John, Mary and Sue are sick?* than the answer *(At least) Sue is sick*, although less relevant than the complete answer to the question that *Only Sue is sick*. These examples suggest that we want to determine a *total* ordering relation between answers and that we should compare answers to one another in a more *quantitative* way. When probability doesn't play a role (or when all worlds have equal probability), this can simply be done by *counting* the numbers of cells of the partition the answers are compatible with, or the number of worlds compatible with what is expressed by the answers. I won't discuss such a proposal further in this paper, and turn in the next section straightaway to probabilities.

Above I have argued that we should define a more fine-grained ordering relation between answers. Something similar also holds for questions. If I want to find out who of John, Mary and Sue are sick, the question *Who of John and Mary are sick?* is felt to be more informative, or relevant, than the question *Is Sue sick?*, although none of the complete answers to the first question will solve the second issue. What this example suggests is that (i) also questions should be compared to each other with respect to a *quantitative* ordering relation, but also that (ii) to compare the usefulness of two questions with each other, we should *relate* the questions to (something like) a *third question*. Later in this paper, this third question, or problem, will show up again as a *decision problem*.

We have suggested to *extend* our *partial* ordering relations between questions and answer to *total* orderings by measuring the informativity and relevance of propositions and questions in a more *quantitative* way. But how can we do that?

4   Information and Communication Theory

4.1   The Amount of Information of a Proposition

There turns out to be a standard way to determine the informativity of propositions that give rise to a total ordering, such that this total ordering is an extension of the ordering induced by entailment. Notice that if one proposition entails another, it is more informative to learn the former than to learn (only) the latter. That is, it would be *more surprising* to find out that the former proposition is true, than to find out that the latter is. But it doesn't seem to be a necessary condition for proposition A to be more surprising than proposition B that A entails B. All what counts, intuitively, is that the *probability* that A is true is smaller or equal than the probability that B is true. Assuming that an information state should be modeled by a probability function, P, we might say that for each proposition A, its measure of surprise can be defined as either $1 - P(A)$, or $1/P(A)$.[7] Both measures will induce the same total ordering of propositions with respect to their informativity. For reasons that will become clear later, however, we will follow Bar-Hillel & Carnap (1953),[8] and define the informativity of proposition A, inf(A), as the logarithm with base 2 of $1/P(A)$, which is the same as the negative logarithm of the probability of A:[9]

$$\text{inf}(A) \quad = \quad \log_2 \left(1/P(A)\right) \quad = \quad -\log_2 P(A).$$

Also in terms of this notion of informativity we can totally order the propositions by means of their informativity, or measure of surprise, and it turns out that the so induced ordering corresponds exactly with the ones suggested earlier.[10]

---

[7] In this paper I will assume that probabilities are assigned to worlds, and not (primarily) to propositions. Thus, a probability function, P, is a function in $[W \to [0,1]]$, such that $\sum_{w \in W} P(w) = 1$. Notice that this allows lots of worlds to have a probability of 0. A proposition, A, is represented by a set of worlds, and the probability of such a proposition, $P(A)$, is defined as $\sum_{w \in A} P(w)$.

[8] Who in turn take over Hartley's (1928) proposal for what he calls the 'surprisal value' of a proposition.

[9] The 'inf'-value of a proposition is a function of its probability; for different probability functions, the 'inf'-value of a proposition might be different. In the text I won't mention, however, the particular probability function used.

[10] To determine this ordering it is also irrelevant what we take as the base of the logarithm. But certainly in our use of the informational value of propositions for determining the informational value of questions, the chosen base 2 will be most appealing.

To explain the 'inf'-notion, let us consider again the state space where the relevant issues are whether John, whether Mary, and whether Sue are sick or not. The three issues together give rise to $2^3 = 8$ relevantly different states of the world, and assuming that it is considered to be equally likely for all of them to be sick or not, and that the issues are independent of one another, it turns out that all 8 states are equally likely to be true. In that case, the informativity of proposition A equals the number of the above 3 binary issues solved by learning A. Thus, in case I learn that John is sick, one of the above three binary issues, i.e. *yes/no*-questions, is solved, and the informativity of the proposition expressed by the sentence *John is sick* $= J$, inf(J), is 1. Notice that proposition J is compatible with 4 of the 8 possible states of nature, and on our assumptions this means that the probability of J, $P(J)$, is $\frac{1}{2}$. To determine the informational value of a proposition, we looked at the negative logarithm of its probability, where this logarithmic function has a base of 2. Recalling from high-school that the logarithm with base 2 of n is simply the power to which 2 must be raised to get n, it indeed is the case that inf(J) = 1, because $-\log P(J) = -\log \frac{1}{2} = 1$, due to the fact that $2^{-1} = \frac{1}{2}$. Learning that both Mary and Sue are sick however, i.e. learning proposition $M \wedge S$, has an informative value of 2, because it would resolve 2 of our binary issues given above. More formally, only in 2 of the 8 cases it holds that both women are sick, and thus we assume that the proposition expressed, $M \wedge S$, has a probability of $\frac{1}{4}$. Because $2^{-2} = \frac{1}{4}$, the amount of information learned by $M \wedge S$, inf($M \wedge S$), is 2.

What if a proposition does not resolve a single one of our binary issues, like the proposition expressed by *At least one of the women is sick*, i.e. $M \vee S$? Also such propositions can be given an informative value, and in accordance with our above explanation the informative value of this proposition will be *less* than 1, because it does not resolve a single of the relevant binary issues. Notice that the proposition is true in 6 of the 8 states, and thus has a probability of $\frac{3}{4}$. Looking in our logarithm-table from high-school again, we can find that $-\log \frac{3}{4} = 0.415$, which is thus also the amount of information expressed by the proposition according to Bar-Hillel & Carnap's proposed measure.

In our above examples we have only looked at the special case where each of the 8 states were equally likely, and thus limited ourselves to a rather specific probability function.[11] But it should be clear that the informative value of a proposition can also be determined in case the states are not equally probable. Bar-Hillel & Carnap prove that their value function has a number of properties, and here I want to mention only the most important ones.

---

[11] The kind of probability function we used is closely related to Carnap's (1950) *objective* probability function, and also used in Bar-Hillel & Carnap (1953), to define an *objective* notion of amount of the *semantic* information of a proposition. But the way they define the informativity of a proposition does obviously not demand the use of such an objective probability function. The informative value of a proposition is always calculated with respect to a particular probability function, and this probability function might well be *subjective* in the sense that it represents the beliefs of a particular agent.

They note that when proposition A is already believed by the agent, i.e. when $P(A) = 1$, the amount of information gained by learning A is 0, $\inf(A) = 0$, which is a natural measure for the lower bound. The higher bound is reached when proposition A is 'learned' of which the agent believes that it cannot be true, $P(A) = 0$. In that case it holds that $\inf(A) = \infty$. The 'inf'-value of all 'contingent' propositions, i.e. of all propositions A such that $0 < P(A) < 1$, will be finite, and higher than 0.

Let us say that two propositions A and B are *independent* with respect to probability function P when $P(A \wedge B) = P(A) \times P(B)$, that is, when $P(B/A) = P(B)$. In that case it holds that $\inf(B/A) = \inf(B)$, where $\inf(B/A)$ measures the amount of information of B given that A holds, and defined as the difference between $\inf(A \wedge B)$ and $\inf(A)$:

$$
\begin{aligned}
\inf(B/A) &= \inf(A \wedge B) - \inf(A) \\
&= -\log_2 P(B/A).
\end{aligned}
$$

When A and B are independent, conjunction behaves informationally *additive*, i.e. $\inf(A \wedge B) = \inf(A) + \inf(B)$. And indeed, in our above example M and S – the propositions that Mary and Sue are sick, respectively – are independent, and both have the same 'inf'-value as J, namely 1. Thus, $\inf(M) + \inf(S) = 2$, which is exactly the 'inf'-value of $M \wedge S$, as we have observed above.

An important property of the 'inf'-function for our purposes is that it is *monotone increasing* with respect to the entailment relation. That is, if $A \subseteq B$, it holds that $\inf(A) \geq \inf(B)$. And indeed, in our above example we saw that $\inf(M \wedge S) \geq \inf(M \vee S)$. Exactly because the 'inf'-function behaves monotone increasing with respect to the entailment relation, the total ordering relation induced by the 'inf'-function has the nice property that it is an *extension* of the partial ordering relation induced by the entailment relation. The entailment relation and the ordering relation induced by the 'inf'-function are even closer related to each other: if with respect to every probability function it holds that $\inf(A) \geq \inf(B)$, then it will be the case that A semantically entails B. What this suggests is that the *semantic entailment* relation is an *abstraction* from the more *pragmatically oriented amount-of-information relation*.[12]

---

[12] Of course, the semantic entailment relation (a partial ordering) is defined in terms of *meaning*, while the total ordering relation is defined in terms of a different kind of concept. Some early proponents of communication theory, however, didn't make a great effort to keep the concepts separate. Norbert Wiener (1950), for instance, takes *amounts of information* and *amount of meaning* to be equivalent. He says, "The amount of meaning can be measured. It turns out that the less probable a message is, the more meaning it carries, which is entirely reasonable from the standpoint of common sense." But, to quote Dretske (1999, p. 42) "It takes only a moment's reflection to realize that this is *not* 'entirely reasonable' from the standpoint of common sense. There is no simple equation between meaning (or amount of meaning) and information (or amount of information) as the latter is understood in the mathematical theory of information. The utterance *There is a gnu in my backyard* does not have *more meaning* than *There is a dog in my backyard* because the former is, statistically, less probable."

4.2 The Entropy of a Question

Now that we have extended the ordering relation between propositions with respect to their information values to a total relation, the question arises whether something similar can be done for *questions*. As before, I will think of questions as semantic objects, and in particular as partitions of the state space.

To determine the informative value of a question, we will again follow the lead of Bar-Hillel & Carnap (1953). They discuss the problem how to determine the *estimated* amount of information conveyed by the outcome of an *experiment* to be made. They equate the value of an experiment with its estimated amount of information, and they assume that the possible outcomes denote propositions such that the set of outcomes are mutually exclusive and jointly exhaust the whole state space. In other words, they assume that the set of possible outcomes *partitions* the set of relevant states. This suggests, obviously, that we can also equate the informative value of a *question* with the estimated amount of information conveyed by its (complete) answers. The estimated amount of information of the answers will simply be the *average* amount of information of the answers. For reasons that will become clear soon, I will denote the informative value of question Q by $E(Q)$, which will be defined as follows:

$$E(Q) \;=\; \sum_{q \in Q} P(q) \times \inf(q).$$

To strengthen our intuitions, let us look again at the case where we have 8 relevantly different states of the world, such that each of the states are equally likely to be true. Consider now the question *Who of John, Mary and Sue are sick?* Notice that any complete answer to this question will reduce our 8 possibilities to 1. Thus, any complete answer, q_i, will have an 'inf'-value of 3, i.e. it will resolve all three of the relevant binary issues. But if each answer to the question has an informative value of 3, the *average* amount of information conveyed by the answers, and thus the informative value of the question, $E(Q)$, should also be 3. And indeed, because each of the complete answers has a probability of $\frac{1}{8}$ to be true, the informative value of the question is according to the above formula equated with $(\frac{1}{8} \times 3) + ... + (\frac{1}{8} \times 3) = 8 \times (\frac{1}{8} \times 3) = 3$. In general it will hold that when we have n mutually exclusive answers to a question, and all the answers are considered to be equally likely true, the informative value of the question can simply be equated with the informative value of each of its answers, which is $-\log_2 \frac{1}{n} = \log_2 n$. The informative value of the question *Will the outcome of the flipping of an unbiased coin be heads?*, for instance, will be 1, because the question has 2 answers, which are by assumption equally likely to be true.

What if not all of the n answers are equally likely to be true? In that case some answers have a higher informative value than $\log_2 n$, and others have a lower one. It turns out, however, that the *average* amount of information conveyed by the answers will in that case be *lower* than in case the answers are equally likely to be true. Consider for instance the flipping of a *biased coin*, whose chance to come up heads after flipping is $\frac{3}{4}$. Because the 'inf'-value of outcome/answer *Heads* is in that case $-\log_2 \frac{3}{4} = 0.415$, and the 'inf'-value of answer *Tails* is $-\log_2 \frac{1}{4} = 2$, the *average* amount of information of the answers is $(\frac{3}{4} \times 0.415) + (\frac{1}{4} \times 2) = 0.811 < 1$.

Thus, although one of the answers has an informative value that is 2 times as high as the informative values of the outcomes/answers in case of an unbiased coin, the average amount of information of the answers turns out to be lower.

This is in general the case; the informative value of question Q defined as above is *maximal* just in case the answers are all equally likely to be true. And this seems to confirm our intuitions. If you want to be sure to find out after 3 *yes/no*-questions which of the 8 states of our toy-example actually obtains, we should ask the three *yes/no*-questions which have maximal E-value. That is, we should ask for each individual separately whether he or she is sick, which all have an 'inf'-value of 1, and we should not ask *risky* questions that might, but need not, convey more information, like *Are John and Mary the ones who are sick?* In fact, we might even define the *risk* of question Q which has n different possible answers, as (a function of) the difference between the E-value of the n-ary question with maximal informative value, i.e. with an E-value of $\log_2 n$, and $E(Q)$.

Having defined when a question has its maximal informative value, we now would like to know under which circumstances it reaches its *minimal* value. Intuitively, a question is (at least) valueless in case you already know the answer to the question. And, unsurprisingly, this is what comes out; $E(Q) = 0$ just in case only one answer has a positive probability (and thus has the probability 1), and for all other cases the question has a value strictly higher than 0.

Our aim was to define a value of questions (partitions) that allows us to extend the partial ordering on questions induced by the '$\sqsubseteq$' relation to a total ordering. We have succeeded in doing that: it always will be the case that when $Q \sqsubseteq Q'$, it will also be the case that $E(Q) \geq E(Q')$. Moreover, as a special case of a theorem stated in section 5 it will be the case that if $E_P(Q) \geq E_P(Q')$ with respect to all probability functions P, it holds that $Q \sqsubseteq Q'$.

We have defined the informative value of questions in the same way as Bar-Hillel & Carnap (1953) defined the value of doing an experiment. As they have noted themselves, the way this value is defined is *formally* exactly analogous to the way the *entropy* of a *source*, i.e. coding system, is defined by Shannon (1948) in his *Communication Theory*. This is why we denoted the informative value of question Q by $E(Q)$, and from now on I will call the informative value of a question simply its entropy. In Communication Theory 'entropy' is the central notion, because engineers are mostly interested in the issue how to device a coding system such that it can transmit *on average* as much as possible information via a particular *channel*. Although we have defined the entropy of a question formally in the same way as Shannon defined the entropy of a source, there is an important difference between Shannon's original use of the formalism within Communication Theory on the one hand, and Bar-Hillel & Carnap's and our application of it on the other: Shannon looked at things from a purely *syntactic* point of view while we interpret notions like 'informativity' and 'entropy' in a *semantic/pragmatic* sense.

4.3   Conditional Entropy, and the Informative Value of Expressions

Although we have followed Bar-Hillel & Carnap in making a different use of the formalism Shannon invented than originally intended, this doesn't mean that we are not allowed to 'borrow' some mathematical results Shannon proved for his theory of entropy. In particular, we can make use of what in Communication Theory is known as *conditional entropy*, and of what is sometimes called *Shannon's inequality*, to determine the estimated reduction of uncertainty due to getting an answer to a question.

To use those notions, we first have to say what the *joint entropy* is of two questions, Q and Q', $E(Q, Q')$, is, where both Q and Q' are as usual represented by partitions:

$$E(Q, Q') \;=\; \sum_{q \in Q} \sum_{q' \in Q'} P(q \cap q') \times \mathrm{Inf}(q \cap q').$$

It should be clear that this joint entropy of Q and Q' is equivalent to the entropy of $Q \sqcap Q'$, $E(Q \sqcap Q')$, where $Q \sqcap Q' \overset{def}{=} \{q \cap q' : q \in Q \ \& \ q' \in Q' \ \& \ q \cap q' \neq \emptyset\}$.

Until now we have defined the entropy of a question with respect to a set of ways the world might be. Notice that the set of worlds consistent with what is believed, $\{w \in W : P(w) > 0\}$, corresponds itself also to a partition, namely the most fine-grained partition $\{\{w\} : P(w) > 0\}$. Calling this latter partition B, also this partition can be thought of as a question that has a certain entropy, $E(B)$.

Let us now assume that the agent learns answer q to question Q. What is then the entropy of B *conditional* on learning q, $E_q(B)$? The definition of this conditional entropy can be easily given:

$$E_q(B) \;=\; \sum_{b \in B} P(b/q) \times \inf(b/q),$$

and measures the entropy of, or uncertainty in, B when it is known that the answer to Q is q. In terms of this notion we might now define the *entropy of B conditional on Q*, $E_Q(B)$. This is defined as the *average* entropy of B conditional on learning an answer to question Q:

$$
\begin{aligned}
E_Q(B) \;&=\; \sum_{q \in Q} P(q) \times E_q(B) \\
&=\; \sum_{q \in Q} P(q) \times \sum_{b \in B} P(b/q) \times \inf(b/q) \\
&=\; \sum_{q \in Q} \sum_{b \in B} P(q \wedge b) \times \inf(b/q).
\end{aligned}
$$

Now it can be shown that for any two partitions X and Y of the same set of worlds, it holds that $E(X, Y) - E(X) = E_X(Y)$:

$$
\begin{aligned}
E(X, Y) - E(X) \;&=\; -\sum_{x \in X} \sum_{y \in Y} P(x \wedge y) \times log\,P(x \wedge y) + \\
&\qquad \sum_{x \in X} P(x) \times log\,P(x) \\
&=\; \sum_{x \in X} \sum_{y \in Y} P(x \wedge y) \times log\,P(x) - \\
&\qquad \sum_{x \in X} \sum_{y \in Y} P(x \wedge y) \times log\,P(x \wedge y) \\
&=\; \sum_{x \in X} \sum_{y \in Y} P(x \wedge y) \times [log\,P(x) - log\,P(x \wedge y)] \\
&=\; \sum_{x \in X} \sum_{y \in Y} P(x \wedge y) \times log\,\tfrac{P(x)}{P(x \wedge y)} \\
&=\; \sum_{x \in X} \sum_{y \in Y} P(x \wedge y) \times inf(y/x) \\
&=\; E_X(Y).
\end{aligned}
$$

A similar calculation shows that $E(X,Y) - E(Y) = E_Y(X)$, and thus that $E_X(Y) + E(X) = E_Y(X) + E(Y)$. Notice that thus in particular it holds for our two partitions Q and B that $E_Q(B) = E(Q,B) - E(Q)$. I will just state, and not show, *Shannon's inequality*, which says that for any two partitions X and Y of the same state space, it holds that

$$E_X(Y) \leq E(Y),$$

where the two values are the same exactly when the two issues are completely *orthogonal* to one another, i.e. when the issues are *independent*. Notice that this means that the entropy of $Q \sqcap Q'$ only equals the entropy of Q plus the entropy of Q' in case the partitions are fully independent. That the entropy of the combined question is only in these special cases equal to the sum of the entropies of the questions separately, conforms to our intuition that *on average* we learn less by getting an answer to the combined question *Who of John and Mary will come to the party?*, than by getting two separate answers to both questions *Will John come to the party?* and *Will Mary come to the party?*, when John only, but *not* if and only, comes when Mary comes.

Shannon's inequality will turn out to be a nice property of what I will call the average information gained from the answer to a question. To define this notion, let us first define what might be called the *Informational Value* of answer q, with respect to partition B, $IV_B(q)$, as the reduction of entropy, or uncertainty, of B when q is learned:[13]

$$IV_B(q) = E(B) - E_q(B).$$

Because learning q might flatten the distribution of the probabilities of the elements of B, it should be clear that $IV_B(q)$ might have a negative value. Still, due to Shannon's inequality, we might reasonably define the informational value of question Q, the *Expected Informational Value* with respect to partition B, $EIV_B(Q)$, as the *average* reduction of entropy of B when an answer to Q is learned:

$$
\begin{aligned}
EIV_B(Q) &= \sum_{q \in Q} P(q) \times IV_B(q) \\
&= \sum_{q \in Q} P(q) \times [E(B) - E_q(B)] \\
&= E(B) - [\sum_{q \in Q} P(q) \times E_q(B)]^{[14]} \\
&= E(B) - E_Q(B)
\end{aligned}
$$

The difference between $E(B)$ and $E_Q(B)$ is also known as the *mutual information* between B and Q, $I(B,Q)$. Shannon's inequality tells us now that our *average* uncertainty about B can never be increased by asking a question, and it remains the same just in case Q and B are orthogonal to each other. In the latter case we might call the question *irrelevant*.

---

[13] A similar notion was used by Lindley (1956) to measure the informational value of a particular result of an experiment.

[14] This step is allowed because the *unconditional* entropy of B, $E(B)$, does not depend on any particular element of Q.

To strengthen our intuitions, let us look at our toy-example again. Recall that 8 worlds were at stake, and all the 8 worlds had the same probability. In that case, learning which element of B obtains, i.e. learning what the actual world is, gives us 3 bits of information, and thus $E(B) = 3$. Remember also that learning the answer to the *yes/no*-question *Is John sick?* will give us 1 bit of information, i.e. $E(\text{Sick(j)?}) = E(Q) = 1$, because each answer to the question is equally likely true, and from both answers we would gain 1 bit of information. It's almost equally easy to see that for both answers to the question, the entropy of B *conditional* on learning this answer q, $E_q(B)$, is also 1, and thus that the *average* reduction of uncertainty due to an answer to Q, $E_B(Q)$ is 1, too. It follows that thus the expected information value, $EIV_B(Q)$, is $E(B) - E_Q(B) = 3 - 1 = 2$. The same result is achieved when we determine $EIV_B(Q)$ by taking the average difference between $E(B)$ and $E_q(B)$ for both answers q, because both answers are equally likely, and for both it holds that $E(B) - E_q(B) = IV_B(q) = 2$.

We have defined the expected informational value of question Q with respect to partition B, $EIV_B(Q)$, as the average reduction of entropy of B when an answer to Q is given, i.e. as the difference between $E(B)$ and the *conditional* entropy $E_Q(B)$. And to make sure that this is always positive, we have made use of Shannon's inequality. But notice that the entropy of B conditional on Q, $E_Q(B)$, is simply the same as the entropy of Q, $E(Q)$, itself. But this means that the expected informational value of Q with respect to B, $EIV_B(Q)$, can also be defined as the difference between the entropy of B and the entropy of Q, $E(B) - E(Q)$. Notice also that we don't have to make use of Shannon's inequality to see that for any question Q it holds that $EIV_B(Q)$ will never be negative. The reason is that for any question Q it holds that $B \sqsubseteq Q$, and we have noted already that in that case it will hold that the entropy of B will be at least as high as the entropy of Q: $E(B) \geq E(Q)$. But if we can assure that the informational value of a question is non-negative without making use of Shannon's inequality, why did we define the value of a question in such a roundabout way via the conditional entropy of B given Q?

4.4   Deciding between Hypotheses

The reason is that we don't want to restrict ourselves to the *special case* where in the end we want to have *total* information about the world, where we have completely reduced all our uncertainty. Remember that partition B was the most fine-grained partition possible; the elements of B were *singleton sets* of worlds. Because the entropy of Q measures the average uncertainty about how the world looks like when we've got an answer to Q, this measure, $E(Q)$, is only the same as the entropy of B conditional on Q, $E_Q(B)$, because the elements of our special partition B correspond one-to-one to the worlds.[15]

But now suppose that we need not to know how *exactly* the world looks like, but rather just want to find out which of the mutually exclusive and exhaustive set of *hypotheses* in the set $H = \{h_1, ..., h_n\}$ is true, where the h_i's denote

---

[15] More in general, it holds that for two partitions Q and Q', if $Q \sqsubseteq Q'$, then $E_Q(Q') = E(Q')$.

arbitrary propositions. The problem now is to determine the value of question Q with respect to this other partition H, $EIV_H(Q)$, and this is in general not equal to $E(H) - E(Q)$. To determine the value $EIV_H(Q)$, we need to make use of the conditional entropy of H given (an answer to) Q.

Notice that Shannon's inequality tells us now also something informative; $EIV_H(Q)$ will never be negative, although it need not be the case that $H \sqsubseteq Q$. And not only for our special partition B, but also for any other issue H we can determine when question Q is informationally relevant. Question Q is *informationally relevant* with respect to a set of hypotheses H just in case the true answer to Q is expected to reduce the uncertainty about what the true hypothesis of H is, i.e. $EIV_H(Q) > 0$.

This notion of 'informational relevance' is important when an agent is fronted with the *decision problem* which of the mutually exclusive hypotheses $\{h_1, ..., h_n\}$ he should choose. In case the agent only cares about the issue which of the hypotheses is true, and that all ways of choosing falsely are equally bad, the risk of choosing depends only on the uncertainty about what the right hypothesis is. It seems natural to advice him in these circumstances always to choose that hypothesis that has the highest prior probability. But this means that the *risk* of choosing depends entirely on the entropy of H, $E(H)$. And indeed, the flatter the distribution of the probabilities of the hypotheses is, the more risky the choice will be.

Notice that asking a question, and thereby expecting to get an answer (that is true), might reduce the entropy of H, i.e. the uncertainty about which hypothesis is true, and thus also the *risk* of the decision, even if all answers to the question are compatible with all hypotheses. But this means that even if no answer to the question will eliminate a single hypothesis, it might still be *useful*, or *relevant*, to ask the question. Indeed, at this point it seems only natural to equate the *usefulness* of question Q with respect to the *decision problem* which of the hypotheses of H should be chosen, with the reduction of uncertainty about H due to Q, i.e. $EIV_H(Q)$. Moreover, we can say that question Q is *relevant* with respect to H just in case $EIV_H(Q)$ is strictly higher than 0.

Thus, instead of the *partial* order between questions induced by the relation '$\sqsubseteq$', we can now determine a *total* order. We say that if $Q \neq Q'$, question Q is *better* than question Q' with respect to hypotheses H, $Q >_H Q'$, just in case the expected information value of Q is higher than the value of Q', or, if both are the same, the former is less fine-grained than the latter:[16]

$$Q >_H Q' \quad \text{iff} \quad \begin{array}{ll} (i) & EIV_H(Q) > EIV_H(Q'), \text{ or} \\ (ii) & EIV_H(Q) = EIV_H(Q') \text{ and } Q \sqsupseteq Q'. \end{array}$$

Just as the usefulness, and relevance, of question Q with respect to decision problem H can be defined in terms of $EIV_H(Q)$, we can also define the usefulness,

---

[16] As before, I assume always a particular probability function. If we don't do that, the following general fact can be proved: Denote the expected utility value of Q with respect to H and probability function P: If $Q \sqsubseteq Q'$, then for all P : $EIV_H^P(Q) \geq EIV_H^P(Q')$. We will see in section 5 what is needed to strengthen this fact to the stronger *if and only if* statement.

and relevance, of *assertion* A with respect to decision problem H in terms of the information values of answers. That is, we can propose to equate the usefulness of assertion A with respect to issue H with $IV_H(A)$, and we can say that assertion A is *relevant* just in case $IV_H(A) > 0$. Notice that the thus defined notion of relevance predicts that many assertions are relevant, although they are (falsely) predicted to be irrelevant according to the *qualitative* notion of relevance used above. Moreover, our newly defined notion of relevance still has the nice property that it can explain why an assertion is felt to be *irrelevant* although it still is *informative*. For instance, if the issue is who of John and Mary are sick, and we look at our toy-example again where the sickness of John, Mary and Sue are independent of each other, the assertion *Sue is sick* is rightly predicted to be irrelevant, although it does eliminate some possible worlds.

Now we can also turn our *partial* order between answers induced by the relation '$>_Q$', to a *total* order (although it is not an extension of it). We say that assertion A is *better* than assertion B with respect to hypotheses H, $A >_H B$, just in case the informational value of A, $IV_H(A)$, is higher than the corresponding value of B, $IV_H(B)$, or, in case both are the same, the former should be less surprising than the latter:

$$A >_H B \quad \text{iff} \quad \begin{array}{ll} (i) & IV_H(A) > IV_H(B), \text{ or} \\ (ii) & IV_H(A) = IV_H(B) \text{ and } inf(A) < inf(B). \end{array}$$

Thus, if A reduces the entropy of H more than B does, it is a better answer to 'question' H than B.

Notice that according to our definition of the relevance of an assertion, an assertion is predicted to be irrelevant when it flattens the probability distribution of the hypotheses. In such cases the assertion indeed has the effect that it doesn't make the decision any easier. Intuitively, however, this doesn't mean that thus the assertion is felt to be irrelevant. The assertion seems to *be* relevant exactly *because* it makes the decision more risky. This wrong prediction can, fortunately, be removed easily. Just say that A is *relevant* with respect to H exactly when the acceptance of A *changes* the probability distribution of the hypotheses, i.e. when $IV_H(A) \neq 0$.

4.5   Limitations of the Analysis in Terms of Entropy

The measure of usefulness and relevance of questions and assertions with respect to a decision problem that we have defined above is, I think, reasonable for some, but also *only* reasonable for some kinds of decision problems. First, in our description of decision problems, we only looked at problems where the choice between a set of *hypotheses* is at stake. We would like to extend the analysis from the choice between hypotheses, to choices between more general kinds of *actions*. Extending our analysis from decisions between hypotheses to decisions between actions need not yet worry us. It doesn't seem to be completely unreasonable to represent actions as propositions; an action is true in a world just in case the result of the action is true in that world. Indeed, in the well respected decision theory of Jeffrey (1965), actions are represented by propositions.

What is more problematic for the way we have analyzed the usefulness of questions and answers in this section is that once we think of a decision problem as consisting of a set of actions, it seems only natural to assume that the decision depends not only on the *probabilities* involved, but also on the *desirabilities*, or *utilities*, of the states that result when the various actions would be chosen. But once desirabilities enter the picture, it is obvious that our analysis of the usefulness of questions and answers can no longer be defined simply in terms of the dependencies between certain probability distributions, i.e. in terms of conditional entropies.

Consider, for instance, the decision problem faced by airpilot Smith who wonders whether he should drop the bomb, with the reasonable chance to trigger a world-war, or not dropping the bomb, and thereby missing an excellent chance to strike a potential future enemy in war, and getting a scolding for this by his commanding officer. It is clear that Smith's desirabilities of the expected outcomes of the relevant actions will heavily influence his decision.

Even if the relevant actions just involve a choice between a set of hypotheses, the most probable hypothesis is not always the one that intuitively is preferable. The reason is that choosing this hypothesis might give rise to very nasty consequences. Consider, for instance, scientist Jones who is facing the dilemma between choosing the generally accepted theory h_1 and working in this framework, or choosing the alternative theory h_2 that he thinks is somewhat more likely to be true, but that has a very bad reputation among his fellow researchers. Because Jones knows that choosing h_2 will turn him into a black sheep of his family whose papers will never be read, even the more purists among us could understand Jones' choice for theory h_1.

Let me give a simple example showing that the reduction of entropy of the relevant set of hypotheses/actions does not always measure the usefulness of questions and assertions in a satisfying way. Consider John, who wonders whether he should go to the party tonight, or not. His decision depends almost entirely on whether Mary will go, because he is secretly in love with Mary, and believes that going to the party is his only chance to meet her. He prefers meeting her tonight, to not meeting her, but if Mary won't go, he prefers to stay home. But going to the party when Mary comes too obviously involves a risk; perhaps Mary will turn him down when he makes his advances. We might say that in this situation 4 different states (worlds) are involved: one world, w_1, where Mary goes to the party, John will go, too, he will try his luck, and is successful; a world, w_2, where Mary goes, John goes, he tries his luck, and is unsuccessful; world w_3, where Mary won't go to the party, and thus neither does John, but where the counterfactual statement holds that when John *would* try his luck, he *would* be successful, and w_4 which is similar to w_3 except that in this world the counterfactual would be false. On the additional assumption that John thinks all worlds are equally likely to come out true, that he doesn't care about what Mary would do if they don't go to the party, and that John has a negative attitude towards taking risks, we might represent his decision problem by the following table:

World	Probability	Desirability
w_1	1/4	12
w_2	1/4	2
w_3	1/4	8
w_4	1/4	8

In this case, it is *relevant*, intuitively, for John to learn that the above mentioned counterfactual statement is true. It is, however, easily seen that learning the proposition expressed by this statement, $\{w_1, w_3\} = A$, does *not* change the entropy of the decision problem that can be represented by $\{\{w_1, w_2\}, \{w_3, w_4\}\} = H$. That is, $IV_H(A) = E(H) - E_A(H) = 0$, because learning A does not change the probability distribution of the elements of H, i.e. both $E(H)$ and $E_A(H)$ have a value of 1.

In a similar way, it also seems relevant for John to know the answer to the question whether he would be successful if he tried, that is, to learn which element of the partition $\{\{w_1, w_3\}, \{w_2, w_4\}\}$ is true. It is straightforward to check, however, that not only the positive answer to the question, A, but also the negative answer, $\neg A$, has no effect on the probability distribution of the elements of H. Representing the question whether the counterfactual is true or not by Q, it is thus predicted that also $EIV_H(Q) = E(H) - E_Q(H) = 0$. We can conclude that the value $EIV_H(Q)$ is at least not always the proper measure to determine the relevance of a question with respect to a decision problem.

What we need, or so it seems, is a measure that not only looks at the probabilities, but also at the desirabilities involved. In the next section we will define such a measure by looking seriously at statistical decision theory.

5   Utility Values of Questions and Answers

5.1   Utilies of Answers and Expected Utilities of Questions

In Savage's (1954) decision theory, actions are taken to be primitives, and if we assume that the utility of performing action a in world w is $U(a, w)$, we can say that the *expected utility* of action a, $EU(a)$, with respect to probability function P is

$$EU(a) \quad = \quad \sum_w P(w) \times U(a, w).$$

Let us now assume that our agent, John, faces a *decision problem*, i.e. he wonders which of the alternative actions in $\mathcal{A}$ he should choose. A decision problem of an agent can be modeled as a triple, $\langle P, U, \mathcal{A} \rangle$, containing (i) the agent's probability function, P, (ii) his utility function, U, and (iii) the alternative actions he considers, $\mathcal{A}$. You might wonder why we call this a decision *problem*; shouldn't the agent simply choose the action with the highest expected utility? Yes, he should, if he *chooses now*. But now suppose that John doesn't have to choose now, but that he has the opportunity to first receive some useful information by *asking question Q*.

Before we can determine the utility of Q, we first have to say how to determine the expected utility of an action conditional on learning some new information. For each action $a \in \mathcal{A}$, its conditional expected utility with respect to new proposition C, $EU(a, C)$ is

$$EU(a, C) \quad = \quad \sum_{w} P(w/C) \times U(a, w).$$

When John learns proposition C, he will of course choose that action in $\mathcal{A}$ which maximizes the above value. Then we can say that the utility value of making an informed decision conditional on learning C, $UV(\text{Learn } C, \text{ choose later})$, is the expected utility conditional on C of the action that has highest expected utility:

$$UV(\text{Learn } C, \text{ choose later}) \quad = \quad max_{a \in \mathcal{A}} EU(a, C).$$

In terms of this notion we can determine the value, or *relevance*, of the assertion C. Referring to a^* as the action that has the highest expected utility according to the original decision problem, $\langle P, U, \mathcal{A} \rangle$, i.e. $max_{a \in \mathcal{A}} EU(a) = EU(a^*)$, we can determine the *utility value* of the *assertion* C, $UV(C)$, as follows:

$$UV(C) \quad = \quad max_{a \in \mathcal{A}} EU(a, C) - EU(a^*, C).$$

This value, which in statistical decision theory (cf. Raiffa & Schlaifer, 1961) is known as the *value of sample information* C, $VSI(C)$, can obviously never be negative. In fact, it predicts that an assertion only has a positive utility value in case it influences the action that John will perform. And indeed, it seems natural to say that a cooperative participant of the dialogue only makes a *relevant* assertion in case it makes John *change* his mind with respect to which action he should take. It also seems not unreasonable to claim that in a cooperative dialogue one assertion, A, is 'better' than another, B, just in case the utility value of the former is higher than the utility value of the latter, $UV(A) > UV(B)$.

In terms of the utility value of assertions/answers, we can now determine the utility values of *questions*. Suppose that question Q is represented by the partition $\{q_1, ..., q_n\}$. Just like in section 4 we defined the informative value, or entropy, of a question as the *expected*, or *average*, informative value of its answers, in this case we can determine the *expected* utility value of a question, $EUV(Q)$ as the *average* utility value of the possible answers:

$$EUV(Q) \quad = \quad \sum_{q \in Q} P(q) \times UV(q).$$

Notice that this value, which in statistical decision theory is known as the *expected value of sample information*, $EVSI$, will never be negative. In fact, the value will only be 0 in case no answer to the question would have the result that the agent will change his mind about which action to perform, i.e. for each answer $q \in Q$ it will be the case that $max_{a \in \mathcal{A}} EU(q, a) = EU(q, a^*)$. In these circumstances the question really seems irrelevant, and it thus seems natural to say that question Q is *relevant* just in case $EUV(Q) > 0$. It should be obvious that this measure function also totally orders all questions with respect to their expected utility value.

It is of some interest to see that we can determine the expected utility value of questions also in another way. According to this alternative way of determining the value of questions, we first have to determine the utility value of *choosing now*. The utility value of choosing now is defined as the expected utility of the action which has the highest expected utility according to the original decision problem, i.e. with respect to the original probability function:

$$UV(\text{Choose now}) \quad = \quad max_{a \in A} EU(a).$$

Now we can determine the expected utility value of choosing after you learn the answer, $EUV(\text{Learn answer, choose later})$, in terms of $UV(\text{Learn q, choose later})$, by averaging over the answers to the question:

$$
\begin{aligned}
EUV(\text{Learn answer, ch. later}) &= \textstyle\sum_{q \in Q} P(q) \times UV(\text{Learn } q, \text{ ch. later}) \\
&= \textstyle\sum_{q \in Q} P(q) \times max_{a \in A} EU(a, q).
\end{aligned}
$$

The expected utility value of question Q, $EUV^\dagger(Q)$, is now defined as the difference between the expected utility value of choosing after you got the answer, and the utility value of choosing now:

$$EUV^\dagger(Q) \quad = \quad EUV(\text{Learn answer, choose later}) - UV(\text{Choose now}).$$

It can be easily shown that the second way of determining the expected utility value of a question gives rise to the same result as determining the expected utility value of a question according to the first way, i.e. $EUV^\dagger(Q) = EUV(Q)$:[17]

$$
\begin{aligned}
EUV^\dagger(Q) &= EUV(\text{Learn answer, choose later}) - UV(\text{Choose now}) \\
&= [\textstyle\sum_{q \in Q} P(q) \times UV(\text{Learn } q, \text{ choose later})] - UV(\text{Choose now}) \\
&= [\textstyle\sum_{q \in Q} P(q) \times max_{a \in A} EU(a, q)] - EU(a^*) \\
&= [\textstyle\sum_{q \in Q} P(q) \times max_{a \in A} EU(a, q)] - [\textstyle\sum_{q \in Q} P(q) \times EU(a^*, q)] \\
&= \textstyle\sum_{q \in Q} P(q) \times [max_{a \in A} EU(a, q) - EU(a^*, q)] \\
&= \textstyle\sum_{q \in Q} P(q) \times UV(q) \\
&= EUV(Q).
\end{aligned}
$$

According to the qualitative comparison method of section 3, one question, Q, is better than another question, Q', just in case the former *entails* the latter, that is, in case the partition Q is *finer* than the partition Q': $\forall q \in Q : \exists q' \in Q' : q \subseteq q'$. We have seen in section 4 that measuring the expected informational value of questions, $EIV_H(Q)$, in terms of reduction of entropy of the set of hypotheses H, accords with the qualitative measurement, in the sense that when Q is a finer partition than Q', it also holds that Q will have a greater expected

---

[17] Where a^* is again the action which maximizes expected utility in the original decision problem.

informational value, $EIV_Q(H) \geq EIV_{Q'}(H)$, whatever the set of hypotheses is. Now we can ask a similar question with respect to the question's expected utility value. Denoting by $EUV_{DP}(Q)$ the expected utility value of Q with respect to decision problem DP, Marschak & Radner (1972) have proved as a special case of Blackwell's (1953) theorem the following strong, but also very appealing theorem:

Theorem $Q \sqsubseteq Q'$ iff $\forall DP : EUV_{DP}(Q) \geq EUV_{DP}(Q')$.

The 'only if' part is natural, and shows that it is never irrational (if collecting evidence is cost free) trying to get more information to solve one's decision problem. This part was already implicitly assumed by Savage (1954) and Raiffa & Schlaifer (1961), and was explicitly proved by Good (1966) to follow from the Bayesian principle of maximizing expected utility.[18]

The 'if' part is more surprising, and it suggests that the *semantic* entailment relation between questions is an *abstraction* from the more *pragmatic* usefulness relation of questions. The proof is based on the idea that when two partitions are qualitatively incomparable, one can always find a pair of decision problems such that the first partition has a higher expected utility value than the second one according to one decision problem, and a lower expected utility value than the second one according to the other decision problem.

Given this result for questions, one might expect that something similar holds for *assertions*. We have seen in section 4.1 that whenever $A \subseteq B$, it also holds that $\inf(A) \geq \inf(B)$. In section 4.3, however, we saw that in such circumstances it still might be that $IV_H(B) > IV_H(A)$, i.e. the informational value of a proposition does not behave monotone increasingly with respect to the (ordering induced by the) classical entailment relation between propositions. Still, it might be the case that stronger propositions always *do* have a higher utility value. But in fact, they do not. The utility value of choosing now, UV(Choose now), might be higher than the utility value of first learning proposition C, and then choosing later, UV(Learn C, choose later), because from learning C I might learn that my worst nightmare has come out true, and that I have to perform an action that I otherwise never would have performed.

If neither the informative value of proposition A, $IV_H(A)$, nor its utility value, $UV(A)$, behaves monotone increasing with respect to the '$\subseteq$'-relation, perhaps they *do* behave monotone increasing with respect to one another. But also that is in general not the case, as it should be according to our argumentation in section 4.5.

First, it might be the case that learning a proposition that doesn't change the entropy, still effects a change of mind. Look at the following matrix for the example discussed in section 4.5, but now for a Savage-style decision theory:

---

[18] But see Skyrms (1990), who traces this result back all the way to an unpublished manuscript of Ramsey.

World		Prob	John goes	doesn't go
Mary comes	w_1	1/4	12	0
Mary comes	w_2	1/4	2	0
Mary doesn't come	w_3	1/4	0	8
Mary doesn't come	w_4	1/4	0	8

Looking at the matrix, we can equate the action *John goes* with the worlds where this action has a higher utility than the alternative action. Thus, the action corresponds in this case with the proposition $\{w_1, w_2\}$. The decision problem which action John should perform can thus be represented by the partition $\{\{w_1, w_2\}, \{w_3, w_4\}\} = H$. Note that due to the fact that all worlds have an equal probability, the informational value of proposition $\{w_1, w_3\} = A$ is 0, $IV_H(A) = 0$. Still, learning the proposition has a positive utility value, i.e. $UV(A) > 0$, because learning the proposition would have the result that John changes his mind. Facing his original decision problem, John would decide not to go to the party, because that action has the highest expected utility, $UV(\text{Choose now}) = max_i EU(a_i) = EU(\text{doesn't go}) = 4$. When he would learn proposition $A = \{w_1, w_3\}$, however, John would change his mind, because $EU(\text{John goes}, A) = 6 > EU(\text{doesn't go}, A) = 4$. Due to this latter inequality, together with the fact that the action *doesn't go* is the one that would originally have been chosen, it follows that also $UV_H(A) = 2 > 0$. This shows that information can be useful with respect to a decision problem, although it doesn't reduce the problem's entropy.

With the help of the same matrix we can also show that a proposition might reduce the entropy of a decision problem, although it doesn't have a positive utility value. We just have to find a proposition that strengthens the choice for the action/hypothesis that would have been chosen anyway, in our case for action/hypothesis $\{w_3, w_4\}$. Of course, any subset of this action/hypothesis will do this trick.

5.2   Decision between Hypotheses

In section 4 our problem was to choose an hypothesis from set H, and base this decision only on the probabilities involved. A decision problem can in such cases be modeled by a pair like $\langle P, H \rangle$. As for all kinds of decision problems, we are interested in two kinds of questions: (i) What is the hypothesis the agent should go for? and (ii) What kind of question should the agent ask to make a better informed decision concerning the hypotheses? The answer to the first question seems obvious; the hypothesis the agent should choose is the hypothesis which is most likely to be true, i.e. the hypothesis with the greatest probability. The second question is somewhat more difficult to answer. Let me now show, following Marschak (1974a), that this is a special case where the decision problem should be modeled by a triple like $\langle P, U, \mathcal{A} \rangle$, as in the previous section.

We have assumed in the previous section that a decision problem partly consists of a set of alternative actions, and that each action $a \in \mathcal{A}$ has a utility in a world w, $U(a, w)$. Let us now assume that the set of alternative actions, $\mathcal{A}$, is such that for each world w there is always exactly one action $a \in \mathcal{A}$ such that $\forall a' \in (\mathcal{A} - \{a\}) : U(a, w) > U(a', w)$. This means that the set of alternative actions *partitions* the set of worlds; to each action $a \in \mathcal{A}$ there corresponds a cell of the partition, and in each world of this cell a is the unique best action to do. This set corresponds of course exactly to a set of mutually exclusive and jointly exhaustive hypotheses, $H = \mathcal{A}^*$, that we used in section 4 to measure the informational values of questions and answers when we define this partition as follows:

$$H = \mathcal{A}^* = \quad \{\{w \in W | \; \forall a' \in (\mathcal{A} - \{a\}) : \; U(a, w) > U(a', w)\} | \; a \in \mathcal{A}\}.$$

For each action $a \in \mathcal{A}$ we will denote the cell corresponding with a by a^*, and this, again, is exactly a hypothesis in the original set H. This shows that choosing a hypothesis can be thought of as a special kind of action.

But to show that a decision problem of the form $\langle P, H \rangle = \langle P, \mathcal{A}^* \rangle$ is a special case of a problem of the form $\langle P, U, \mathcal{A} \rangle$, we also have to eliminate the utility function in a natural way. The most natural way in which this can be done is to assume that for this case the utility function is the utility function of someone who cares about the truth, and nothing but the truth.

Suppose that there are only two units of utilities, u_1 and u_2, such that u_1 is strictly higher than u_2. In combination with the foregoing assumption this means that the actions taken in a world can be counted as being either wrong or right, i.e. having a utility of either 1 or 0; action a has utility 1 in a world iff hypothesis a^* is true in that world, and has utility 0 otherwise. Thus, the utility function is nothing else but a truth-value function. The utility value of choosing now is in these special circumstances the same as the probability value of the hypothesis with the highest utility:

$$\begin{aligned} UV_H(\text{Choose now}) &= max_{a \in \mathcal{A}} EU(a) \\ &= max_{a \in \mathcal{A}} \textstyle\sum_w P(w) \times U(a, w) \\ &= max_{a^* \in \mathcal{A}^*} [(\textstyle\sum_{w \in a^*} P(w) \times 1) + (\textstyle\sum_{w \notin a^*} P(w) \times 0)] \\ &= max_{a^* \in \mathcal{A}^*} \textstyle\sum_{w \in a^*} P(w) \\ &= max_{a^* \in \mathcal{A}^*} P(a^*). \end{aligned}$$

Now we can determine for each action $a \in \mathcal{A}$ its conditional expected utility with respect to new proposition C:

$$\begin{aligned} EU(a, C) &= \textstyle\sum_w P(w/C) \times U(a, w) \\ &= P(a^*/C). \end{aligned}$$

Thus, in these special cases the expected utility of action a after learning C is the same as the probability of a^* conditional on C. As a result it also follows that the action a which maximizes the expected utility conditional on learning new proposition C, is the proposition a^* which has the highest probability conditional on C. Now we can also determine the utility value of choosing after learning C:

$$\begin{aligned} UV_H(\text{Learn } C, \text{ choose later}) &= max_{a \in \mathcal{A}} EU(a, C) \\ &= max_{a^* \in \mathcal{A}^*} P(a^*/C). \end{aligned}$$

In terms of this notion we can define a utility value of learning proposition C, $UV_{\mathcal{A}^*}^*(C)$, that slightly differs from the one defined in the previous section, $UV(C)$, in that according to the new function we immediately subtract the utility value of choosing now.

$$UV_{\mathcal{A}^*}^*(C) = UV_H(\text{Learn } C, \text{ choose later}) - UV_{\mathcal{A}^*}(\text{Choose now})$$
$$= max_{a \in \mathcal{A}} EU(a, C) - max_{a \in \mathcal{A}} EU(a)$$
$$= max_{a^* \in \mathcal{A}^*} P(a^*/C) - max_{a^* \in \mathcal{A}^*} P(a^*).$$

Thinking of $\mathcal{A}^*$ again as the set of hypothesis H, one can see that $UV_H^*(C)$, in distinction with $UV_H(C)$, can have a negative value, but also a positive one in case C only strengthens the initially already preferred hypothesis.

Given our new definition of the utility value of *assertions*, $UV_H^*(C)$ it is, under the special circumstances sketched in this subsection, true that

$$UV_H^*(A) \geq UV_H^*(B) \quad \text{iff} \quad max_{h \in H} P(h/A) \geq max_{h \in H} P(h/B).$$

Thus, we have shown the utility value of an assertion is the larger, according to this measure function, the larger the probability of the hypothesis that has maximal posterior probability derived from it.

Notice that when $max_{h \in H} P(h/A) \geq max_{h \in H} P(h/B)$, it also holds that learning A reduces the entropy of H more than B does, in case H consists of 2 hypotheses, because in these cases $E_A(\{h, \neg h\}) \leq E_B(\{h, \neg h\})$. We can conclude that at least in these very special cases, utility values of assertions behave similar to their informational values: $UV_H^*(A) \geq UV_H^*(B)$ iff $IV_H(A) \geq IV_H(B)$. However, when H contains more than 2 hypotheses the result doesn't go through anymore. The reason is, intuitively, that to determine $UV_H^*(A)$ we only look at the optimal hypothesis, while to determine $IV_H(A)$ we also look at the various sub-optimal hypotheses.

Let us now, finally, look at proposition C that completely resolves the issue. That is, let us look at the case where for each $h \in H$, it either holds that $C = h$, or $C \cap h = \emptyset$. Notice that in that case the value $max_{h \in H} P(h/C)$ will always be 1, and the utility value of C, $UV_H^*(C)$, depends only on the prior probability of h^*.[19] Let us now look at the question that completely corresponds with decision problem H, i.e. let us look at question H itself. We might evaluate the expected gain from this question, $EUV_H^*(H)$, by averaging over the corresponding expected values of the answers:

$$EUV_H^*(H) = \sum_{h \in H} P(h) \times UV_H^*(h),$$

because for each $h \in H$ it holds that $UV_H^*(h) = -P(h^*)$, we can conclude that for these special cases the expected gain from question H, $EUV_H^*(H)$, decreases as the prior probability of the least surprising message, i.e. h^*, increases.

---

[19] Of course, this does not mean that the utility values of propositions are thus always independent of the propositions themselves. This is only the case when we only compare the utility values of different propositions that all *fully* resolve the issue.

5.3   Questioning Procedures

We ended section 5.1 with a negative result: even if we can represent the actions of a decision problem by a set of propositions, i.e. by a partition like H, there still exists *in general* no connection between the *informational value* of a proposition A, $IV_H(A)$, and its *utility value*, $UV_H(A)$. Something similar is the case for questions: $EIV_Q(Q')$ is in general no special case of $EUV(Q)$. Let us assume that for every action a_i of our decision problem $\mathcal{A}$ there corresponds a set of worlds a_i^* in which a_i is the unique best action to perform. Assume $\mathcal{A} = \{a_1, ..., a_5\}$ and that the corresponding $\mathcal{A}^* = \{a_1^*, ..., a_5^*\}$ partitions the set of worlds compatible with what our agent believes. According to the prior probability function, all 'worlds' a_i^* are equally likely. Suppose, moreover that the utility function is as follows:

$$U(a_i, a_j^*) = 1, \text{ if } i = j, 0 \text{ otherwise.}$$

In this case one should pick the a_i whose corresponding proposition has the maximal probability. Suppose we have two questions, $Q = \{q_1, q_2\}$ and $Q' = \{q_1', q_2'\}$. The following table gives the probabilities of a_i^* given that we learn an answer to one of these questions:

$$EIV_{\mathcal{A}^*}(Q) \neq EUV_{\mathcal{A}}(Q):$$

	q_1	q_2	q_1'	q_2'
a_1^*	0.4	0.3	0.5	0.2
a_2^*	0.3	0.4	0.2	0.5
a_3^*	0.1	0.1	0.1	0.1
a_4^*	0.1	0.1	0.1	0.1
a_5^*	0.1	0.1	0.1	0.1

Because $max_{a \in \mathcal{A}} EU(a, q_i') = 0.5 > 0.4 = max_{a \in \mathcal{A}} EU(a, q_i)$, it is obviously the case that $EUV_{\mathcal{A}}(Q) < EUV_{\mathcal{A}}(Q')$. However, it turns out that $E_Q(\mathcal{A}^*) < E_{Q'}(\mathcal{A}^*)$ and thus that $EIV_{\mathcal{A}^*}(Q) > EIV_{\mathcal{A}^*}(Q')$. Thus, in general $EUV(Q)$ and $EIV(Q)$ do not behave monotone increasing with respect to one another.

However, as shown by Sneed (1967), in the following special case they do.[20] Suppose our agent wants to know which of the elements of $X_0 = \{x_1, ...x_N\}$ is true. Our agent may partition X_0 into $n \leq N$ disjoint, non-void subsets.

$$X_1^1, X_2^1, ..., X_n^1.$$

Now he is given the choice to pay a fee r to be told which member of the partition contains the true member of X. Say he is told X_1^1. If $N(X_1^1) \geq n$ he may then partition X_1^1 and pay r to be reliably told which member of this new partition contains the true member of X_0. The agent can go on in this way until every answer to a new question contains only elements of one of the elements of X_0.

For any number n and N there is a finite number v of different questioning procedures of this sort that the agent could employ in attempting to discover which member of X is true. Call these n-ary questioning procedures for X at constant rate r. Let

---

[20] For other special cases, see van Rooij (2004a).

$$QP_X = \{p_1, p_2, ..., p_v\}$$

be the mutually exclusive and jointly exhaustive propositions describing the employment of these different n-ary questioning procedures to discover which member of X_0 is true. The decision problem is now which questioning procedure to follow: $\mathcal{A} = QP_X$.

To determine the utility of new information with respect to a questioning procedure, we have to determine the utility of a questioning procedure for the remaining set of possibilities X'. We will assume that this depends completely on the *costs* of the questioning procedure, $C(p)$, and that this is measured in terms of the number of n-ary questions of procedure p that still has to be asked before the member of X_0 can be determined for certain. But this means that in the optimal case $max_{a \in \mathcal{A}} EU(a) = -min_{p \in QP_X} C(p) = -E^n(X)$ and $max_{a \in \mathcal{A}} EU(a, q) = -min_{p \in QP_X} C(p, q) = -E_q^n(X)$.[21]

To make life easier, we will make use of a *decision rule* that assigns a unique action to every possible answer to Q. Because Q is a partition, $Q(w)$ is simply the element of Q that has to be answered if w is the case. Now we can determine the utility value of the decision rule d with respect to question Q, $EU(d, Q) = \sum_w P(w) \times U(d(Q(w)), w)$. In terms of the utility of a decision rule, we can now show in a simple way that the expected utility value of a question with respect to the decision problem which questioning procedure to adopt if you want to know which member of X is true reduces to the expected informativity value of this question with respect to 'question' X:

$$
\begin{aligned}
EUV_{QP_X}(Q) &= max_d EU(d, Q) - max_{p \in QP_X} EU(p) \\
&= -min_d C(d, Q) - -min_{p \in QP_X} C(p) \\
&= -\sum_{q \subset Q} P(q) \times E_q(X) - -E(X) \\
&= E(X) - E_Q(X) \\
&= EIV_X(Q).
\end{aligned}
$$

6   Conclusions and Outlook

In this paper I have shown how we can measure the usefulness, or *relevance*, of questions and answers using Stochastic Communication Theory, Inductive Logic and Statistical Decision Theory, and I have suggested that some of these measures are of greater value than others. In other papers I have used these notions for linguistic purposes to account for (i) the meaning of questions and assertions (van Rooij, 2003a,b); (ii) conversational implicatures (van Rooij, 2003c), and (iii) the licensing of polarity items (van Rooij, 2003d). In Van Rooij (2003a), for instance, I argue that measuring the relevance, or value, of questions and answers is of importance for linguistic theory, because it helps the answerer to determine what is actually expressed by an interrogative sentence, and the questioner to calculate which proposition is expressed by a declarative answer. What is expressed

---

[21] From Shannon's noiseless coding theorem it follows that in general $E^n(X) \leq min_{p \in DP_X} C(p) < (E^n(X) + 1)$.

by interrogative and declaratively used sentences is very context-dependent, and depends heavily on the decision problem of the questioner. Assuming that both participants know what the decision problem of the questioner is, I propose that what is expressed by an interrogative sentence is that question that would be *most relevant* with respect to the questioner's decision problem.

In this paper I have implicitly assumed that the participants of a dialogue are always *cooperative*. In particular, that it can never do any harm for the questioner to make her decision problem public, and that the answerer will always help the questioner as much as he can to solve her decision problem by giving complete answers. Although cooperativity is standardly assumed in Gricean (1989) pragmatics, the participants of a dialogue do not always behave accordingly. It has been argued by Merin (1999), for instance, that for linguistic purposes we should base our notion of *relevance* on the assumption that the two participants of a dialogue try to win an argument. Adopting Anscombre & Ducrot's (1983) conjecture that by making assertions we always want to *argue* for particular hypotheses, he suggests to measure the relevance of an assertion in terms of its argumentative function. Assuming that the two participants of a dialogue always argue for mutually exclusive hypotheses, he proposes to determine the relevance of assertion A with respect to hypothesis h in terms of Good's (1950) measure of the weight of evidence: $r_h(A) = \log(P(h/A)/P(\neg h/A))$. It is, perhaps, reassuring that adopting such a radically non-cooperative view on language use doesn't make our whole investigation useless. It turns out that $r_h(A)$ can also be defined as the difference between $\inf(A/\neg h)$ and $\inf(A/h)$, i.e. $r_h(A) = \inf(A/\neg h) - \inf(A/h)$, and it is easily seen that $r_h(A) = 0$ just in case the informative value of A with respect to *yes/no*-question $\{h, \neg h\}$, $IV_{\{h,\neg h\}}(A) = E(\{h, \neg h\}) - E_A(\{h, \neg h\})$, is 0, too. Thus, Merin (1999) takes a proposition to be a relevant argument with respect to an hypothesis, just in case we (in section 4) say it is relevant with respect to the corresponding *yes/no*-question. This doesn't mean that our notions of relevance are, thus, the same. It might well be that $r_h(A) < 0$ although $IV_{\{h,\neg h\}}(A) > 0$, and the other way around, due to the fact that Merin measures the relevance of assertions with respect single *hypotheses*, while we measure them with respect to *questions*, or *decision problems*.

Only very recently it has become clear that an analysis of relevance in terms of the hearer's decision problem is not quite appropriate to account for conversational implicatures: the speaker's beliefs and preferences should be taken into account as well. The proper way to do this would be to embed our information- and decision theoretic analyses into a more general game theoretic one. It would be beyond the scope of the present paper to discuss this embedding, though.

References

1. Anscombre, J.C., Ducrot, O.: L'Argumentation dans la Langue. Mardaga, Brussels (1983)
2. Bar-Hillel, Y.: An examination of Information Theory. Philosophy of Science 22, 86–105 (1955)

3. Bar-Hillel, Y., Carnap, R.: Semantic Information. In: Proceedings of the Symposium on Applications of Communication Theory. Butterworth Scientific Publications, London (1953)
4. Blackwell, D.: Equivalent comparisons of experiments. Ann. Math. Statist. 24, 265–272 (1953)
5. Carnap, R.: Logical Foundations of Probability. University of Chicago Press, Chicago (1950)
6. Dretske, F.I.: Knowledge and the Flow of Information. MIT Press, Cambridge (1981)
7. Good, I.J.: Probability and the Weighing of Evidence. Griffin, London (1950)
8. Good, I.J.: On the principle of total evidence. British Journal for the Philosophy of Science 17, 319–321 (1966)
9. Grice, H.P.: Studies in the Way of Words. Harvard University Press (1989)
10. Groenendijk, J.: The logic of interrogation. In: Matthews, T., Strolowitch, D.L. (eds.) The Proceedings of the Ninth Conference on Semantics and Linguistic Theory. CLC Publications, Stanford (1999)
11. Groenendijk, J., Stokhof, M.: Studies in the Semantics of Questions and the Pragmatics of Answers, Ph.D. thesis, University of Amsterdam (1984)
12. Groenendijk, J., Stokhof, M.: Questions. In: van Benthem, J., ter Meulen, A. (eds.) Handbook of Logic and Language, pp. 1055–1124. Elsevier, Amsterdam (1997)
13. Hamblin, C.L.: Questions. Australian Journal of Philosophy 36, 159–168 (1958)
14. Hartley, R.V.L.: Transmission of information. Bell System Technical Journal (1928)
15. Jeffrey, R.: The Logic of Decision. McGraw-Hill, New York (1965)
16. Lehmann, E.L.: Testing Statistical Hypotheses. Wiley, New York (1959)
17. Lewis, D.: Relevant implication. Theoria 54, 161–174 (1988)
18. Lindley, D.V.: On a measure of information provided by an experiment. Ann. Math. Stat. 29, 986–1005 (1956)
19. Marschak, J., Radner, M.: Economic Theory of Teams. Yale University Press, New Haven (1972)
20. Marschak, J.: Prior and posterior probabilities and semantic information. In: Menges, G. (ed.) Information, Inference and Decision, pp. 167–180. D. Reidel, Dordrecht (1974a)
21. Marschak, J.: Information, decision, and the scientist. In: Cherry, C. (ed.) Pragmatic Aspects of Human Communication, pp. 145–178. D. Reidel, Dordrecht (1974b)
22. Merin, A.: Information, relevance, and social decisionmaking. In: Moss, L., Ginzburg, J., de Rijke, M. (eds.) Logic, Language, and Computation, Stanford, vol. 2, pp. 179–221 (1999)
23. Raiffa, H., Schlaifer, R.: Applied Statistical Decision Theory. MIT Press, Cambridge (1961)
24. van Rooij, R.: Questioning to resolve decision problems. Linguistics and Philosophy 26, 727–763 (2003a)
25. van Rooij, R.: Asserting to resolve decision problems. Journal of Pragmatics 35, 1161–1179 (2003b)
26. van Rooij, R.: Conversational implicatures and communication theory. In: van Kuppevelt, J., Smith, R. (eds.) Current and New Directions in Discourse and Dialogue, pp. 283–303. Kluwer, Dordrecht (2003c)
27. van Rooij, R.: Negative polarity items in questions. Journal of Semantics 20, 239–273 (2000d)
28. Rooij, R., van Rooij, R.: Utility, informativity and protocols. Journal of Philosophical Logic 33, 389–419 (2004a)

29. van Rooij, R.: Utility of mention-some questions. Research on Language and Computation 2, 401–416 (2004b)
30. Rosenkrantz, R.: Experimentation as communication with nature. In: Hintikka, J., Suppes, P. (eds.) Information and Inference, pp. 58–93. Reidel, Dordrecht (1970)
31. Rosenkrantz, R.: Inference, Method and Decision. D. Reidel, Dordrecht (1977)
32. Savage, L.J.: The Foundations of Statistics. Wiley, New York (1954)
33. Shannon, C.: The Mathematical Theory of Communication. Bell System Technical Journal 27, 379–423, 623-656 (1948); The paper also appeared as the second chapter of Shannon, C., Weaver, W. (eds.) The Mathematical Theory of Communication. University of Illinois Press (1949)
34. Skyrms, B.: The Dynamics of Rational Deliberation. Harvard University Press, Cambridge (1990)
35. Sneed, J.S.: Entropy, information and decision. Synthese 17, 392–407 (1967)
36. Stalnaker, R.: Assertion. In: Cole, P. (ed.) Syntax and Semantics. Pragmatics, vol. 9, pp. 315–332. Academic Press, New York (1978)
37. Wiener, N.: The human use of human beings. Houghton Mifflin Company, Boston (1950)

Channels: From Logic to Probability

Jeremy Seligman

Philosophy Department, The University of Auckland, New Zealand
j.seligman@auckland.ac.nz

Information arises in conditions of uncertainty. When we are unsure of what has happened, our uncertainty is reduced by gaining information. Central to any mathematical model of information is the representation of a state of uncertainty and the change of state brought about by the acquisition of a new piece of information.

Probability theory provides such a model, starting with a set of possible outcomes, and representing an event as the set of outcomes compatible with the event's occurrence. A state of uncertainty is represented by a function measuring the probability each event's occurrence. When we are fairly sure what has happened, some events will have a high probability and others a low probability; when we are very uncertain, the probabilities are more evenly distributed. The acquisition of new information results in a change in the probability distribution and, possibly, a reduction of uncertainty. I'll call this cluster of modelling assumptions 'Information Via Probability.'

Formal logic supplies an alternative. The possible configurations of an unknown situation are represented by relational structures and a state of uncertainty is represented by a set of such structures. Information is modelled by expressions of a formal language. Each formal proposition (formula) represents the information that the unknown situation has a structure satisfying the formula. Formal reasoning can then be shown to be related to information content: if formula ψ is a deductive consequence of formula φ then the information represented by ψ is contained in the information represented by φ. I'll call this approach 'Information Via Logic.'

Shannon's groundbreaking work on the mathematics of communication, [1], showed how Information Via Probability can be used to give a precise measure of uncertainty and to model the movement of information in a system of communication 'channels'. Each channel connects an informational 'source' to a 'receiver'. From the perspective of the receiver, the state of the source is uncertain but information transmitted along the channel reduces this uncertainty by a quantifiable amount. When channels are connected in a network, the movement of information is measured by tracking the relative reduction in uncertainty between different points in the network.[1]

---

[1] More precisely, Shannon's theory tracks the *average* reduction in uncertainty, which is all that is required for determining 'channel capacity', and other technologically significant quantities. It has proved more difficult to adapt the model to provide an account of the informational content of individual events, but as Shannon famously noted, this 'is irrelevant to the engineering problem',[1, p.3].

G. Sommaruga (Ed.): Formal Theories of Information, LNCS 5363, pp. 193–233, 2009.
© Springer-Verlag Berlin Heidelberg 2009

In [2], Jon Barwise and I developed an account of information flow using a more abstract model of channels, which I'll call the Barwise-Seligman framework. Although inspired by Information Via Logic, we made no use of details concerning the structure of formal languages or deductive calculi.[2] Our aim was to provide an algebraic model of information flow that captured general features common to any system of representation, including linguistic and diagrammatic forms, and even the non-representational information flow of natural signs such as the information carried by a thermometer about the temperature and the information carried by the sight of smoke pouring out of the window of a house. Ultimately, this is a contribution to the mathematical foundations of a semantics and epistemology of the kind proposed by Dretske in [6] and developed by Barwise, Perry, and others as 'situation semantics' and 'situation theory'.

The present article aims to consider various problems involved in adapting the Barwise-Seligman framework to provide a similarly abstract account of Information Via Probability. Dretske framed his definition of information flow using probability rather than logic and returning to this point has been a 10-year lacuna in the development of the theory. Section 1 is an introduction to models of channels using logic and probability, and introducing the central concepts of the Barwise-Seligman framework from [2]: classifications, infomorphisms and channels. Section 2 looks in detail at the problem of modelling Dretske's account of information flow using Shannon channels and comparing these to an Information-Via-Logic analogue (Tarski channels). Particular attention is given to three problems involved in applying these models to information flow among real-world events: the Strength Problem, the Modality Problem and the Context Problem. The concept of a link is introduced as a parameter for solutions to these problems. Section 3 investigates how links are determined by theories derived from logic (Tarski theories) and probability (Dretske theories) and compares them to links derived from a more general class of theories that encompasses both (Gentzen theories). In Section 4, the three kinds of theory are related to ideas from situation semantics, culminating in the equivalence of Gentzen theories with certain models of situation semantics (Barwise structures). Section 5 takes steps toward a philosophical and mathematical analysis of Dretske theories. From an epistemological perspective, they are characterised by properties concerning the existence of answers to questions and the degrees of coherence. From a mathematical perspective, the same properties are shown to relate closely to recent work in algebraic measure theory. Finally, Section 6 introduces the account of information flow from [2], which is quite different from Dretske's account, and

---

[2] Early formulations of Information Via Logic by Bar-Hillel and Carnap [3] focused on the definition of information content rather than the dynamics of information flow. Dynamic logic, especially dynamic epistemic logic [4], also provides a well-developed framework for modelling information flow but one that preserves a feature of Information Via Logic that we drop: the use of expressions in a formal language to represent content. Connections between the two approaches are explored in [5].

shows how they are related. Open problems and directions for future research are indicated.

1   Classifications, Infomorphisms, and Channels

The Barwise-Seligman approach to modelling information and information flow uses structures called 'classifications,' maps between them, called 'infomorphisms,' and various constructions, one of which is the 'channel.' This section will introduce these three concepts with examples drawn from formal logic, abstract probability theory and applications to more concrete epistemological situations.

Definition 1.1. *A* classification *A* *consists of a binary relation* $\vDash_A$ *between sets* $\text{tok}(A)$ *and* $\text{typ}(A)$. *The elements of* $\text{tok}(A)$ *are called 'tokens' and the elements of* $\text{typ}(A)$ *are called 'types'; a* $\vDash_A \alpha$ *represents a being of type* α. *Types are* equivalent *if they have the same tokens; tokens are* equivalent *if they are of the same types. A classification is* type-extensional *if there are no two distinct equivalent types and is* token-extensional *if there are no two distinct equivalent tokens.*

The paradigm example of a classification is given by a formal language L whose formulas are evaluated in semantic structures according to a relation of satisfaction. Define $\text{typ}(L)$ to be the set of formulas of L, $\text{tok}(L)$ to be the set of semantic structures of L, and let $m \vDash_L \varphi$ iff m satisfies φ.[3] For formulas to be equivalent as types is for them to be logically equivalent. For structures to be equivalent as tokens is for them to satisfy the same formulas.[4] Typically, L is neither type-extensional (because of logically equivalent but syntactically distinct formulas) nor token-extensional (because of equivalent but distinct structures).

The paradigm example from probability theory is the classification of possible outcomes by events. A probability space $P = \langle \Omega, \Sigma, p \rangle$ consists of a set Ω, whose elements represent 'possible outcomes,' a σ-algebra Σ on Ω, whose elements represent 'events,' and a probability measure p on Σ representing the probability of the occurrence each event.[5] So define $\text{tok}(P)$ to be Ω, $\text{typ}(P)$ to be Σ and

---

[3] The class of semantic structures for a formal language is usually taken to be a proper class not a set, but this is not essential. If the number of propositions is κ then there are at most 2^κ non-equivalent structures and so for most purposes the class of structures can be restricted a set of representative members. Keeping track of caveats concerning size is an unnecessary headache and so here it will be assumed from the outset that the class of structures is a set.

[4] With languages of first-order predicate logic, for example, token equivalence is elementary equivalence; with languages of modal logic, it is bisimilarity.

[5] A σ-*algebra* over Ω is a set Σ of subsets of Ω such that $\emptyset \in \Sigma$, $\Omega - e \in \Sigma$ for each $e \in \Sigma$, and $\bigcup E \in \Sigma$ for each countable set $E \subseteq \Sigma$. p is a *probability measure* on Σ iff it satisfies the Kolmogorov axioms: $p(\emptyset) = 0$, $p(\Omega - e) = 1 - p(e)$, and $p(\bigcup E) = \sum_{e \in E} p(e)$ if E is countable and $p(e_1 \cap e_2) = 0$ for all $e_1 \neq e_2 \in E$.

let $\omega \vDash_P e$ iff $\omega \in e$. The classification is type-extensional because the types are just sets of tokens, and typically it is also token-extensional.[6]

More concretely, classifications arise whenever specific objects or events are classified into kinds or types. The classification of stars by astronomical predicates ('white dwarf', 'red-shifted' etc.), people by height, daily events by diary entries, and thermometers by temperature readings ('22° C') are all appropriately modelled as classifications. When classifying events, the distinction between tokens and types is the usual distinction between token events and event types. A token event, such as my going for a swim this morning, is something that occurs at a particular time in a particular place, whereas the event type of my going for a morning swim may occur more than once or not at all. Information flow, as considered by Shannon and Dretske, concerns a relationship between events and so event classifications will be the primary focus of this article.

The concept of an 'infomorphism' is best illustrated by considering the relationship between event classifications and the more abstract classifications used to reason about them. Consider a game of chess, observed and analysed by a group of experts. The observations can be represented by an event classification G in which the actual moves of the game are classified by the experts into types of varying precision: "White now has a one-pawn advantage," "Black has control of the centre," "White's queen side is looking weak," and so on. A more abstract classification C, representing a theoretical conception of the games of chess, can be defined by taking the tokens to be abstract representations of each possible configuration of pieces on the board, classified into three types: 'W' if there is a winning strategy for white, 'B' if there is a winning strategy for black, and 'D' if either side can force a draw.[7]

The two classifications are related by asking how to classify an actual move m of the game (token of G) by a theoretically ideal outcome (type of C). There are two approaches to answering this question. From an epistemic perspective, in which the final outcome of the game is unknown, our only hope is to associate the outcome types W, B, and D with observational descriptions $f^\wedge(W)$, $f^\wedge(B)$ and $f^\wedge(D)$ and then classify an actual move m as, for example, a predicted win for white if $m \vDash_G f^\wedge(W)$.[8] From a metaphysical perspective – or, for the

---

[6] Token-extensionality corresponds to the topological concept of T_0-separability, which is satisfied by all the spaces usually considered by probability theorists. In fact, most ordinary probability spaces satisfy the stronger separation property, known as T_2 (Hausdorff), which is equivalent to the condition that any two tokens are of two distinct types: if $a \neq b$ then there are α and β such that $a \vDash \alpha$, $a \nvDash \alpha$, $b \vDash \beta$, and $b \nvDash \beta$. The algebra of classifications, in a much more general form than considered here, was first studied as generalisations of topological spaces in [7], where they are known as Chu spaces. They have since been applied widely in theoretical computer science. A good introduction is Pratt's [8] and material on the website [9].

[7] In a seminal work on the mathematics of chess in 1913, Zermelo proved that in chess either white can force a win, or black can force a win, or both sides can force at least a draw. See [10].

[8] For this to work, the observations in G should be assumed to be closed under some form of aggregation. Boolean conjunction is sufficient but not necessary.

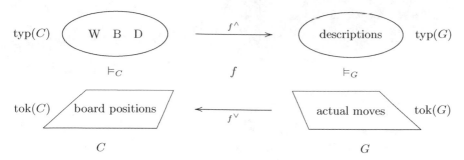

Fig. 1. An infomorphism $f\colon G \rightleftarrows C$

metaphysically squeamish, from a less epistemically disadvantaged perspective, in which all possible outcomes of the game are computed – each move m can be mapped to the resulting board configuration $f^\vee(m)$ and then classified as a win for white if $f^\vee(m) \vDash_C W$. When these two ways of constructing the classification agree, the pair of functions $\langle f^\wedge, f^\vee \rangle$ is an informational coupling between the two classifications, called an 'infomorphism.'

Definition 1.2. *An* infomorphism $f\colon A \rightleftarrows B$ *from classification A to classification B consists of functions $f^\wedge\colon \mathrm{typ}(A) \to \mathrm{typ}(B)$ and $f^\vee\colon \mathrm{tok}(B) \to \mathrm{tok}(A)$ such that for each $\alpha \in \mathrm{typ}(A)$ and $b \in \mathrm{tok}(B)$, $b \vDash_B f^\wedge(\alpha)$ iff $f^\vee(b) \vDash_A \alpha$.*

The existence of an infomorphism represents an ideal informational situation for chess commentary. At each stage of the game, the expert's descriptions in G determine precisely who has the advantage, in the theoretically precise sense given by classification C. No human or computer expert is currently able to classify *all* actual moves in such a way that an infomorphism to C exists but the existence of an infomorphism from the expert classification of some parts of some games is feasible. Excluding the opening moves from the classification and observing games between players of unmatched ability increase the chances of their being an infomorphism (and, consequently, the reputation of the commentator).

An infomorphism between formal language classifications is an interpretation of one language in the other. To take an example familiar to logicians, let A be the language of arithmetic, and let Z be the language of set theory. Each model V of set theory determines a model of arithmetic V_A with domain V_ω and with addition and multiplication suitably defined. Each formula φ of L_A can be mapped to an equivalent formula φ_Z of the language of set theory. The sense in which φ is 'equivalent' to φ_Z is given by the infomorphism condition: $V \vDash_Z \varphi_Z$ iff $V_A \vDash_A \varphi$.

An infomorphism between probability spaces is a continuous function, in the topological sense.[9] Not every continuous function preserves probability measure, and so infomorphisms are too crude to capture the measure-theoretic structure

---

[9] The condition for f to be an infomorphism between spaces P_1 and P_2 is equivalent to the statement that $f^\wedge$ is the set-inverse of $f^\vee$, i.e., that $f^\wedge(e) = \{\omega | f^\vee(\omega) \in e\}$. This ensures that $f^\wedge$ is uniquely determined by $f^\vee$ and exists iff $f^\vee$ is continuous.

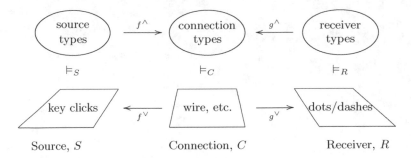

Fig. 2. The telegraph channel

of probability. Nonetheless, the boundaries of what can be expressed in terms of classifications and infomorphisms is less clear-cut than first appearances suggest. For example, not every classification can be assigned a sensible measure of probability.[10]

The third concept to be introduced is the channel. The paradigm example is given by a telegraph wire connecting two telegraph stations, depicted in Figure 2. The source consists of the events occurring when the operator at one station types out a message. Events at the source are represented by an event classification, S. The receiver consists of the events occurring as the message is received at the other station, and this is represented by another event classification, R. A third classification, C, represents the chain of events connecting the source to the receiver, from the click of a key at the source to the sound of the dots and dashes at the receiver and including all the electrical activity along the wire that runs between the two. Each particular connection event c begins with a particular source event $f^\vee(c)$ and ends with a particular receiver event $g^\vee(c)$. Each source event type α determines a connection event type $f^\wedge(\alpha)$, which can be understood as the type of connection event that starts with key clicks of type α, and likewise, each receiver event type β determines a connection event type $g^\wedge(\beta)$, which can be understood as the type of connection event that ends with dots and dashes of type β. This interpretation of f and g is justified by the conditions required for f and g to be infomorphisms. More generally, any pair of infomorphisms into a common classification is called a 'channel'.

Definition 1.3. *A (binary) channel* from classification S to classification R *consists of a classification C and infomorphisms $f: C \rightleftarrows S$ and $g: C \rightleftarrows R$. Classification C is called the* core *of the channel.*

Shannon [1] models communication channels using a probability space C whose outcomes are pairs $\langle s, r \rangle$ consisting of the state s of the source and the state r of

---

[10] A real-valued function on $\mathrm{typ}(A)$ can be considered a 'sensible' measure of probability if it obeys the Kolmogorov axioms with respect to the corresponding sets $\mathrm{tok}(\alpha) = \{a \in \mathrm{tok}(A) \mid a \vDash_A \alpha\}$ of tokens. Classifications that can be assigned probability in this way have been studied by Allwein *et al* in [11]. But if $\mathrm{tok}(A)$ is infinite and each subset $X \subseteq \mathrm{tok}(A)$ has a type α, with $\mathrm{tok}(\alpha) = X$, then there is no such function.

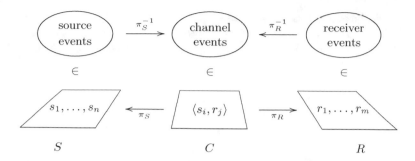

Fig. 3. A finite Shannon channel

the receiver. In the examples Shannon considers, C determines probability spaces S on the set of states of the source and R on the set states of the receiver.[11] The projection functions $\langle s, r \rangle \mapsto s$ and $\langle s, r \rangle \mapsto r$ are continuous, and so provide infomorphisms from S and R to C, respectively. I'll call a channel of this kind a *Shannon channel*. A distinctive property of Shannon channels is that every token of S is connected to every token of R. This certainly does not hold in concrete event channels such as the telegraph channel shown in Figure 2. Although a Shannon channel can be used to model telegraphic communication, it is a model of something slightly different than the telegraph channel because its tokens represent possible states of the source and receiver rather than the actual events occurring in the two places.

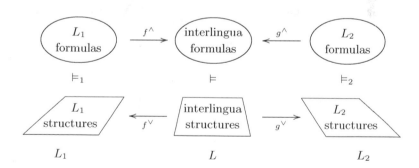

Fig. 4. A Tarski channel

---

[11] In the finite case, this is straightforward because Σ_S and Σ_R can be taken to consist of inverse-images of events in C, with $p_S(e) = p(e \times \Omega_R)$ and $p_S(e) = p(e \times \Omega_S)$. In the infinite case more care is needed because not every σ-algebra on C can be projected to σ-algebras on S and R. Even when this can be done, further conditions are required for the analysis of information flow. Shannon's definition of uncertainty requires that the probability spaces are sufficiently well-behaved to have a probability density function over a suitable measure, as given, for example, by the conditions of the Radon-Nikodym theorem. See, for example, Fremlin [12, Vol. 2, Ch.23].

Channels between formal language classifications are given by interpreting them in an 'interlingua'. At the level of tokens, structures satisfying interlingua formulas are mapped to corresponding structures for the component languages. The language and models of Zermelo-Fraenkel set theory provide the canonical example of an interlingua classification for interpreting between different branches of mathematics. So long as any new branch of mathematics is interpretable in set theory, its practitioners are able to communicate with other mathematicians and use results from other fields (in principle if not in practice). I'll call a channel of this kind a *Tarski channel*. Tarski channels differ from Shannon channels (and are similar to event channels) in that not every structure of one language is 'connected' to every other structure of the other language. But they are similar to Shannon channels (and differ from event channels) in that their tokens represent possible configurations rather than actual occurrences.

When channels are seen as pairings of infomorphisms, it is natural to ask for interpretations of other ways of combining infomorphisms. There are clearly many possibilities, but for one example, I'll return to the chess game commentary. The classification C of board configurations by the outcome types W, B, and D is an idealised one in that it provides no insight into the way in which these outcomes can be inferred from the board configuration. So suppose instead that there is a classification A of abstract representations of entire games of chess, using a more sophisticated language, of the kind used to reason mathematically about chess strategy. Among the sentences of the language will be translations of $g^\wedge(W)$ of W, $g^\wedge(B)$ of B and $g^\wedge(D)$ of D. Likewise, each abstract representation a of a game of chess can be mapped to its initial board configuration $g^\vee(a)$. The correctness of the model is captured by the condition that $g = \langle g^\wedge, g^\vee \rangle$ is an infomorphism. Not all board configurations need be represented, and so models of chess end-game or of the theory of particular openings can be represented in this way. Combining the two infomorphisms we get a complex of a slightly different kind.

The reversal of the direction of the arrows make $\langle f, g \rangle$ a simple *co*-channel rather than a channel. The relationship between a theoretical model and a series of observations can often be seen as forming a co-channel, in which the mediating classification represents a common underlying subject matter. An analysis of this

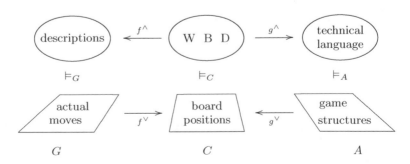

Fig. 5. The Chess co-channel

situation using probability spaces can also be given. If P is a probability space over chess game structures with an appropriate σ-algebra of events then we can define $h^\vee = g^\vee$ and $h^\wedge(W) = \{a|g^\vee(a) \vDash W\}$ (and similarly for B and D), to get a co-channel $\langle f, h \rangle$ from G to P.

More generally, the informational coupling of a variety of concrete and more abstract classifications, in large networks and using a variety of representational systems, can be represented by combinations of infomorphisms. The category (in the sense of Category Theory) of classifications and infomorphism is well behaved, allowing many constructions with classifications, channels and co-channels. Channels can be linked together in chains: a channel from A_1 to A_2 and a channel from A_2 to A_3 can be combined in sequence to produce a channel from A_1 to A_3, which behaves in the expected way. More generally, any network of infomorphisms between classifications has a 'limit' classification which minimally combines all the information in the network into one classification with 'projection' infomorphisms to each of the component classifications. The construction of the composition of channels is a special case of such a limit. Dually, there is also a 'co-limit' classification with infomorphisms from each of the component classifications, which contains all of the information common to different parts of the network.[12]

2   Information Flow via Probability and via Logic

Shannon's analysis of information flow in communication channels was inspiration for Dretske's information-based analysis of knowledge and belief in [6]. This section will review first Shannon's and then Dretske's accounts and show how the latter can be modelled in the Barwise-Seligman framework, at least in outline. I'll identify three problems for filling in the outline, and show how solutions to these problems require sensitivity to a type/token ambiguity in discussions of information flow.[13]

Information flow in Shannon channels is defined as the reduction of uncertainty, which is inversely related to probability. The more likely an event is to occur, the less uncertain we are about it occurring. Logarithms are taken to ensure that the resulting measure is additive. Thus our uncertainty about an event of type e is $\log(1/p(e))$, which is just $-\log p(e)$.[14] In the finite case, uncertainty

---

[12] An account of the category of classifications and infomorphisms is given in [2]. Much more of the mathematics, in greater detail, can be found in the literature on Chu spaces, [8,9], and the 'institutions' of Goguen and Burstall, in [13] but also in [14], which explains the connection to the Barwise-Seligman framework. The category of classifications with infomorphisms is mathematically identical to the category Chu(2) of binary matrices, which has been used in computer science as a model for concurrent processes and, in logic, as an algebraic semantics for linear logic.

[13] The different but related account of information flow given in [2], is postponed to Section 6.

[14] $-\log p(e)$ is also known as the 'surprisal' or 'self-information' of e.

about a source is defined as the sum of our uncertainties about the source states, weighted according to their probability:[15]

$$H(S) = -\sum_{i=1}^{n} p(\{s_i\}) \log p(\{s_i\})$$

This quantity is reduced by observation of the receiver. If we know that an event of type e has occurred at the receiver then our uncertainty about whether an event of type e' has occurred at the source is reduced from $-\log p(e')$ to $-\log p(e'|e)$.[16] The new level of uncertainty about the source is

$$H(S|e) = -\sum_{i=1}^{n} p(\{s_i\}|e) \log p(\{s_i\})$$

and so the amount by which our uncertainty is reduced is

$$I(S;e) = H(S) - H(S|e)$$

The idea that information gain is uncertainty reduction is fundamental to Shannon's analysis but the quantity $I(S;e)$, which measures the reduction of uncertainty brought about by an occurrence of event e, cannot be interpreted as the information content of e.[17] Its value to the qualitative analysis of information flow is less clear, largely because of the possibility that $I(S;e)$ is negative, indicating information loss. Observation of an event at the receiver may make us *more* uncertain about the source, despite telling us something new.[18]

Dretske proposes to identify the information *content* of a receiver event, not by the reduction of uncertainty but simply by the resulting probability of source events. For information to be carried, this must be sufficiently high. His requirement can be stated as follows:

> *Threshold Condition*: For a receiver event of type e to carry the information that a source event of type e' has occurred, the conditional probability $p(e'|e)$ must reach a threshold θ.

---

[15] Uncertainty is also widely known as 'entropy' or 'Shannon entropy' because of the formal similarity with Boltzmann's definition of entropy in thermodynamics.

[16] The expression $p(e'|e)$ is an abuse of notation because e' and e are in different spaces — that of the source and receiver, respectively – and so must first be mapped into the connection space. The correct expression is $p_C(\pi_S^{-1}(e')|\pi_R^{-1}(e))$, where π_S and π_R are the projection functions from C to S and to R.

[17] Shannon's analysis does not use $I(S;e)$ but rather the expected value of this quantity, which is the weighted average of $I(S|\{r_j\})$ for each receiver state r_j, written $I(S;R)$, and known as the 'mutual information' of the channel because it is symmetrical: $I(S;R) = I(R;S)$. Unlike $I(S;e)$, mutual information is non-negative and zero if and only if S and R are probabilistically independent.

[18] For example, if the source has 8 states, one with a very high probability (0.93) and the others with equal low probability (0.01), then we are fairly sure that it is in the high-probability state (uncertainty 0.53). But if we learn that it is *not* in the high-probability state, we are highly uncertain about which of the remaining states the source is in (uncertainty 2.81) and so there is a large information loss (of 2.28).

He argues that the threshold θ can be no less than 1 by considering the possibility of constructing a sequence of channels and events $e_1, \ldots, e_n$ such that e_i is deemed to carry the information that e_{i+1}, for each i. The requirement that $p(e_{i+1}|e_i) \geq \theta$ only ensures that $p(e_n|e_1) \geq \theta^n$ which is arbitrarily small for large n. Dretske takes this to be unacceptable and proposes the following "regulative principle:"

> "*Xerox Principle*: If A carries the information that B, and B carries the information that C, then A carries the information that C." [6, p.57]

The Xerox Principle requires that information flow works like an ideal photocopying machine, in which the quality of the copies is so good that repeatedly copying copies will never spoil the image. If the Threshold Condition is taken to be both necessary and sufficient for information flow, the Xerox Principle implies that θ is either 1 or 0, with the latter value clearly absurd.[19] His final definition of information flow is slightly more complicated:

> "*Informational content*: A signal r carries the information that s is F
> = The conditional probability of s's being F, given r (and k) is 1 (but, given k alone, less than 1)" [6, p.65]

Here, the Threshold Condition (with $\theta = 1$) is conjoined with a requirement that excludes information about events that are certain. The parameter k, represents an agent's knowledge.[20]

In the simplest case, this can be taken to be knowledge of the occurrence of an event and Dretske's conception of information flow can be modelled using a

---

[19] Caveat: Dretske's arguments to justify his definition are more wide-ranging than this simplified summary suggests, and he does not explicitly state the Threshold Condition, which stands here in lieu of his other arguments.

[20] The inclusion of k in the definition raises significant questions about the relationship between the agent's knowledge and the sense in which 'probability' is used by Dretske. Whereas Shannon's model is abstract, in the mathematical sense of applying to whatever satisfies Kolmogorov's axioms, Dretske needs a more substantial interpretation of probability if he is to reach his philosophical goal of explaining epistemology in ultimately non-epistemic terms. The bearing of an agent's knowledge on the determination of channel probabilities would have to be part of such an interpretation. Dretske does not say much about this, and it remains a significant gap in his project which some – such as Loewer in [15] – take to be a fatal flaw. Ultimately, I believe that for an information-based epistemology to work, it must be based on a subjective conception of probability, resulting in a perspectival epistemology, according to which we can know about what an agent knows in the same sense as we can know about the stars and human history, but that there is always the possibility of shifting to a more (or a less) demanding epistemic perspective. Dretske's remarks on skepticism and channel conditions in [6, pp. 111-133] suggest that he would agree with something like this, although it is difficult to square with the naturalistic tone of the rest of the book, and quite out of tune with later developments in situation semantics, which tend to be proudly realist in orientation. A good but highly critical study of Dretske's project is Chater's [16].

Shannon channel C with projection π_S into source S and π_R into receiver R, as follows:[21]

> *Information flow in a Shannon channel*: Given C, the occurrence of e_1 in R carries the information that e_2 has occurred in S, given the knowledge that k has occurred, iff $p(\pi_S^{-1}(e_2)|\pi_R^{-1}(e_1) \cap k) = 1$ but $p(\pi_S^{-1}(e_2)|k) < 1$ and $p(\pi_R^{-1}(e_1) \cap k) > 0$.[22]

That C, the channel, is a parameter of the definition is important. In [6, pp. 111-133], as part of a response to skepticism, Dretske claims that all channels have prerequisite "channel conditions," which must be met if information is to flow. For example, and simplifying a little, when an electrical instrument such as an ammeter is used to make inferences about current flow in some particular circuit component, a particular reading of the meter, '200' say, carries the information that the current flowing through the component is 200mA only if the conditional probability of a 200mA current given the meter reading is 1 (given k). But this condition presupposes the accuracy of a probabilistic model of the channel connecting the ammeter and component. If there is a poor connection between the ammeter and the circuit, or the needle is sticky, or any of a wide variety of other calamities occurs, the meter reading may not carry information about the current at all.

Dretske claims that factual knowledge is an absolute notion, not admitting of degrees, and he "traces the absolute character of knowledge to the absolute character of the information on which it depends" [6, p. 108]. But to retain the absolute character of information, the conditions on which a channel depends must be recognised as a parameter of the account. Skeptical doubts about a specific claim to knowledge can then be diverted as doubts about whether the conditions of the underlying channels are met.[23]

---

[21] Slightly less simply, the agent's knowledge can be represented as a σ-subalgebra of the probability space, adapting the definition of conditional probability accordingly. See, for example, Fremlin [12, Vol. 2,§233].

[22] It is unclear how to interpret Dretske's definition in the case that $p(\pi_R^{-1}(e_1) \cap k) = 0$ because the definition of conditional probability presupposes that it doesn't happen. I have assumed that in this case, no information flows, although not much depends on this decision and the theory could easily be developed with the opposite assumption.

[23] The possibility of a mismatch between model and reality is one that faces any application of mathematics. Every model presupposes conditions under which the model is expected to work and is excused when those conditions are not met. But a model suitable for Dretske's project has to reflect these conditions in the model itself. A Shannon channel or even a network of Shannon channels is not sufficient for this purpose. There must be some explanation within the model of how channels fail to operate when their conditions are not met. A proper treatment of channel conditions eluded many attempts to formalise Dretske's ideas in the theory of situations developed by Barwise, Perry and others in the 1990s. In the literature, this is known as the problem of modelling 'conditional constraints,' which are regularities between situations that do not hold universally. The topic is introduced in [17] and [18] gives a more extensive vision of the role of constraints. [19] gives a good sense of the state of range of questions and options and [20] argues for the need to model 'exceptions' to regularities explicitly using an approach that is close to the Barwise-Seligman framework of [2].

Following Information Via Logic instead of Information Via Probability, a similar model of information can be stated using Tarski channels. In a Tarski channel, information flows from one language to the other via interpretation in a common language (the interlingua). Given a message in the language of L_2 we can learn about events as described in L_1 by translating between the two languages. To determine what information is carried by a formula φ_2 of L_2, it can be translated into L_1. The information expressed by formulas φ_1 of L_1 is carried if it is entailed by a translation of φ_2. Translation in a Tarski channel occurs using the interlingua, by interpreting both φ_2 and φ_1 into the interlingua and then seeing whether there is an entailment between them. By analogy with Dretske's definition, consideration can be given to the agent's knowledge, which can be represented as a set K of interlingua formulas, and the borderline cases in which φ_2 is already known or φ_1 contradicts current knowledge can both be deemed not to count as cases of information flow. Summarising: in a Tarski channel, formula φ_2 of language L_2 can be said to carry the information expressed by formula φ_1 (given K) if and only if the interlingua translation of φ_1 is entailed by K together with the translation of φ_2, excluding the case that φ_1 is already a consequence of K and the case that φ_2 contradicts K.

The role played by probability in a Shannon channel is now played by entailment, which is defined semantically: φ entails ψ iff every structure m of type φ is also of type ψ. The definition of entailment depends only on the satisfaction relation $\vDash$, and so can be formulated in the language of classifications alone.

Definition 2.1. *A* constraint *of classification A is a pair of subsets $\langle \Gamma, \Delta \rangle$ of* $\mathrm{typ}(A)$. *Token $a \in \mathrm{tok}(A)$* violates $\langle \Gamma, \Delta \rangle$ *iff every type of a is in Γ and no type of a is in Δ; otherwise, a* respects *the constraint. The set of constraints respected by every token of A is a relation $\vdash_A$ between subsets of $\mathrm{typ}(A)$, called the 'theory of A': $\Gamma \vdash_A \Delta$ iff every token of A respects $\langle \Gamma, \Delta \rangle$.*

A constraint in a formal language classification is a sequent, in the sense of Gentzen's sequent calculus, and the relation $\vdash_L$ is Gentzen's entailment relation, defined semantically in the Tarskian manner: $\Gamma \vdash_L \Delta$ iff every token that satisfies every formula in Γ also satisfies at least one formula in Δ.[24] With this notation, an information flow in a Tarski channel with interpretations f and g from languages L_1 and L_2, each into interlingua L, can also be characterised as follows:

> *Information flow in a Tarski channel* Formula φ_2 of language L_2 carries the information expressed by formula φ_1 of L_1 (given K) iff $K, g^\wedge(\varphi_2) \vdash_L f^\wedge(\varphi_1)$ but neither $K \vdash_L f^\wedge(\varphi_1)$ nor $Kg^\wedge(\varphi_2) \vdash_L$.

Information flow in Shannon and Tarski channels share a structural similarity. To make this explicit, I'll extend Dretske's definition of information carrying to apply to sets of events.

---

[24] Logicians' standard abuse of notation is permitted. We write $\Gamma_1, \Gamma_2 \vdash_A \Delta, \alpha$ for $\Gamma_1 \cup \Gamma_2 \vdash_A \Delta \cup \{\alpha\}$ and $\vdash_A \alpha, \beta$ for $\vdash_A \{\alpha\} \cup \{\beta\}$.

Definition 2.2. *A constraint $\langle \Gamma, \Delta \rangle$ in the classification of a probability space P is* violated *by an event type e iff $e \subseteq \bigcap \Gamma - \bigcup \Delta$; otherwise, e* respects *the constraint. The set of constraints respected by every non-null event type is a relation $\vdash_P$ between sets of event types, called the 'Dretske theory of P': $\Gamma \vdash_P \Delta$ iff every non-null event type respects $\langle \Gamma, \Delta \rangle$.*

Theorem 2.1. *In Shannon channel C with projections π_R to R and π_S to S, the occurrence of e_2 in R carries the information that e_1 has occurred in S, given the knowledge that events in the set K have also occurred, iff $K, \pi_R^{-1}(e_2) \vdash_P \pi_S^{-1}(e_1)$ but neither $K \vdash_P \pi_S^{-1}(e_1)$ nor $K, \pi_R^{-1}(e_2) \vdash_P$.*

Theorem 2.1 shows a clear parallel between information flow in Shannon and Tarski channels. What is not clear is whether the common structure can be formulated in the Barwise-Seligman framework and whether any of this can be adapted to model information flow in concrete event channels, of the kind required for Dretske's epistemological project. One feature that distinguishes information flow in a Tarski channel is that it is defined in terms of its classification relation alone and so can be generalised to arbitrary classifications, including concrete event classifications. So in any channel C with infomorphisms $f \colon C \rightleftarrows A$ and $g \colon C \rightleftarrows B$, a form of information flow can be defined:

Definition 2.3. *Type β of classification B* strongly indicates *type α of classification A (given a subset $K \subseteq \mathrm{typ}(C)$) iff $K, g^\wedge(\beta) \vdash_C f^\wedge(\alpha)$ but neither $K \vdash_C f^\wedge(\alpha)$ nor $K, g^\wedge(\beta) \vdash_C$.*

If information flow in Tarski and Shannon channels were both forms of strong indication, I would have a unified account of the two kinds of flow within the Barwise-Seligman framework. The following can thus be seen as a partial success.

Theorem 2.2. *In a Tarski channel with interlingua L, formula φ_2 carries the information expressed by formula φ_1 (given K) iff φ_2 strongly indicates formula φ_1 (given K). In a finite Shannon channel C, the occurrence of e carries the information that e' occurs in (given k) iff e strongly indicates e' (given $\{k\}$) in the channel C^+, which is obtained from C by removing all states of probability zero.*[25]

But the theorem leaves a gap containing the infinite Shannon channels, which include channels built from quite ordinary probability spaces, such as the uniform distribution on $[0, 1]$. The strategy of excluding states of probability zero does not work for such spaces because all states have probability zero. Strong indication is therefore too strong to account for flow in all Shannon channels. I'll call this the Strength Problem.

One response to the Strength Problem is simply to ignore it. Although Dretske's account of information flow is framed in the language of probability, he does not

---

[25] The part concerning Tarski channels is obvious. For the second part, it is enough to notice that the following are equivalent: $p(B|A) = 1$; $p(A - B) = 0$; no outcome in $A - B$ is also in C^+; every outcome in C^+ respects the constraint $\langle A, B \rangle$; $A \vdash_{C^+} B$.

explicitly require the resources of continuous mathematics and it may turn out that his epistemological project can be completed using only finite or countable probability spaces. I will not be ignoring the problem, which is addressed in Sections 3 to 5, but I'll put it aside for now because there are other problems to consider when applying the definition of strong information flow to concrete event classifications. The tokens of event classifications are actual events whereas the tokens of formal language classifications and probability spaces represent mere possibilities. Whereas actual events are expected to respect logical or probabilistic constraints between types, they cannot be used to define them if, as we should expect, there are unactualised possibilities. I'll call this the Modality Problem.[26]

A third but related problem is that information flow in concrete event channels depends on conditions being just right. Every channel presupposes that its 'channel conditions' are met. But actual events can violate these presuppositions. If the telegraph wire is cut, events will continue to occur at the source and receiver but they will not respect the same constraints as are presupposed in a model of a working telegraph channel. Strong information flow requires all connection events to respect all constraints, and so cannot allow these anomalies. I'll call this the Context Problem.

A solution to the Modality Problem requires the provision of a suitable relation $\Rightarrow$ between sets of types in the core of a concrete channel to model the regularities on which information flow depends. In general, $\alpha \Rightarrow \beta$ is not implied by $\alpha \vdash_C \beta$ because it may have counterexamples that have not actually occurred or are simply not part of the channel. Likewise, a solution of the Context Problem requires the provision of a suitable set N of tokens to characterise the contextual connections between particular events. These will be called the 'normal' tokens, allowing some tokens to be 'abnormal'. Only normal tokens are required to respect $\Rightarrow$ and so, in general, $\alpha \Rightarrow \beta$ does not imply $\alpha \vdash_C \beta$. The desiderata are captured in the following definition:

Definition 2.4. *A link $\langle N, \Rightarrow \rangle$ on a classification A consists of a subset N of* tok(A) *and a binary relation $\Rightarrow$ between types of A such that if $\alpha \Rightarrow \beta$ then every token of N respects the constraint $\langle \alpha, \beta \rangle$.*

To model information flow in a concrete channel, a link on the channel core is needed. This is shown in Figure 6. In general, the Modality and Strength Problems prevents defining $\Rightarrow$ in terms of $\vdash_C$ (for different reasons) and the Context Problem prevents the taking N to be the whole of tok(C). But schematically, at least, a link on the channel core provides an account of information flow in concrete event channels:

---

[26] David Lewis's modal realism can be seen as one solution to the Modality Problem in which possibilities are taken to be multiple actualities ('possible worlds') and counterfactual constraints are defined by quantifying over them. This solution is not available for Dretske because the events occurring in a concrete communication channel are localised. Information flow in a concrete channel depends on contextual factors which are eliminated when moving to the global perspective of possible worlds. The contrast is one of the motivations for the term 'local logic' in [2], and to be reintroduced below.

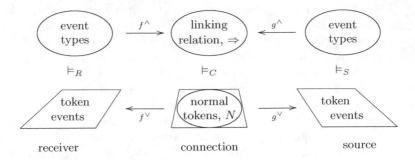

Fig. 6. Information flow in a concrete channel

Information flow relative to a link For a receiver event r to carry information about a source event s there must be a (normal) connection event c that connects them ($f^{\vee}(c) = r$ and $g^{\vee}(c) = s$). Then r's being of type α carries the information that s is of type β if the type of connection correlated with α is linked to the type of connection correlated with β, i.e., $f^{\wedge}(\alpha) \Rightarrow g^{\wedge}(\beta)$.

To illustrate the analysis, I'll take Dretske's example of his experience of hearing a ringing doorbell carrying information that there is someone on the porch pressing the button.[27] The receiver consists of Fred's auditory perceptions when inside his house, classified into experiences of the sound of a doorbell ringing and the rest. Source events occur on the porch outside the front door, classified by whether or not there is a person standing there. Simultaneous events in the two locations are connected by a chain of events loosely described as 'the doorbell' and classified according to various electrical properties such as the state of the button switch (open or closed) and the battery, whether there is current flowing through the bell, and so forth. Not every perception is connected to a porch event by a doorbell event. Fred may have the stereo on too loud or there may be a person on the porch who hasn't yet pushed the button. For there to be a 'doorbell event' the chain linking the porch event to his perception must be complete. But this is not to assume that the doorbell itself is working properly. When it is, the doorbell event is 'normal' and so in the set N of normal connections. A linking relation $\Rightarrow$ describes the workings of the doorbell, so that, for example, there being current flowing through the bell is linked to the switch being closed. The pair $\langle N, \Rightarrow \rangle$ is required to be a link, which implies that every normal doorbell event respects the constraints of a working doorbell. Abnormal doorbell events involving shorts or breaks in the electrical circuit are not required to respect these constraints, although they may do 'accidentally' as when a break in the circuit occurs while the switch is open.

The channel modelled by infomorphisms f and g relates Fred's perceptual events to those on the porch via the doorbell events that connect them. Each

---

[27] The doorbell is a running example in [6] and is used to make a number of points. Here I am using it only as an example of information flow in a concrete event channel.

such doorbell event is mapped by $f^\vee$ and $g^\vee$ to the simultaneous perception and porch event, respectively. At the level of types, $f^\wedge$ maps his hearing a doorbell ringing to the current flowing through the bell and the presence of a person on the porch to the closure of the switch. The infomorphism conditions capture the requirement that the connecting events are genuine connections. If the stereo is on too loud or the person on the porch doesn't know how to use a doorbell, the simultaneous doorbell event is prevented from being in the set of token connections by the infomorphism condition of f or of g.[28]

Information flows between perception and porch event when they are connected by a normal doorbell event and then Fred's hearing a ringing bell indicates there really is someone at the door. The doorbell constraints, modelled by $\Rightarrow$, include the constraint linking current flowing through the bell (which is correlated with Fred's hearing a bell) to the switch being closed (which is correlated with there being someone on the porch). But information can fail to flow for a variety of reasons. The doorbell connection may be abnormal (when there is a short or a circuit break) or there may be no connection between Fred's perception and the simultaneous event on the porch (when the stereo is too loud or the person hasn't pressed the button). Failure will also occur if the linking relation is too weak. A reasonable linking relation may include the constraint linking current not flowing through the bell and the battery being fully charged to the switch being open but may not include a constraint linking the lack of current directly to the switch being open. Some condition implying that the battery isn't flat is needed. This prevents information flowing from Fred's not hearing a bell to their not being a person on the porch. But now suppose that all normal doorbell events happen to be ones in which the battery is not flat. This is not enough for information to flow. In this way, the model incorporates an epistemic element. The doorbell linking relation represents what must be known about the doorbell channel for information to flow.

To get further insight into the model, it is helpful to introduce terms to distinguish between what happens at the level of types and tokens.[29]

Definition 2.5. *In a channel C with infomorphisms $f\colon C \rightleftarrows A$ and $g\colon C \rightleftarrows B$, and a link $\langle N, \Rightarrow \rangle$ on C,*

1. *for types $\alpha \in \mathrm{typ}(A)$ and $\beta \in \mathrm{typ}(B)$, α indicates β if $f^\wedge(\alpha) \Rightarrow g^\wedge(\beta)$;*
2. *for tokens $a \in \mathrm{tok}(A)$ and $b \in \mathrm{tok}(B)$, a signals b if there is a connection $c \in N$ such that $f^\vee(c) = a$ and $g^\vee(c) = b$*

That one event carries information about another event can now be analysed into the two components of indicating and signalling. Fred's hearing a bell

---

[28] The infomorphism conditions also limit how many types can be used to describe Fred's perceptions and the events on the porch. To include more discriminating types, such as those identifying the person on the porch, an extra channel would be required.

[29] The metaphysics of information flow is difficult to pin down in a way directly related to the notorious confusion between tokens and types in discussions of events. See [20] for further discussion.

indicates that someone is at the door (event type) because it is a hearing of a bell and signals someone being at the door (event token) because it is appropriately connected to that particular event on the porch. Both signalling and indicating must occur for information to flow. That someone is on the porch may be indicated without there being anyone there and Fred's perception may signal someone being at the door without his gaining the information that there is someone there.[30]

Both signalling and indicating satisfy Dretske's Xerox Principle in the following sense:

Theorem 2.3. *Given two composable channels C_1 and C_2, tokens a_1 (in the receiver of C_1), a_2 (in the source of C_1, which is also the receiver of C_2) and a_3 (in the source of C_2), and types α_1 (in the receiver of C_1), α_2 (in the source of C_1, which is also the receiver of C_2) and α_3 (in the source of C_2),*

1. *if a_1 signals a_2 in C_1 and a_2 signals a_3 in C_3 then a_1 signals a_3 in $C_1; C_2$,*
2. *if α_1 indicates α_2 in C_1 and α_2 indicates α_3 in C_3 then α_1 indicates α_3 in $C_1; C_2$.*

3   Deriving Links from Theories

Section 2's analysis of information flow in concrete channels is schematic in that it does not provide a particular solution to the Strength, Modality and Context Problems. Instead, the concept of a link was introduced as a parameter, whose value may be determined by further considerations. In this section, more will be said about how to assign this value using ideas from both Information Via Logic and Information Via Probability.

In Tarski channels and Shannon channels, the link defining information flow is constructed from the core of the channel using either $\vdash_L$ or $\vDash_P$. To see what these constructions have in common requires a closer look at the two relations and what they have in common. Given a set Σ, a relation $\vdash$ between subsets of Σ is a *theory* on Σ. Given a classification A, a theory on the $\mathrm{typ}(A)$ determines a link $\langle N_\vdash, \Rightarrow_\vdash \rangle$ on A with $N_\vdash$ defined to be the set of tokens that respect $\vdash$ and $\Rightarrow_\vdash$ defined by: $\alpha \Rightarrow_\vdash \beta$ iff $\alpha \vdash \beta$ but neither $\alpha \vdash$ nor $\vdash \beta$. To account for the role of background knowledge, a theory $\vdash$ can be *conditioned* by a set of types K to get a new theory $\vdash_K$ defined by $\Gamma \vdash_{|K} \Delta$ iff $\Gamma, K \vdash \Delta$.

Theorem 3.1. *Information flow in a Tarski channel C (given K) is information flow relative to the link determined by $\vdash_{C|K}$.[31] Information flow in a Shannon channel C (given K) is information flow relative to the link determined by $\vDash_{C|K}$.*

---

[30] Perhaps the best way of understanding this situation is that the concept of information flow involves an ambiguity of mutually presupposing assertions. To say that event a of type α carries the information that event b is of type β is either to assert that α indicates β presupposing that a signals b, or to assert that a signals b presupposing that α indicates β.

[31] More precisely, φ_2 carries the information that φ_1 given K in C iff φ_2 carries the information that φ_1 given K in the link $\langle N_{\vdash_{C|K}}, \Rightarrow_{\vdash_{C|K}} \rangle$.

Theorem 3.1 shows what is common between information flow in Tarski and Shannon channels and suggests a generalisation to information flow relative to the link determined by *any* theory. The ultimate philosophical objective is to find such a theory, while avoiding the various problems indicated in the previous section. But by way of groundwork, more can be done to understand the possible theories to choose from. In particular, it is useful to try to understand what it is that theories on Tarski and Shannon channels have in common, some of which can be characterised axiomatically.

Definition 3.1. *A theory* $\vdash$ *on* Σ *is a* Gentzen theory *if for all* $\alpha \in \Sigma$ *and* $\Gamma, \Gamma', \Delta, \Delta' \subseteq \Sigma$,

    (Identity) $\alpha \vdash \alpha$
    (Weakening) *if* $\Gamma \vdash \Delta$ *then* $\Gamma, \Gamma' \vdash \Delta, \Delta'$
    (Cut)   *if* $\Gamma \vdash \alpha, \Delta$ *and* $\Gamma, \alpha \vdash \Delta$ *then* $\Gamma \vdash \Delta$

Both $\vdash_L$ and $\vDash_P$ are easily shown to be Gentzen theories, so a better generalisation of information flow in Tarski and Shannon channels is to say that both are information flow relative to links determined by Gentzen theories.

Further progress can be made by finding more properties that theories of the form $\vdash_L$ and $\vDash_P$ have in common. To do this, I'll first define classes of theories that behave in ways similar to those of these two forms.

Definition 3.2. *Given a theory* $\vdash_1$ *on* Σ_1 *and a theory* $\vdash_2$ *on* Σ_2, *a function* $f \colon \Sigma_1 \to \Sigma_2$ *is an* interpretation *of* $\vdash_1$ *in* $\vdash_2$ *if for all* $\Gamma, \Delta \subseteq \Sigma_1$, $\Gamma \vdash_1 \Delta$ *iff* $f[\Gamma] \vdash_2 f[\Delta]$.[32] *Theory* $\vdash_1$ *is* interpretable *in theory* $\vdash_2$ *iff there is an interpretation of* $\vdash_1$ *in* $\vdash_2$.

Theory interpretations are discussed in more detail in [2, Ch. 9]. Here they are needed merely to define the following kinds of theory:

Definition 3.3. *A theory is a* Tarski theory *iff it is interpretable in the theory* $\vdash_L$ *of some formal language classification* L. *A theory is a* Dretske theory *iff it is interpretable in the theory* $\vDash_P$ *of some probability space* P.

Theorem 3.2. *The classes of Gentzen, Tarski, and Dretske theories are closed under interpretability and conditioning. In other words, any theory that is interpretable in a Gentzen/Tarski/Dretske theory is a Gentzen/Tarski/Dretske and the result of conditioning a Gentzen/Tarski/Dretske theory is again a Gentzen/-Tarski/Dretske theory.*[33]

Theorem 3.3. *Every Tarski theory and every Dretske theory is a Gentzen theory.*[34]

---

[32] $f[\Gamma]$ is the image of Γ under f, i.e., the set $\{f(\alpha) \mid \alpha \in \Gamma\}$.

[33] The proof is straightforward. For conditioning, use conditional probability.

[34] It is easy to show that if f is an interpretation of $\vdash$ in a Gentzen theory, then $\vdash$ is also a Gentzen theory. But $\vDash_P$ and $\vdash_L$ are Gentzen theories and so the theorem follows.

Theorem 3.3 is the only positive thing that can be said about inclusion between these three classes of theories. It can be shown that there are Gentzen theories that are neither Tarski theories nor Dretske theories, and that there are Tarski theories that are not Dretske theories and Dretske theories that are not Tarski theories. There are also theories that are both Dretske theories and Tarski theories. Most of these results are consequences of the following characterisation of Tarski theories:

Theorem 3.4. *Suppose* $\vdash$ *is a Gentzen theory on* Σ. *The following are equivalent:*[35]

1. $\vdash$ *is a Tarski theory*
2. $\vdash$ *is the theory* $\vdash_A$ *of some classification* A
3. $\vdash$ *satisfies the Partition rule:*[36]
     (Partition) *if* $\Gamma' \vdash \Delta'$ *for each partition* $\langle \Gamma, \Delta \rangle$ *such that* $\Gamma' \supseteq \Gamma$ *and* $\Delta' \supseteq \Gamma$ *then* $\Gamma \vdash \Delta$.

The characterisation of Dretske theories is more difficult and will take up much of the next two sections. First note that there are Dretske theories that are also Tarski theories. The theory of any finite probability space is an example.[37] There are also Dretske theories that do not satisfy Partition (and so are not Tarski theories). An example is given by uniform Borel measure on $[0, 1]$.[38] Some light is shed on this difference by showing that Partition lies at the far end of a hierarchy of principles, one for each cardinal κ and one with $\kappa = \infty$, involving no restriction of size:

(κ-Cut Left) If Θ is of cardinality $\leq \kappa$ and $\Gamma \vdash \alpha, \Delta$ for each $\alpha \in \Theta$ and $\Gamma, \Theta \vdash \Delta$ then $\Gamma \vdash \Delta$
(κ-Cut Right) If Θ is of cardinality $\leq \kappa$ and $\Gamma \vdash \Theta, \Delta$ and $\Gamma, \alpha \vdash \Delta$ for each $\alpha \in \Theta$ then $\Gamma \vdash \Delta$

If $\kappa_1 \leq \kappa_2$ then κ_2-Cut implies κ_1-Cut so the weakest of these principles is 1-Cut, which is the ordinary Cut principle. The strongest is ∞-Cut but even this is weaker than Partition, which can be restated in a form that shows its resemblance to the other principles in the hierarchy:

---

[35] Tarski theories are known as 'regular theories' in [2] and the theorem follows from results proved there, especially Proposition 9.5, p. 119 and Theorem 9.33, p.130. The construction giving the implication from 3 to 2 is worth noting here because it will be adapted to situation semantics in Section 4. A Gentzen theory $\vdash$ on Σ that satisfies Partition can be used to construct a classification A whose tokens are the partitions $\langle \Gamma, \Delta \rangle$ such that $\Gamma \nvdash \Delta$, and with types Σ and $\langle \Gamma, \Delta \rangle \vDash_A \alpha$ iff $\alpha \in \Gamma$ (or, equivalently, $\alpha \notin \Delta$).

[36] $\langle \Gamma, \Delta \rangle$ is a *partition* of Σ if $\Gamma \cap \Delta = \emptyset$ and $\Gamma \cup \Delta = \Sigma$.

[37] See Theorem 2.2.

[38] This is because for every partition $\langle \Gamma, \Delta \rangle$, $\bigcap \Gamma - \bigcup \Delta$ is either empty or a singleton, which is has zero probability, and so $\Gamma \vdash \Delta$. Yet $[0, 1] \nvdash [0, \frac{1}{2}]$ and so Partition cannot hold. In general, $\vdash_P$ is a Tarski theory iff P is atomic; most useful non-discrete spaces are not.

(Global Cut) if $\Gamma, \Sigma_0 \vdash \Delta, \Sigma_1$ for each partition $\langle \Sigma_0, \Sigma_1 \rangle$ of Θ, then $\Gamma \vdash \Delta$.[39]

The strongest cut principle known to apply to all Dretske theories is $\aleph_0$-Cut, otherwise known as 'Countable Cut'.[40] But there are clearly Gentzen theories that don't satisfy any of the cut principles other than Cut, and these cannot therefore be either Tarski theories or Dretske theories.

Before going any further in trying to characterise Dretske theories, it is worth pausing to ask whether the relationship between Gentzen, Tarski and Dretske theories is duplicated with respect to the links they determine. Surprisingly, the answer is no. The reason this is possible is that a link can be determined by more than one theory. For a simple finite example of this, consider the set $\{2, 3, 6\}$ with $m \Rightarrow n$ iff m is divisible by n. This can be extended to two Tarski theories $\vdash_1$ and $\vdash_2$ such that $2, 3 \vdash_1 6$ but $2, 3 \nvdash_2 6$.[41] In fact, the links determined by Dretske theories are all also determined by Tarski theories, but not conversely. To see this, some properties of links derived from Dretske theories must be identified.

Theorem 3.5. *The link determined by the Dretske theory $\vdash_P$ has the following properties:*

(Transitivity)   *if $\alpha \Rightarrow \beta$ and $\beta \Rightarrow \gamma$ then $\alpha \Rightarrow \gamma$*
(Quasi-reflexivity) *if $\alpha \Rightarrow \beta$ then $\alpha \Rightarrow \alpha$ and $\beta \Rightarrow \beta$*

And there is a function $\alpha \mapsto \neg\alpha$, such that

(Involution) $\neg\neg\alpha = \alpha$
(Contraposition) *if $\alpha \Rightarrow \beta$ then $\neg\beta \Rightarrow \neg\alpha$*
(Contingency) $\alpha \nRightarrow \neg\alpha$

Any relation $\Rightarrow$ with these properties can be used to construct a classification A with the same types as are related by $\Rightarrow$, tokens of the form $\langle \alpha, \beta \rangle$ for which $\alpha \nRightarrow \beta$, and a classification relation defined by

$$\langle \alpha, \beta \rangle \vDash \sigma \text{ iff either } \alpha \Rightarrow \sigma \text{ or } \neg\beta \Rightarrow \sigma$$

---

[39] Taking $\Theta = \Sigma$ gives Partition, the strongest; taking $\Theta = \{\alpha\}$ gives Cut; and size restrictions on Θ give the others. See [2, 9.1,9.2] for more on these principles, such as their relationship to compactness.

[40] For example, with Borel measure on $[0, 1]$, take $\Delta = \{\{x\} | x \in [0, 1]\}$, then $\vdash \Delta$ and $x \vdash$ for each $\{x\} \in \Delta$, but $\emptyset \nvdash \emptyset$ and so $\vdash$ does not satisfy $2^{\aleph_0}$-Cut. This is enough, assuming the Continuum Hypothesis.

[41] For example, the first can be given by $\Gamma \vdash_1 \Delta$ iff the greatest common multiple of the numbers in Γ is divisible by the lowest common divisor of the numbers in Δ. This corresponds to a classification with two tokens, one of type 2 and the other of type 3. The other relation can given by $\Gamma \vdash_2 \Delta$ iff at least one of the numbers in Γ is divisible by at least one of the numbers in Δ. This corresponds to a classification with three tokens: the two previous tokens plus one that is both of type 2 and of type 3.

The Tarski theory $\vdash_A$ of this classification can then be shown to determine a link with linking relation $\Rightarrow$.[42] This proves:

Theorem 3.6. *Every link determined by a Dretske theory is also determined by a Tarski theory.*

Only the first two of the listed properties (transitivity and quasi-reflexivity) are shared by all links determined by Tarski theories. The existence of a 'negation' with the remaining properties is a special feature of the link determined by a Dretske theory of the form $\vdash_P$, and is not even present in all links determined by Dretske theories. Moreover, not all links determined by Tarski theories are also determined by Dretske theories. A simple counterexample is obtained from the theory of the classification of $[0, 1]$ by arbitrary subsets.[43]

4   Modelling Dretske Theories with Situation Semantics

Theorem 3.4 gives a complete characterisation of Tarski theories, both axiomatically, and in terms of the Barwise-Seligman framework. Tarski theories are just the theories of classifications. The picture of Dretske theories is much more sketchy. Axiomatically, the previous section shows that they are Gentzen theories satisfying Countable Cut, but no more powerful cut principles. They arc also known (by definition) to be interpretable in theories of the form $\vdash_P$, but this falls short of a characterisation in the Barwise-Seligman framework, because it involves reference to probability measures. The aim of this section and the next is to understand Dretske theories from the perspective of Information Via Logic, to the fullest extent possible. This section explores (non-probabilistic) models for Dretske theories and the next examines various axioms.

The theory of any classification is a Tarski theory and not all Dretske theories are Tarski theories. So models of Dretske theories will have to be found elsewhere. I'll start by showing how to model any Gentzen theory (because all Dretske theories are Gentzen theories) and then look for properties of Dretske theories that narrow down the field. The inspiration for classifications and their theories comes from semantics and logic. The inspiration for these new models again comes from semantics, but this time from situation semantics, principally [17] and [21].

---

[42] It must be shown that $\alpha \Rightarrow \beta$ iff $\alpha \vdash_A \beta$ and neither $\alpha \vdash_A$ nor $\vdash_A \beta$. There is no type β for which $\vdash_A \beta$ so this part can be ignored. Also note that $(\star)$ $\alpha \vdash_A$ iff $\alpha \not\Rightarrow \alpha$. Now if $\alpha \Rightarrow \beta$ then $\alpha \Rightarrow \alpha$ by quasi-reflexivity, and so $\alpha \not\vdash_A$. And for any token $\langle \alpha_1, \beta_1 \rangle$ of type α, either $\alpha_1 \Rightarrow \alpha$ or $\neg\beta_1 \Rightarrow \alpha$, so by transitivity, either $\alpha_1 \Rightarrow \beta$ or $\neg\beta_1 \Rightarrow \beta$, and so $\langle \alpha_1, \beta_1 \rangle$ is also of type β. Thus $\alpha \vdash_A \beta$, which is enough to show that α is linked to β by the link determined by $\vdash_A$. Conversely, if $\alpha \not\Rightarrow \beta$ then $\langle \alpha, \beta \rangle$ is a token of A so by $(\star)$, either $\alpha \vdash_A$ or $\alpha \Rightarrow \alpha$ and so $\langle \alpha, \beta \rangle \vDash_A \alpha$. But in the latter case, $\langle \alpha, \beta \rangle \not\vDash_A \beta$ because neither $\alpha \Rightarrow \beta$ (assumption) nor $\neg\beta \Rightarrow \beta$ (contingency). Thus either $\alpha \vdash_A$ or $\alpha \not\vdash_A \beta$, and in either case, α is not linked to β by in the link determined by $\vdash_A$.

[43] There is no measure p with the required property that $p(\beta - \alpha) = 0$ iff $\alpha \subseteq \beta$. For a counterexample, just pick α and β to differ by any nonempty set of measure zero.

A specification of the vocabulary of a formal language L provides a definition of what counts as a structure for evaluating the truth of sentences of L. If for example, L has just a unary predicate letter P and a binary relation symbol R, an L-structure is a triple $\langle m, \mathsf{P}^m, \mathsf{R}^m \rangle$ of a set m together with a subset $\mathsf{P}^m \subseteq m$ and a subset $\mathsf{R}^m \subseteq m^2$. The same information can be represented using 'infons'. An L-*infon* is a triple $\langle\!\langle r, \boldsymbol{a}, p \rangle\!\rangle$ consisting of a predicate symbol r (in this case 'P' or 'R') a sequence $\boldsymbol{a}$ whose length depends on r (in this case, 1 if r is 'P' and 2 if r is 'R') and p, known as a 'polarity', which is either '$+$' or '$-$'. An L-*situation* is a set of L-infons. The L-structure m can be represented by the following situation, s_m:

$\langle\!\langle \mathsf{P}, a, + \rangle\!\rangle$ for each $a \in \mathsf{P}^m$
$\langle\!\langle \mathsf{P}, a, - \rangle\!\rangle$ for each $a \in m - \mathsf{P}^m$
$\langle\!\langle \mathsf{R}, \langle a, b \rangle, + \rangle\!\rangle$ for each $\langle a, b \rangle \in \mathsf{R}^m$
$\langle\!\langle \mathsf{R}, \langle a, b \rangle, - \rangle\!\rangle$ for each $\langle a, b \rangle \in m^2 - \mathsf{R}^m$

A definition of satisfaction for formulas of L can be given in terms of L-situations instead of L-structures, but this requires a separation of the satisfaction relation $\vDash$ into a positive $\vDash^+$ and a negative $\vDash^-$ part. For atomic formulas (with assignment function g) the definition is as follows:

$s \vDash^+ \mathsf{P}x$ iff $\langle\!\langle \mathsf{P}, g(\mathsf{x}), + \rangle\!\rangle \in s$
$s \vDash^- \mathsf{P}x$ iff $\langle\!\langle \mathsf{P}, g(\mathsf{x}), - \rangle\!\rangle \in s$
$s \vDash^+ \mathsf{R}xy$ iff $\langle\!\langle \mathsf{R}, \langle g(\mathsf{x}), g(\mathsf{y}) \rangle, + \rangle\!\rangle \in s$
$s \vDash^- \mathsf{R}xy$ iff $\langle\!\langle \mathsf{R}, \langle g(\mathsf{x}), g(\mathsf{y}) \rangle, - \rangle\!\rangle \in s$

The relations $\vDash^+$ and $\vDash^-$ are extended to the rest of the language according to the rules of a system of partial logic, of which there are a number of candidates.[44] But if s_m is the situation representing structure m, then the definitions agree:

$$m \vDash \varphi \quad \text{iff} \quad s_m \vDash^+ \varphi \quad \text{iff} \quad s_m \not\vDash^- \varphi$$

Of course, there are many situations that do not represent L-structures. Some of these are incomplete, in that for some φ neither $s \vDash^+ \varphi$ nor $s \vDash^- \varphi$ and some are incoherent, in that for some φ both $s \vDash^+ \varphi$ and $s \vDash^- \varphi$. A relation of entailment is defined by isolating a set S of situations that are sufficiently well-behaved.[45] Again, there are a number of choices here but the definition I'll use is as follows:

$\Gamma \vdash_S \Delta$ iff there is no situation $s \in S$ such that $s \vDash^+ \varphi$ for all $\varphi \in \Gamma$ and $s \vDash^- \varphi$ for all $\varphi \in \Delta$

If S_0 is the set of situations of the form s_m for some L-structure m then this agrees with the ordinary Tarskian entailment on the classification of situations in S by sentences that they satisfy. But if S_0 contains an incomplete situation,

---

[44] For example, Kleene's strong 3-valued logic. The details are of interest to situation semantics but are of minor concern here. All that matters is the correspondence shown.

[45] As before, I'm making the simplifying assumption that the class of L-structures is a set.

there will be a φ for which $\nvDash_S \varphi, \neg\varphi$, and if it contains an incoherent situation, there will be a φ for which $\varphi, \neg\varphi \nvDash_S$. Situations are therefore suitable to model semantic theories of truth-value gaps and gluts.[46]

A problem arises in saying what it is for one situation to be part of another. As sets of infons, situations are naturally ordered by $\subseteq$ but this has some unfortunate properties. Say that a formula φ is 'persistent' if $s \vDash^+ \varphi$ implies $s' \vDash^+ \varphi$ and $s \vDash^- \varphi$ implies $s' \vDash^- \varphi$ for all $s \subseteq s'$. In a first-order language with standard semantics, atomic formulas and their Boolean combinations are all persistent, but universally quantified formulas are problematic because a situation does not contain the information that there are no more elements in its domain. Consequently, the semantic theory of quantification has a choice: either assume that the domain of s is just the set of elements that occur in infons in s or allow the possibility that there may be more. The second option implies that universally quantified formulas are never $+$-vely satisfied (and existentially quantified formulas are never $-$-vely satisfied) whereas the first option makes quantified formulas non-persistent. A lack of persistence is an indication that the $\subseteq$ ordering is not really representing an information ordering.

This can be redressed by adding such an ordering to the model, and using that ordering in the semantic theory. An *L-situation order* is a partial order (reflexive, transitive and antisymmetric relation) $\preceq$ on a set S together with an L-situation s^* for each element s of S such that for all s_1, s_2 in S, if $s_1 \preceq s_2$ then $s_1^* \subseteq s_2^*$. Now define $s \vDash^+ \varphi$ iff there is no s' such that $s \preceq s'$ and $s'^* \vDash^- \varphi$. Likewise $s \vDash^- \varphi$ iff there is no s' such that $s \preceq s'$ and $s'^* \vDash^+ \varphi$.[47] The new model ensures that all formulas are persistent. It allows a distinction between a complete situation s_m with a closed domain (so nothing above it in the $\preceq$ ordering) and a situation with exactly the same infons but the possibility that the domain can be enlarged.

Just as ordinary Tarskian semantics was generalised to give classifications and their regular entailment relations, so situation semantics can be generalised to provide a representation of Dretske theories.

Definition 4.1. *A situation structure S consists of binary relations $\vDash_S^+$ and $\vDash_S^-$, each between set $\mathrm{tok}(S)$ and $\mathrm{typ}(S)$, together with a partial order $\preceq_S$ on $\mathrm{tok}(S)$ such that*

> (Persistence) *for $s_1, s_2 \in \mathrm{tok}(S)$ if $s_1 \preceq_S s_2$ then for all $\sigma \in \mathrm{typ}(S)$, if $s_1 \vDash_S^+ \sigma$ then $s_2 \vDash_S^+ \sigma$ and if $s_1 \vDash_S^- \sigma$ then $s_2 \vDash_S^- \sigma$*

S is *extensional* if the converse of Persistence also holds, and in this case, $\preceq$ is definable from $\vDash_S^+$ and $\vDash_S^-$. Tokens of S are referred to as 'situations' and types as 'infons'. A situation s *determines* σ if either $s \vDash_S^+ \sigma$ or $s \vDash_S^- \sigma$, and

---

[46] In the literature on situation semantics, situations are typically required to be coherent, but the literature on relevant and paraconsistent logic makes use of incoherent situation.

[47] Essentially, this is how Kripke models for intuitionistic predicate logic are constructed.

over-determines σ if both $s \vDash_S^+ \sigma$ or $s \vdash_S^- \sigma$. It is *coherent* if it does not over-determine any infon. The following three conditions are of interest:

(Coherence)   Every situation in S is coherent.
(Completeness) Every situation in S determines every infon.
(No Mystery)   For every situation s in S and every infon σ there is a
                situation s' that determines σ and $s \preceq s'$.

The situation structure determined by an L-situation order satisfies No Mystery if it satisfies Coherence, so with the standard assumption of Coherence in situation semantics, No Mystery is implied. Completeness, however, is only satisfied in the case that all situations are equivalent to ordinary L-structures. The closest generalisation of situation semantics is therefore to the following situation structures:

Definition 4.2. *A* Barwise *structure is a situation structure satisfying Coherence and No Mystery.*

Now the situation semantic definition of entailment can be carried over directly to the more general setting of situation structures:

$\Gamma \vdash_S \Delta$ iff there is no situation $s \in S$ such that $s \vDash^+ \varphi$ for all $\varphi \in \Gamma$
and $s \vDash^- \varphi$ for all $\varphi \in \Delta$

A situation structure satisfying Coherence and Completeness is essentially just a classification because $s \vDash^- \sigma$ iff $s \nvDash^+ \sigma$, and in this case $\vdash_S$ agrees with the Tarskian definition. So every Tarski theory is the theory of a Barwise structure. More generally, there is a correspondence between conditions on situation structures and conditions on their theories:

Theorem 4.1. *For any situation structure S,*

1. $\vdash_S$ *satisfies Weakening*
2. $\vdash_S$ *satisfies Identity iff S satisfies Coherence*
3. $\vdash_S$ *satisfies Cut if S satisfies No Mystery*
4. $\vdash_S$ *satisfies No Mystery if S is extensional and satisfies Cut.*[48]

The theorem gets us almost half way to the following characterisation of Barwise structures.

Theorem 4.2. *A theory is a Gentzen theory iff it is the theory of a Barwise structure.*

---

[48] The proof is straightforward but as an illustration of the method consider the case of showing that Cut follows from No Mystery. Suppose that $\Gamma \vdash_S \Delta, \sigma$ and $\sigma, \Gamma \vdash_S \Delta$. If $\Gamma \nvdash_S \Delta$ then there is a situation $s \in S$ such that $s \vDash^+ \varphi$ for all $\varphi \in \Gamma$ and $s \vDash^- \varphi$ for all $\varphi \in \Delta$. By No Mystery, there is a situation s' that determines σ and $s \preceq s'$. Then by Persistence, $s' \vDash^+ \varphi$ for all $\varphi \in \Gamma$ and $s' \vDash^- \varphi$ for all $\varphi \in \Delta$, which contradicts either $\Gamma \vdash_S \Delta, \sigma$ or $\sigma, \Gamma \vdash_S \Delta$.

Proof. That theory of a Barwise structure is a Gentzen theory is given by Theorem 4.1. In the other direction, the construction is similar to the one used for Theorem 3.6. Define a Barwise structure S to have as infons the elements of X and as situations the pairs $\langle \Gamma, \Delta \rangle$ of subsets of X for which $\Gamma \nvdash \Delta$. Define $\langle \Gamma, \Delta \rangle \vDash^+ \alpha$ iff $\alpha \in \Gamma$ and $\langle \Gamma, \Delta \rangle \vDash^- \alpha$ iff $\alpha \in \Delta$. Finally, let $\langle \Gamma_1, \Delta_1 \rangle \preceq \langle \Gamma_2, \Delta_2 \rangle$ iff $\Gamma_1 \subseteq \Gamma_2$ and $\Delta_1 \subseteq \Delta_2$. Weakening ensures that $\Gamma \vdash \Delta$ iff $\Gamma \vdash_S \Delta$ and Identity and Cut ensure that S satisfies Coherence and No Mystery (by Theorem 4.1, since the constructed situation structure is extensional).   □

Now one of the ways in which Dretske theories are distinguished from other Gentzen theories is by satisfying Countable Cut. This, and the whole hierarchy of cut principles, corresponds to a suitable generalisation of the No Mystery condition on Barwise structure:

(No κ-Mystery, +ve) For every situation s in S and every set Γ of infons of cardinality $\leq \kappa$, there is a situation s' such that $s \preceq s'$ and either $s' \vDash^+ \sigma$ for all $\sigma \in \Gamma$ or $s' \vDash^- \sigma$ for some $\sigma \in \Gamma$.

(No κ-Mystery, -ve) For every situation s in S and every set Γ of infons of cardinality $\leq \kappa$, there is a situation s' such that $s \preceq s'$ and either $s' \vDash^- \sigma$ for all $\sigma \in \Gamma$ or $s' \vDash^+ \sigma$ for some $\sigma \in \Gamma$.

An easy adaption of the argument for Cut and No Mystery shows the correspondence between κ-Cut and No κ-Mystery. It follows that the extensional Barwise structures representing Dretske theories also satisfy the principle of No Countable Mystery.

5   Characterising Dretske with Questions and Coherence

To progress further, two new concepts are required.[49]

The first new concept comes from a generalisation of structures already implicit in Shannon's understanding of information flow. Recall that uncertainty of a probability space was defined for discrete spaces. The standard approach to extending this concept to non-discrete spaces applied only in those cases in which a probability density function can be defined. It does not depend on general concepts from measure theory or algebra. Another approach is to look at ways of reducing the space to a discrete space.

Suppose that P is a non-discrete space and P_0 is a discrete space of outcomes $\omega_1, \omega_2, \ldots$, related to P by a continuous, measure-preserving transformation, f. The uncertainty $H(P_0)$ of the resulting space P_0 can be used to measure of our uncertainty about P. This quantity depends on which the transformation applied, and so is not an invariant of P, but it is an invariant of P of certain

---

[49] The new concepts arose from close consideration of relevant results in algebraic measure theory, especially Kelley's seminal paper, [22], and Fremlin's treatment of Kelley's work in [12]. Despite the purely technical origin of these concepts, their application to information theory is well-motivated and yields a surprising connection between Information Flow Via Probability and Information Flow Via Logic.

question about P: which of the events $f^{-1}[\omega_n]$ occurred? This question, and our uncertainty about its answer depends only on the event types $f^{-1}[\omega_n]$ and P, not on f. In this way Shannon's concept of uncertainty can be applied straightforwardly to any scenario in which a question is posed about a stochastic system, so long as the question admits of at most a countable number of mutually exclusive and exhaustive answers.[50]

The conditions of exhaustivity and mutual disjointness, imposed by the structure of the function f, can be relaxed a bit. It is only necessary that the probability of getting exactly one answer is 1. More precisely, say that a *question* in a probability space is a countable set Q of non-null events, called 'answers', such . that $\Omega - \bigcup Q$ and the intersection of any two distinct answers is negligible.[51] The *uncertainty* of Q is defined by

$$H(Q) = -\sum_{e \in Q} p(e) \log p(e)$$

The uncertainty of a question with just one answer is zero, and the uncertainty of a question of size 2, such as whether a particular event e occurs, is at most 1, with this maximum value achieved only for events of probability $\frac{1}{2}$. More generally, question Q with $\#(Q) = n$ has a maximum uncertainty of $\log n$ when each answer has probability $\frac{1}{n}$.[52] A countably infinite question ($\#(Q) = \aleph_0$) with probabilities $\frac{1}{2^n}$ has uncertainty 2.

Shannon's model of communication between a finite source and receiver can now be framed in arbitrary probability spaces, as a relationship between two questions. The conditional uncertainty of question Q_2 given question Q_1 is defined by

$$H(Q_2|Q_1) = -\sum_{e_1 \in Q_1} \sum_{e_2 \in Q_2} p(e_2|e_1) \log p(e_2|e_1)$$

This is the expected uncertainty about Q_2 given that Q_1 is answered. The mutual information $I(Q_1; Q_2)$ is calculated as $H(Q_2) - H(Q_2|Q_1)$, the amount by which the uncertainty of Q_1 is expected to be reduced. It is symmetric: $I(Q_1; Q_2) = I(Q_2; Q_1)$.

Now suppose $\vdash$ is the Dretske theory given by an interpretation f into a probability space. Then $f[Q]$ is a question iff the following conditions all hold:

(Exhaustive Answers) $\vdash Q$
(Coherent Answers) $\alpha \nvdash$ for each α in Q
(Incompatible Answers) $\alpha_1, \alpha_2 \nvdash$ for each distinct pair α_1, α_2 in Q

---

[50] Any set Q of mutually disjoint and exhaustive event types in P, determines a continuous, measure-preserving transformation f from P to a discrete space P_0 such that Q is the set of inverse-images of the outcomes of P_0.

[51] A subset of Ω is *null* if it has probability zero; it is *negligible* if it is contained in a null set. In complete measures, all negligible sets are null, but not all useful measures are complete (although they can be completed).

[52] $\#(Q)$ is the cardinality of Q, i.e., its number of answers.

These three conditions make no mention of probability or countability, because all the probability required is already packaged into the definition of $\vdash$ and every set Q satisfying these conditions is countable. This motivates a purely formal definition of questions for theories in general and a specific property of Dretske theories.

Definition 5.1. *A question is a set of types with Exhaustive, Coherent, and Incompatible Answers.*

Theorem 5.1. *Every question in a Dretske theory is countable.*[53]

Questions can also be understood semantically, using classifications (for Tarski theories) and Barwise structures (for Gentzen theories). For a classification, A, say a function f from tokens to types is an 'experiment' if maps each token a to an 'observed' type $f(a)$, i.e., $a \vDash f(a)$. Then a set Q is a question in the Tarski theory $\vdash_A$ iff Q is the range of some experiment in A. In a Barwise structure, S, say that a situation s 'answers' Q with α if $s \vDash^+ \alpha$ for some $\alpha \in Q$ and $s \vDash^- \beta$ for all $\beta \neq \alpha \in Q$. A set Q is a question in the theory of S iff each α in Q is resolved by some situation and each situation can be extended to resolve Q. Although the concept of mutual information has no direct correlate for theories other than Dretske theories, it can be approximated in various useful ways. First, note that $I(Q_1; Q_2)$ is maximal (for a given Q_2) when $H(f[Q_2]|f[Q_1]) = 0$, and in the Dretske theory given by interpretation f,

$$H(f[Q_2]|f[Q_1]) = 0 \text{ iff for each } \alpha_1 \in Q_1 \text{ there is an } \alpha_2 \in Q_2 \text{ such that } \alpha_1 \vdash \alpha_2.$$

So here is another invariant of Dretske theories: the maximal information about one question that can be obtained by answering another.

Definition 5.2. *Question Q_1 is a* strong refinement *of question Q_2 if Q_2 is non-trivial, i.e., $\#(Q_2) > 1$, and for every answer α_1 in Q_1 there is an α_2 in Q_2 such that $\alpha_1 \vdash \alpha_2$.*[54]

The above remarks establish that Q_1 is a strong refinement of Q_2 iff $I(Q; Q_2)$ is maximal when $Q = Q_1$. The Dretske theory of a probability space, has various nice properties that can be expressed using the concept of strong refinement.

Theorem 5.2. *Given a probability space, P,*

---

[53] The proof of countability is easy. Let $Q_n = \{\alpha \in Q \,|\, p(f\alpha) > \frac{1}{n}\}$. The event types in $f[Q_n]$ are pairwise incompatible, so the probability of its union is $\sum_{\alpha \in Q_n} p(f\alpha)$ which is greater than $\#(Q_n)/n$ and less than 1; and so Q_n is finite. As the countable union of finite sets, Q is therefore countable. Questions in Tarski theories and other non-Dretske Gentzen theories are not all countable.

[54] If $\#(Q_2) = 1$ then it already has an answer.

1. *Every non-null event type e is an answer to some question of $\vdash_P$.*[55]
2. *For any two questions of $\vdash_P$, there is a question that is a strong refinement of both.*[56]
3. *If $Q = \bigcup_{n \in \mathbb{N}} Q_n$ is question of $\vdash_P$ with the Q_n disjoint, then there is a question $\{e_n\}_{n \in \mathbb{N}}$ strongly refined by Q in such a way that $e \vdash_P e_n$ for each $e \in Q_n$.*[57]

Despite part 2 of Theorem 5.2, there may not be a strong refinement of an infinite set of questions, even if that set is countable, and even if each question in the set is finite.[58] One diagnosis is that $H(Q_2|Q_1) = 0$ is too demanding. A better notion of refinement is obtained by staying closer to Shannon's definition of mutual information, which allows a question to make some progress toward answering another question without guaranteeing to provide a definite answer in all cases. Define the *restriction* $Q|\beta$ of question Q to β to be the set of answers to Q remaining after discovering β, i.e.,

$$Q|\beta = \{\alpha \in Q \mid \alpha, \beta \nvdash\}$$

Now question Q_1 is guaranteed to make progress by reducing the size of Q_2 if each answer α to Q_1 results restricts Q_2 to a smaller question $Q_2|\alpha$. [59]

Definition 5.3. *Question Q_1 is a* refinement *of question Q_2 if for every answer α to Q_1, $\#(Q_2|\alpha) < \#(Q_2)$.*

Refinements are easier to achieve than strong refinements, and this is reflected in the following result:

Theorem 5.3. *For every countable set K of non-trivial questions in a probability space, there is a question Q that refines every question in K.*[60]

---

[55] Specifically, $\{e, \Omega - e\}$ if $\nvdash_P e$, and the trivial question $\{e\}$ if $\vdash_P e$.

[56] This fact relies on the countability of questions in a Dretske theory. To see why, let $Q_1 \cap Q_2 = \{e_1 \cap e_2 \mid e_1 \in Q_1, e_2 \in Q_2\}$ and let $Q = \{e \in Q_1 \cap Q_2 \mid e \nvdash_P\}$. Now for each $e_1 \in Q_1$ and $e_2 \in Q_2$, $e_1, e_2 \vdash_P e_1 \cap e_2$ so $e_1, e_2 \vdash_P Q_1 \cap Q_2$ by Weakening. But for each $e \in Q_1 \cap Q_2 - Q$, $e \vdash_P$, so by Countable Cut $e_1, e_2 \vdash_P Q$ for each $e_1 \in Q_1$ and $e_2 \in Q_2$. Now $\vdash_P Q_1$ and $\vdash_P Q_2$ so by two more application of Countable Cut, $\vdash_P Q$. The argument does not work for Gentzen theories, which conform only to the ordinary finite Cut. It also doesn't work for Tarski theories, which can have uncountable questions, but which also have stronger versions of Cut. The existence of a finite conjunction is also required.

[57] Take $e_n = \bigcup Q_n$.

[58] Consider for example, a random variable X marking the position of a point on the interval $[0, 1]$ with uniform probability, and measure the position of X with a series of rulers of ever greater precision. The nth ruler has marks at $\frac{1}{2^n}, \ldots, \frac{2^n - 1}{2^n}$. There is no refinement of the sequence of questions, "between which two marks of the nth ruler is X?" The answers to any putative refinement would have to be of the form "X=r" so as to inform us of the answers to all the questions, but $p(X = r) = 0$.

[59] $Q|\alpha$ is not a question in the theory $\vdash$ (unless it is equal to Q, in which case no progress is made) but it is a question in the theory $\vdash_{|\beta}$, which is also a Dretske theory, by Theorem 3.2.

[60] The proof of Theorem 5.3 is in the Appendix.

A quick application of Theorem 5.3 shows that the concept of refinement can be extended to sets that are not questions, in the following sense:

Corollary 5.1. *Given any countable set Γ of questions in a probability space P, there is a question Q such that for every $e \in \Gamma$ and every answer e' in Q, either $e' \vdash_P e$ or $e', e \vdash_P$.*[61]

Theorem 5.3 gives a distinctive and epistemologically significant property of theories of the form $\vdash_P$ that is not shared by all Gentzen theories, nor by all Tarski theories. Yet, for current purposes, it is less than ideal because it applies only to those Dretske theories of the form $\vdash_P$ not to all Dretske theories.

The second new concept for the study of Dretske theories is coherence. The reliability of information sources can be bounded by their coherence. Suppose that a source provides answers to questions in a language L, whose logic is given by a theory $\vdash$. If the source is 100% reliable, then any set of Γ of answers obtained from the source must at least be coherent, i.e., $\Gamma \nvdash$. A newspaper with the editorial line 'we only print the truth' fails to meet its own standards if there is any set Γ of printed sentences for which $\Gamma \vdash$. But an editorial claim that 'most of what we print is true' is less easily undermined. It must be shown that there is a set Γ of printed sentences for which there is no coherent subset containing a majority of the sentences. If the paper passes the test then I'll say that Γ has a 'coherence bound' of 50%. More generally, the bounds on coherence are defined as follows:

Definition 5.4. *δ is a* coherence bound *of a finite set Γ if there is a coherent subset Γ_0 of Γ for which $\#(\Gamma_0)/\#(\Gamma) \geq \delta$.*

For example, Γ has a coherence bound of 90% if it has a coherent subset containing 90% of the elements of Γ. Information sources providing an indefinitely large amount of information, such as daily newspaper presumed to continuation publication into the future, can be represented by infinite sets, whose coherence can be bounded as follows:

Definition 5.5. *δ is a* uniform coherence bound *of a set Γ if δ is a coherence bound of every finite subset of Γ.*

For a newspaper to have a uniform coherence bound of 90%, any sample of sentences taken from (possibly different editions of) the paper must have a coherence bound of 90%; that is, it must be possible to remove no more than 10% of the sample and obtain a coherent set.[62]

In a Dretske theory, coherence bounds are closely related to probability.

---

[61] To prove this, let K be the set of questions $\{e, \Omega - e\}$ for each $e \in \Gamma$. The Theorem 5.3 gives the question Q which refines each of the questions in K. Since each has size 2, Q must be a strong refinement, and so either $e' \vdash_P e$ or $e' \vdash_P \Omega - e$ (and so $e', e \vdash_P$) for each $e' \in Q$.

[62] Every uniform coherence bound of a finite set is a coherence bound but not all coherence bounds are uniform. For example, using propositional logic, $\{p, q, \neg p\}$ has a coherence bound of 2/3 but this is not a uniform coherence bound because the largest coherence bound of the subset $\{p, \neg p\}$ is only $\frac{1}{2}$.

Theorem 5.4. *If Γ is a set of events, each of which has probability at least δ, then δ is a uniform coherence bound on Γ.*[63]

The existence of a probability measure allows us to compare events that are unrelated logically, according to how likely they are to occur. This extra-logical structure casts a shadow on the theories defined on a probability space, by allowing partitions of the set of all events into classes ranked by their probability. For example, take $\Sigma_n = \{e \in \Sigma \mid \frac{1}{n+1} < p(e) \le \frac{1}{n}\}$, together with Σ_∞ the set of null events. More generally, coherence bounds are of little use in comparing individual events, but they can be used to describe the possibility of dividing the set of all events into ranked classes.

Definition 5.6. *A theory is has uniform coherence ranks if there is a partition of the set of all types into classes $\Sigma_1, \ldots, \Sigma_\infty$ such that Σ_n has uniform coherence bound of $\frac{1}{n}$ and all types in Σ_∞ are incoherent.*

Then Theorem 5.4 can be applied to show:

Corollary 5.2. *The Dretske theory of a probability space has uniform coherence ranks.*

The condition given by Corollary 5.2 is the last in a series of conditions satisfied by Dretske theories of probability spaces that are jointly sufficient to characterise the class, up to equivalence of theories.[64]

Theorem 5.5. *A Gentzen theory with the following properties is equivalent to the Dretske theory of a probability space:*

1. *For countable Σ,*
     if $\Gamma \vdash e, \Delta$ for each $e \in \Sigma$ and $\Gamma, \Sigma \vdash \Delta$ then $\Gamma \vdash \Delta$
     if $\Gamma \vdash \Sigma, \Delta$ and $\Gamma, e \vdash \Delta$ for each $e \in \Sigma$ then $\Gamma \vdash \Delta$
2. *Every question is countable.*
3. *Every coherent event is an answer to some question.*
4. *If $Q = \bigcup_{n \in \mathbb{N}} Q_n$ is question with the Q_n disjoint, then there is a question $\{\alpha_n\}_{n \in \mathbb{N}}$ strongly refined by Q in such a way that $\alpha \vdash \alpha_n$ for each $\alpha \in Q_n$.*
5. *For every countable set K of non-trivial questions, there is a question Q that refines every question in K. (And if K is finite or consists of finite questions, then Q strongly refines every question in K.)*
6. *The theory has uniform coherence ranks.*

Theorem 5.5, which is proved in the Appendix, gives sufficient but not necessary conditions for a Barwise theory to be a Dretske theory. Conditions 1,2 and 6 are necessary but those that require the existence of particular questions (conditions 3,4 and 5) are not. A precise characterisation of Dretske theories is therefore still an open problem. Moreover, recent work in algebraic measure theory (especially [23] and [24]) suggest alternative approaches.

---

[63] The proof of Theorem 5.4 is in the Appendix.

[64] Theories $\vdash_1$ and $\vdash_2$ are equivalent if there are functions f and g such that $\alpha \vdash_1 g\beta$ iff $f\alpha \vdash_2 \beta$ and $g\alpha \vdash_1 \beta$ iff $\alpha \vdash_2 f\beta$. Essentially, this amounts to an isomorphism between the quotients of the two theories under equivalence ($e \vdash e'$ and $e' \vdash e$).

6   Logics on the Move

In this final section, I'll return to the model of information flow introduced in Section 2 and reflect on an alternative approach used in [2]. The general conditions for information to flow in a concrete channel were depicted in Figure 6, repeated here:

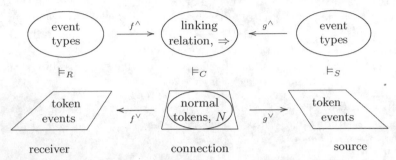

Dretske's definition of information flow was separated into two components. At the level of types, a receiver event type α 'indicates' a source event type β if $f^\wedge(\alpha) \Rightarrow g^\wedge(\beta)$, and at the level of tokens, a particular receiver event r 'signals' a source event s if there is a 'normal' connection event c such that $f^\vee(c) = r$ and $g^\vee(c) = s$. This provides the closest analog to Dretske's account of information flow within the Barwise-Seligman framework, although it differs from the account offered in [2] in several respects. Firstly, the linking relation $\Rightarrow$ was distinguished from the theory that determines it, in order to allow comparison with Dretske's definition of information flow, which deems that what is logically necessary or already known is not information. This is a minor point. Secondly, the concept of a link $\langle N, \Rightarrow \rangle$ was introduced to provide an umbrella for different ways of deriving the linking relation, especially from theories that are not Tarski theories such as theories of continuous probability spaces and theories arising from situation semantics (such as the theories of Barwise structures). And lastly, the account of information flow in [2] has a quite different view of what happens when information flows. It is this point that I will now address.

The Tarski theory $\vdash_A$ of a classification A represents information about the structure of A by specifying all the constraints respected by tokens of a. But this is not all the information that can be had about A. Unless A is completely uniform (all tokens having the same types), there will be constraints that are satisfied by some tokens but not by others. This information can be represented by restricting attention to a subset of 'normal' tokens.

Definition 6.1. *Given a classification A, a local (Tarski) logic L on A is a pair $\langle N_L, \vdash_L \rangle$ consisting of a set $N_L \subseteq \mathrm{tok}(A)$ of tokens, all of which respect the constraints of the (Tarski) theory $\vdash_L$.* [65]

The set N_L of normal tokens represents the part of the the classification that the logic is 'about'–its focus. The theory $\vdash_L$ represents patterns in the distribution

---

[65] The 'local logic' of [2, p. 150] is a local Tarski logic, and I'll continue to use 'local logic' to refer to local Tarski logics throughout this section. The generalisation to local Dretske logics and local Gentzen logics, is an interesting project for future research.

of types among these tokens. Some terminology for describing local logics carries over directly from formal logic classifications, in which $\vdash_L$ is a deductively presented theory and N_L is a class of structures under consideration. A local logic is *complete* if all constraints satisfied by the normal tokens are in the theory ($\vdash_{N_L} \subseteq \vdash_L$), and is *co-complete* if every token that satisfies all the constraints of the theory is normal ($\vdash_L \subseteq \vdash_{N_L}$).[66] It is *sound* if every token is normal (implying $\vdash_L \subseteq \vdash_A$), and so sound logics are also co-complete.

In [2], information flow is modelled not as a relation between individual types (or tokens) in the source and receiver, but as a movement of local logics around a network of classifications. The logics represent information content that is local to a particular classification in the network but in a way that is portable: the effect of acquiring that information on other classifications in the network can be calculated.

Local logics on a classification represent information about the regularities within it. But the information may be partial in two respects. First, the set of normal tokens may satisfy constraints not represented in the theory. A logic is complete if this doesn't happen. Second, the set of normal tokens may be smaller or larger. A complete logic L with just one normal token a represents total information about a, so that $a \models \alpha$ iff $\vdash_L \alpha$. But such a logic provides no information about other tokens. At the other extreme, a logic that is both sound and complete has information about all the tokens but only with respect to what they have in common. Except in trivial cases, there is no local logic that allows the classification relation to be fully recovered. An ordering relation on local logics that conforms to these remarks about partiality is the following:

Definition 6.2. *If L_1 and L_2 are local logics on the same classification then L_1 is contained in L_2, written $L_1 \leq L_2$, iff $N_{L_1} \subseteq N_{L_2}$ and for all Γ, Δ, if $\Gamma \vdash_{L_1} \Delta$ then $\Gamma \vdash_{L_2} \Delta$.*[67]

A local logic is maximal among local logics in the $\leq$ order iff it is both complete and co-complete. In general, there are many maximal local logics but only one for each set of normal tokens. A maximal local logic L represents all the regularities exhibited by tokens in the set N_L. Each set of logics has a *meet* (greatest lower bound) but has a *join* (least upper bound) only if it is bounded above.[68] The

---

[66] The distinction between complete logics that are co-complete and those that are not is important in modal logic. For example, if K is the class of transitive and converse well-founded frames and $\vdash_L$ is the modal logic generated by $(\Box(\Box\varphi \equiv \varphi) \supset \Box\varphi)$, then $\langle K, \vdash_L \rangle$ is co-complete but not complete; by contrast, if $\vdash_{GL}$ is the modal logic generated by $(\Box(\Box\varphi \supset \varphi) \supset \Box\varphi)$, then $\langle K, \vdash_{GL} \rangle$ is both co-complete and complete. Incomplete modal logics were first discovered in the 1970s. Until then, it was implicitly assumed that co-complete logics were also complete.

[67] The order $\leq$ is not the same as the order $\sqsubseteq$ defined in [2] (p. 158), in which the inclusion between the sets of normal tokens is reversed.

[68] Given a set S of logics, $\bigwedge S$ has normal tokens $\bigcap\{N_L \mid L \in S\}$ and $\Gamma \vdash_\Delta$ iff $\Gamma \vdash_L \Delta$ for each $L \in S$, and $\bigvee S$ has normal tokens $\bigcup\{N_L \mid L \in S\}$ and $\Gamma \vdash_\Delta$ iff $\Gamma \vdash_L \Delta$ for some $L \in S$. But $\bigvee S$ is a local logic only if S is bounded above by a local logic, which is the case iff every token that is normal in at least one of the logics satisfies the constraints of all of the logics.

existence of meets is very useful. For example, it shows that for any link $\langle N, \Rightarrow \rangle$, there is a smallest local logic L such that $N \subseteq N_L$ and $\alpha \Rightarrow \beta$ implies $\alpha \vdash_L \beta$. Call this the local logic *extension* of the link. Also any logic L can be *focused* on a set X of tokens to give a logic $L|X$, which is the largest local logic contained in L whose normal tokens are all in X.[69] $L|X$ represents the part of L that is about X, and this provides a way of combining logics even when they lack a join. For any two logics, L_1 and L_2, the logics $L_1|N_{L_2}$ and $L_2|N_{L_1}$ always have an upper bound and so can be joined. This is called the *merge* of L_1 and L_2, written $L_1 \sqcup L_2$.[70]

Now for any infomorphism $f \colon A \rightleftarrows B$, say that a logic L_1 on A is *f-equivalent* to logic L_2 on B if they agree on what counts as a normal token and what is entailed by what; that is,

1. $b \in N_{L_2}$ iff $f^\vee(b) \in N_{L_1}$
2. $\Gamma \vdash_{L_1} \Delta$ iff $f^\wedge[\Gamma] \vdash_{L_1} f^\wedge[\Delta]$

The basic idea for moving local logics across infomorphisms is to move a logic from one classification to an equivalent logic in the other classification, but there may be more than one equivalent logic and so the smallest is selected.[71]

Definition 6.3. *Suppose $f \colon A \rightleftarrows B$ is an infomorphism. The* image *$f[L]$ of local logic L on A is the smallest f-equivalent local logic on B. The* inverse image

---

[69] I choose these operations to indicate a connection with Kohlas's information algebra in [25] and this volume. For any local logic, L, the set of local logics contained in L form an information algebra whose domain is the powerset of $\mathrm{tok}(L)$ (ordered by $\subseteq$), with the focusing operation just defined, and with monoid operation given by join. Mathematically, the construction is not particularly interesting. It is of a very general kind, in which a lattice can be regarded as an information algebra with itself as domain, using meet as focusing. (To make this work precisely, a set X of tokens must be represented by the complete local logic with X as its set of normal tokens.) Nonetheless, the interpretation of normal tokens as being what a logic is 'about' coincides well with the motivating intuitions of information algebra. Another example of this kind can be obtained by 'focusing' a question Q to $Q|e$. The construction cannot be extended to *all* local logics on a classification because only directed joins exist.

[70] $L_1 \sqcup L_2$ has normal tokens $N_{L_1} \cap N_{L_2}$ and a theory that is the smallest regular theory containing both $\vdash_{L_1}$ and $\vdash_{L_2}$. It is the join in the $\sqsubseteq$ ordering, and also defines an information algebra, with focusing, as above.

[71] There may be more than one f-equivalent logic when either A has too many tokens or B has too many types. For example, if $f \colon A \rightleftarrows B$ is given by $f^\vee(b) = b$ and $f^\wedge(\alpha) = \alpha$ where

$$
\begin{array}{c|c} A & \alpha \\ \hline a & 1 \\ b & 1 \end{array}
\qquad
\begin{array}{c|cc} B & \alpha & \beta \\ \hline b & 1 & 1 \end{array}
$$

Let $\vdash_1$ be the smallest regular theory on $\mathrm{typ}(A)$ with $\vdash_1 \alpha$, and let $\vdash_2$ be the smallest regular theory on $\mathrm{typ}(B)$ with $\vdash_2 \alpha$, and let $\vdash_{2'}$ be the smallest regular theory on $\mathrm{typ}(B)$ with $\vdash_{2'} \alpha$ and $\vdash_{2'} \alpha$. Then both $\langle \{b\}, \vdash_1 \rangle$ and $\langle \{a, b\}, \vdash_1 \rangle$ on A are f-equivalent to both $\langle \{b\}, \vdash_2 \} \rangle$ and $\langle \{b\}, \vdash_{2'} \} \rangle$ on B.

$f^{-1}[L]$ of local logic L on B is the smallest local logic on A that is f-equivalent to B.[72]

Movement of local logics provides a much more coherent model of information flow in concrete channels than the two-part signalling/indicating analysis of Section 2, which can be seen as a special case. In the concrete channel depicted in Figure 6, the information provided by the observation that a particular receiver event r is of type α is represented by the local logic L_r^α whose set of normal tokens is the singleton $\{r\}$ and whose entailment relation is the smallest regular theory for which $\vdash \alpha$. The logic L_r can be moved along the infomorphism f to get the logic $f[L_r^\alpha]$ on the core of the channel. The image logic $f[L_r^\alpha]$ can then be merged with L_C, a conservative extension of the link on the core, and moved to the source classification to get $g^{-1}[f[L_r^\alpha] \sqcup L_C]$. An observation about the source, that an event s occurs and is of type β is represented as the local logic L_s^β. Then the logic-movement model of information flow and the signalling/indicating model are related as follows:

Theorem 6.1. r signals s and α indicates β iff

$$L_s^\beta \le g^{-1}[f[L_r^\alpha] \sqcup L_C]$$

But the logic $g^{-1}[f[L_r^\alpha] \sqcup L_C]$ captures more about the classification B than just observations about particular events at the source. To see this, define the *signalling-and-indicating* content of the observation that r is of type α to be the logic[73]

$$\bigvee \{L_s^\beta \mid r \text{ signals } s \text{ and } \alpha \text{ indicates } \beta\}$$

That this is no greater than the moved logic $g^{-1}[f[L_r^\alpha] \sqcup L_C]$ is given by Theorem 6.1. Yet it is possible for the signalling-and-indicating content to be strictly smaller. Suppose, for example, that $g\beta_1, f\alpha \vdash_{L_C} g\beta_2$ but $f\alpha \nvdash_{L_C} g\beta_2$. Then $\beta_1 \vdash \beta_2$ holds in the moved logic but it is not in the signalling-and-indicating content. The two models coincide when the core logic is the smallest logic conservatively extending the link.

Theorem 6.2. If L_C is the smallest conservative extension of $\langle N, \Rightarrow \rangle$ then

$$\bigvee \{L_s^\beta \mid r \text{ signals } s \text{ and } \alpha \text{ indicates } \beta\} = g^{-1}[f[L_r^\alpha] \sqcup L_C]$$

---

[72] The image and inverse image are defined by explicit construction in [2] but the concepts presented there and here are the same. With respect to the $\sqsubseteq$ ordering used in [2], the inverse image $f^{-1}L$ is the *largest* such logic on. See Theorem 13.7, p.167. This reveals a degree of incoherence in taking $\sqsubseteq$ to be modelling information containment while claiming that movement of local logics preserves information content. It is resolved by using the $\le$ order, which gives a better account of information content in concrete classifications.

[73] Thereom 6.1 shows that set $\{L_s^\beta \mid r \text{ signals } s \text{ and } \alpha \text{ indicates } \beta\}$ is bounded above and so that this join exists.

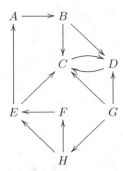

Fig. 7. A network of infomorphisms

The logic movement model of information flow and aggregation can easily be extended to networks of infomorphisms. Given the network of infomorphisms shown in Figure 7, logics L_1 on classification A and L_2 on classification G may be combined by moving them to a common classification. For example, L_1 could be moved to C along the path ABC and L_2 could be moved to C via GC.[74] At C, the images of L_1 and L_2 can be merged. Many other paths are possible, however, and there is no guarantee that the same result will be achieved.

When the results of moving logics is independent of the path taken, it is possible to construct a classification M and a logic K on it, together with canonical infomorphisms from each of the classifications of the network into M. This gives a binary channel between any two classifications in the network, with core M, so that the result of moving a logic from any one classification to any other is done in the same way as considered above. For example, if $f : A \rightleftarrows M$ and $g : G \rightleftarrows M$ are the canonical infomorphisms from A and G into M, then a logic L on A is moved to G as $g^{-1}[f[L] \sqcup K]$.[75]

To extend the movement model to apply other classes of Gentzen theories, especially Dretske theories, the conditions that enable movement and combination of content must be separated from the assumption that content is represented by local logics. This is one direction for further research, related closely to the provision of a full axiomatisation of Dretske theories, extending the results of Section 5. A semantic characterisation of Dretske theories using Barwise structures could be used to similar effect, as well as being of independent interest in building a bridge between early work on situation semantics and the Barwise-Seligman framework of [2]. The 1980s fascination with building an epistemology and semantics based on information and information flow, today seems grandly ambitious. The details of Dretske's account have been widely criticised, and the somewhat amorphous collection of ideas known as 'situation semantics' is no longer in vogue.[76] But information-based approaches in analytic philosophy

---

[74] Specifying movements by listing the classifications on the path works in this example but is not sufficient if there is more than one infomorphism between any two classifications.

[75] This and similar constructions are given in detail in Ch. 15 of [2].

[76] See [26] for a review and prospects.

continue to attract attention, and still lack a precise mathematical toolbox of the kind provided by Kripke semantics in the analysis of modality. There is good reason to continue to look for one.

Acknowledgements. Many thanks to Giovanni Sommaruga for organising the Muenchenwiler Workshop, which was the genesis of the project reported here, and subsequently for his patient work as editor of this volume. Many thanks also to Fred and Sheryl Kroon for providing a calm and supportive environment in which to complete the project. Jon Barwise, who died eight years ago, was very much in my thoughts while writing this paper, and my intellectual debt to him will be obvious from the numerous references to his work both with me and with others.

References

1. Shannon, C.E., Weaver, W.: The Mathematical Theory of Communication. University of Illinois Press (1949)
2. Barwise, J., Seligman, J.: Information Flow: the Logic of Distributed Systems. Cambridge University Press, Cambridge (1997)
3. Bar-Hillel, Y., Carnap, R.: An outline of a theory of semantic information. Technical Report 247, Research Laboratory for Electronics, MIT
4. van Ditmarsch, H., van der Hoek, W., Kooi, P.: Dynamic Epistemic Logic. Synthese Library Series, vol. 337. Springer, Heidelberg (2007)
5. van Benthem, J.: Two logical concepts of information: Correlation vs range. In: Moss, L. (ed.) In Memory of Jon Barwise (2008)
6. Dretske, F.: Knowledge and the Flow of Information. Basil Blackwell, Malden (1981)
7. Barr, M.: *-Autonomous categories. Lecture Notes on Mathematics, vol. 752. Springer, Heidelberg (1979)
8. Pratt, V.R.: Chu spaces. Notes for the School on Category Theory and Applications, University of Coimbra (July 1999)
9. Pratt, V.: The Chu spaces website, http://chu.stanford.edu/
10. Schwalbe, U., Walker, P.: Zermelo and the early history of game theory. Games and Economic Behaviour 34, 123–137 (1997)
11. Allwein, G.T., Moskowitz, I.S., Chang, L.W.: A new framework for Shannon information theory. Technical Report A801024, Naval Research Laboratory (2004)
12. Fremlin, F.H.: Measure theory. Torres Fremlin (2002)
13. Goguen, J.A., Burstall, R.M.: Institutions: abstract model theory for specification and programming. J. ACM 39(1), 95–146 (1992)
14. Goguen, J.: Information integration in institutions. In: Moss, L. (ed.) In Memory of Jon Barwise (2008)
15. Loewer, B.: Review of Knowledge and the Flow of Information. Philosophy of Science 49(2), 297–300 (1982)
16. Chater, N.: Information and Information Processing. PhD thesis, Centre for Cognitive Science, University of Edinburgh (1989)
17. Barwise, J., Perry, J.: Situations and Attitudes. MIT Press, Cambridge (1983)
18. Israel, D., Perry, J.: Information and architecture. In: Barwise, J., Gawron, J.M., Plotkin, G., Tutiya, S. (eds.) Situation Theory and Its Applications, vol. 2. CSLI Press, Stanford (1991)

19. Barwise, J.: Constraints, channels, and the flow of information. In: Aczel, P., Israel, D., Katagiri, Y., Peters, S., Israel, D. (eds.) Situation Theory and Its Applications, vol. 3, Center for the Study of Language and Information, Stanford, CA (1993)
20. Barwise, J., Seligman, J.: The rights and wrongs of natural regularity. Philosophical Perspectives 8, 331 (1994)
21. Barwise, J.: The Situation in Logic. CSLI, Stanford (1989)
22. Kelley, J.L.: Measures in Boolean algebras. Pacific J. Math. 9(11), 1165–1177 (1959)
23. Balcar, B., Jech, T.: Weak distributivity, a problem of von Neumann and the mystery of measurability. The Bulletin of Symbolic Logic 12(2), 241–266 (2006)
24. Jech, T.: Algebraic characterizations of measure algebras. Proceedings of the American Mathematical Society 136(4), 1285–1294 (2008)
25. Kohlas, J.: Information Algebras: Generic Structures for Inference. Springer, New York (2003)
26. Perry, J.: Situation semantics. In: Routledge Encyclopedia of Philosophy, Routledge (1999)
27. Galvin, F., Prikry, K.: On Kelley's intersection numbers. Proceedings of the American Mathematical Society 129(2), 315–323 (2001)

APPENDIX

Some details of the proofs of theorems from Section 5 are given here.

Theorem 5.3 *For every countable set K of non-trivial questions in a probability space, there is a question Q that refines every question in K.*

Proof. By cases of increasing difficulty. Let $P = \langle \Omega, \Sigma, p \rangle$ be the probability space.

1. K is finite. Any two questions of $\vdash_P$ have a common refinement, by Theorem 5.2, and the 'is a refinement of' relation is transitive, so there is a question Q that strongly refines every question in K. Any strong refinement is also a refinement.

2. Every question in K is finite and they can be arranged in a sequence Q_1, $Q_2, \ldots$ such that Q_{n+1} is a strong refinement of Q_n. Define a sequence $\beta_1, \beta_2, \ldots$ as follows. Pick $\beta_1 \in Q_1$ so that $p(\beta_1) \leq \frac{1}{2}$. (This can be done because Q_1 is non-trivial). Pick β_{n+1} such that $\beta_{n+1} \in Q_{n+1}|\beta_n$ and either $p(\beta_{n+1}) = p(\beta_n)$ or $2p(\beta_{n+1}) \leq p(\beta_n)$.[77] Now let $\alpha_1 = \neg\beta_1$ and $\alpha_{n+1} = \neg\beta_{n+1} - \alpha_n$ if $2p(\beta_{n+1}) \leq p(\beta_n)$, otherwise $\alpha_{n+1} = \alpha_n$. Finally, let Q be the set of all the α_n. For each n it can be shown (by induction) that $\{\beta_n, \alpha_1, \ldots, \alpha_n\}$ is a question. Now consider the limit δ of $p(\beta_n)$ (a non-increasing sequence) as n increases. If $\delta = 0$ then the limit of $p(\alpha_1 \cup \ldots \cup \alpha_n)$ is

---

[77] This can be done because $\beta_n \not\vdash_P$ and $\vdash_P Q_{n+1}$ so by Countable Cut, $Q_{n+1}|\beta_n$ is nonempty. Moreover, each $\beta \in Q_{n+1}|\beta_n$ has probability $\leq p(\beta_n)$ because $\beta \vdash_P \beta_n$ (strong refinement). If it has more than one element then choose β_{n+1} to be one of those with smallest probability so that $p(\beta_{n+1}) \leq p(\beta_n)/2$. If it has only one element then choose this to be β_{n+1} and $p(\beta_n) = p(\beta_{n+1})$ because $\beta_n \vdash_P \beta_{n+1}$ (Countable Cut, again, noting that $\beta_n, \beta \vdash_P$ for all $\beta \in Q_{n+1} - \{\beta_{n+1}\}$).

1 and $p(\bigcup Q) = 1$. If $\delta > 0$ then $p(\beta_n)$ is constant for n greater than some sufficiently large N. But then $\alpha_n = \alpha_N$ for $n > N$ and so $Q = \{\alpha_1, \ldots, \alpha_N\}$ and again, $p(\bigcup Q) = 1$. Thus $\vdash_P Q$ and so Q is a question. Q refines each Q_n because $Q_n|\alpha_n$ does not contain β_n, which is in Q_n, and so $\#(Q_n|\alpha_n) < \#(Q_n)$.

3. Every question in K is finite (but maybe case 2 doesn't apply). Enumerate the questions in K as $P_1, P_2, \ldots$. Let $Q_1 = P_1$ and let Q_{n+1} be a strong refinement of Q_n and P_{n+1}. Now $Q_1, Q_2, \ldots$ is a sequence of the kind covered by case 2 and so has a refinement Q. By transitivity of refinement, Q is also a refinement of each P_n.

4. If the previous cases do not apply, then K is a countably infinite set of countably infinite questions. Enumerate K as $Q_1, Q_2, \ldots$. The following lemma is required:

> For any question Q of $\vdash_P$ and any $\epsilon > 0$, there is an event e such that $Q|e$ is finite and $p(e) > 1 - \epsilon$.[78]

Then, for each n and m, there is an event e_m^n such that $p(e_m^n) > 1 - \frac{1}{m2^{n+2}}$ and $Q_n|e_m^n$ is finite. Let $e_m = \bigcap_{n \in \mathbb{N}} e_m^n$. Then $p(e_m) > 1 - \frac{1}{m}$ and $Q_n|e_m$ is finite for each n. Without loss of generality, assume that the e_m are ordered by strictly increasing probability, removing elements of equal probability from the sequence if necessary. Now let $Q = \{\beta_1, \beta_2, \ldots\}$ where $\beta_1 = e_1$ and $\beta_{m+1} = e_{m+1} - (\beta_1 \cup \ldots \cup \beta_m)$. Then (1) $\beta_1 \not\vdash_P$ because $p(\beta_1) > 0$ and $\beta_{m+1} \not\vdash_P$ for each m because $p(e_{m+1}) > p(e_m)$. Also (2) $\beta_m, \beta_k \not\vdash_P$ for $m \neq k$. And (3) $\vdash_P Q$ because $p(\bigcup Q) = p(\bigcup_{m \in \mathbb{N}} \beta_m) = p(\bigcup_{m \in \mathbb{N}} e_m) \geq \lim_{m \to \infty} p(e_m) = 1$. Thus, Q is a question. Finally, Q is a refinement of each Q_n in K because the restricted question $Q_n|\beta_m$ is a subset of $Q_n|e_m$, which is finite, and so strictly smaller than the countably infinite question Q_n. $\square$

Theorem 5.4 *If Γ is a set of events, each of which has probability at least δ, then δ is a uniform coherence bound on Γ.*

Proof. Suppose $\{e_1, \ldots, e_n\}$ is a finite subset of Γ and let Q be the question that determines the status of Γ, as given by Corollary 5.1, so that for each $e \in Q$ and e_i either $e \vdash e_i$ or $e, e_i \vdash$. For each $e \in Q$, let m_e be the number of elements of Γ entailed by e, i.e.,

$$m_e = \#(\{e_i \mid e \vdash e_i \; i \leq n\})$$

For each e_i, $Q = \{e \in Q \mid e \vdash e_i\} \cup \{e \in Q \mid e, e_i \vdash\}$ and $\vdash Q$, so by Cut, $e_i \vdash \{e \in Q \mid e \vdash e_i\}$. Thus

$$p(e_i) \leq p(\bigcup \{e \in Q \mid e \vdash e_i\})$$

Q is mutually exclusive so $p(\bigcup \{e \in Q \mid e \vdash e_i\}) = \sum_{\substack{e \in Q \\ e \vdash e_i}} p(e)$, and so

$$\sum_{i=1}^{n} p(e_i) \leq \sum_{i=1}^{n} \sum_{\substack{e \in Q \\ e \vdash e_i}} p(e) = \sum_{e \in Q} \sum_{e \vdash e_i} p(e) = \sum_{e \in Q} m_e p(e)$$

---

[78] Proof: Q is countable and so can be enumerated as $\alpha_1, \alpha_2, \ldots$. Now $Q = \bigcup_{n \in \mathbb{N}} \{\alpha_1, \ldots, \alpha_n\}$ so $\lim_{n \to \infty} p(\alpha_1 \cup \ldots \cup \alpha_n) = p \bigcup Q = 1$ and so there is an n such that $p(\alpha_1 \cup \ldots \cup \alpha_n) > 1 - \epsilon$. Take $e = \alpha_1 \cup \ldots \cup \alpha_n$ so that $Q|e = \{\alpha_1, \ldots, \alpha_n\}$.

Now let m be the size of the largest coherent subset of $\{e_1, \ldots, e_n\}$. Then m is the maximum value of m_e, and so $\sum_{e \in Q} m_e p(e) \leq m \sum_{e \in Q} p(e)$. But Q is a question, so $\sum_{e \in Q} p(e) = 1$, and so

$$\sum_{i=1}^{n} p(e_i) \leq m$$

Finally, $\delta \leq p(e_i)$, so $n\delta \leq \sum_{i=1}^{n} p(e_i)$, and so $\delta \leq m/n$, as required. $\quad\square$

Theorem 5.5 *A Gentzen theory with the following properties is a Dretske theory:*

1. *For countable Σ,*

   if $\Gamma \vdash \alpha, \Delta$ for each $\alpha \in \Sigma$ and $\Gamma, \Sigma \vdash \Delta$ then $\Gamma \vdash \Delta$
   if $\Gamma \vdash \Sigma, \Delta$ and $\Gamma, \alpha \vdash \Delta$ for each $\alpha \in \Sigma$ then $\Gamma \vdash \Delta$

2. *Every question is countable.*
3. *Every coherent event is an answer to some question.*
4. *If $Q = \bigcup_{n \in \mathbb{N}} Q_n$ is question with the Q_n disjoint, then there is a question $\{\alpha_n\}_{n \in \mathbb{N}}$ strongly refined by Q in such a way that $\alpha \vdash \alpha_n$ for each $\alpha \in Q_n$.*
5. *For every countable set K of non-trivial questions, there is a question Q that refines every question in K. (And if K is finite or consists of finite questions, then Q strongly refines every question in K.)*
6. *The theory has uniform coherence ranks.*

Proof. First map each type α to its equivalence class $[\alpha] = \{\alpha' \mid \alpha \vdash \alpha', \alpha' \vdash \alpha\}$. Let B be the set of equivalence classes, ordered by $[\alpha_1] \leq [\alpha_2]$ iff $\alpha_1 \vdash \alpha_2$. B is a complete Boolean algebra. It is enough to show how to construct complements and joins.

- (Complement) Suppose $b = [\alpha]$. there is a question Q containing α (by 3). Let $Q' = Q - \{\alpha\}$. Then (by 4), there is a question $\{\alpha_1, \alpha_2\}$ such that $\alpha \vdash \alpha_1$ and $\alpha' \vdash \alpha_2$ for each $\alpha' \in Q'$. Q' is countable (by 2), so $\alpha, \alpha_2 \vdash$ and $\vdash \alpha, \alpha_2$ (by 1). Define $\neg b = [\alpha_2]$.

- (Countable Join) Suppose $b_1, b_2, \ldots$ is a countable sequence in B and $b_n = [\alpha_n]$. Let $\neg b_n = [\beta_n]$ so that $Q_n = \{\alpha_n, \beta_n\}$ is a question. There is a question Q that strongly refines each Q_n (by 5). Let $P_n = \{\alpha \in Q \mid \alpha \vdash \alpha_n\}$. By various cuts, $\alpha_n \vdash P_n$ for each n.[79] Let $Q' = \bigcup_{n \in \mathbb{N}} P_n$ and let $Q'' = Q - Q'$. Then (by 4) there is a question $\{\gamma_1, \gamma_2\}$ such that $\alpha \vdash \gamma_1$ for each $\alpha \in Q'$ and $\alpha \vdash \gamma_2$ for each $\alpha \in Q''$. But $P_n \subseteq Q'$ so (by 1), $\alpha_n \vdash \gamma_1$ for each n, and by a few more cuts $\gamma_1 \vdash \{\alpha_n\}_{n \in \mathbb{N}}$.[80] Define $\bigvee_{n \in \mathbb{N}} b_n = [\gamma_1]$.

- (Complete) B has no uncountable antichains (by 2) and all countable joins (above) and so is complete: all joins exist.

---

[79] For each $\alpha \in Q - P_n$, $\alpha \vdash \beta_n$ because Q strongly refines Q_n, and so $\alpha_n, \alpha \vdash$ (by Cut) because $\alpha_n, \beta_n \vdash$. But $\vdash P_n, Q - P_n$ so $\alpha_n \vdash P_n$ (by 1).

[80] $\vdash Q', Q''$ and $\alpha \vdash \{\alpha_n\}_{n \in \mathbb{N}}$ for each $\alpha \in Q'$, so (by 1) $\vdash \{\alpha_n\}_{n \in \mathbb{N}}, Q''$. But $\alpha \vdash \gamma_2$ for $\alpha \in Q''$, so (also by 1), $\vdash \{\alpha_n\}_{n \in \mathbb{N}}, \gamma_2$. Then since $\gamma_1, \gamma_2 \vdash$ (by Cut), $\gamma_1 \vdash \{\alpha_n\}_{n \in \mathbb{N}}$.

Now to complete the proof, we need a probability measure on B.[81] This ensures that the Stone space of B is a probability space. We can then define $f(\alpha)$ to be the event$[\alpha]^*$ in the Stone space corresponding to $[\alpha]$ so that $\Gamma \vdash \Delta$ iff $\bigwedge[\Gamma] \leq \bigvee[\Delta]$ iff $p(\bigcap[\Gamma]^* - \bigcup[\Delta]^*) = 0$. And so $\vdash$ is a Dretske theory.[82]

Necessary and sufficient conditions for the existence of a probability measure on a Boolean algebra were first given by Kelley (in [22]). They are the following:

– (Weakly Distributive) For any sequence $\{P_n\}_{n \in \mathbb{N}}$ of maximal antichains in B, there is a maximal antichain P such that $\{a \in P_n \mid a \wedge b \neq 0\}$ is finite for every $n \in \mathbb{N}$ and $b \in P$.[83]
– (Kelley's Intersection Condition) B^+, the set of non-zero elements of B is a countable union of sets, each of which have a positive intersection number. The intersection number of subset A of B is defined as follows.[84] For each finite sequence $\boldsymbol{a} = \langle a_1, \ldots, n_n \rangle$ of members of A, let $n(\boldsymbol{a}) = n$, the length of the sequence, and let $i(\boldsymbol{a})$ be the maximum number of elements of the sequence with a non-void intersection, i.e.,

$$i(\boldsymbol{a}) = \max\{\#(J) \mid J \subseteq \{1, \ldots, n\}; \bigwedge_{j \in J} a_j \neq 0\}$$

Then the *intersection number* $I(A)$ is defined to be

$$\inf\{i(\boldsymbol{a})/n(\boldsymbol{a}) \mid \boldsymbol{a} \text{ is a finite sequence in } A\}$$

Galvin and Prikry (in [27]) have shown an equivalent condition is obtained by when the intersection number $I(A)$ is replaced by the *weak intersection number* $W(A)$ defined as[85]

$$\inf\{i(\boldsymbol{a})/n(\boldsymbol{a}) \mid \boldsymbol{a} \text{ is a finite sequence in } A \text{ with no repeats}\}$$

It only remains to show that B satisfies these two conditions. That B is weakly distributive follows from 5, given that $\{[\alpha] \mid \alpha \in A\}$ is a maximal antichain in B iff A is a question, and that if Q refines P then $\{\alpha \in P \mid \alpha, \beta \nvdash\}$ is finite for all $\beta \in Q$. For Kelley's condition, note that a finite set $\{\alpha_1, \ldots, \alpha_n\}$ has coherence bound δ iff $i(\langle [\alpha_1], \ldots, [\alpha_n] \rangle) \geq n\delta$ and so that a set Γ has uniform coherence bound δ if $W(\{[\alpha] \mid \alpha \in \Gamma\}) \geq \delta$. Moreover, α is coherent iff $[\alpha] \in B^+$, and so Kelley's condition follows from 6. □

---

[81] A probability measure on a Boolean algebra B is a function $m \colon B \to [0, 1]$ such that $m(0) = 0$, $m(b) > 0$ if $b \neq 0$, $m(1) = 1$, and $m(\bigvee_{n \in \mathbb{N}} b_n) = \sum_{n \in \mathbb{N}} m(b_n)$ whenever $b_n \wedge b_m = 0$ for all $n \neq m$.

[82] Where $[\Gamma] = \{[\alpha] \mid \alpha \in \Gamma\}$ and $[\Gamma]^* = \{[\alpha]^* \mid \alpha \in \Gamma\}$.

[83] There are many equivalent ways of defining weak distributivity. This one is from Fremlin [12, §316G, p.63].

[84] Following [22, p.1166].

[85] $W(A)$ and $I(A)$ are not, in general, equal B^+ is a countable union of sets A with positive $W(A)$ iff B^+ is a countable union of sets A with positive $I(A)$.

Modeling Real Reasoning

Keith Devlin[*]

CSLI, Stanford University
devlin@csli.stanford.edu

This paper is dedicated to my former colleague and good friend, the logician Kenneth Jon Barwise (1942–2000). The work presented here is very much in the spirit of his approach to logic, a theme I pick up in my closing remarks.

1 Introduction

In this article we set out to develop a mathematical model of real-life human reasoning. The most successful attempt to do this, classical formal logic, achieved its success by restricting attention on formal reasoning within pure mathematics; more precisely, the process of proving theorems in axiomatic systems. Within the framework of mathematical logic, a *logical proof* consists of a finite sequence σ_1, σ_2, ..., σ_n of statements, such that for each $i = 1, \ldots, n$, σ_i is either an assumption for the argument (possibly an axiom), or else follows from one or more of σ_1, ..., σ_{i-1} by a rule of logic.

The importance of formal logic in mathematics is not that mathematicians write proofs in the system. To do so would in general be far too cumbersome. Rather, the theory provides a framework for analyzing the notion of mathematical proof. This has led to several benefits. One is a deeper understanding of mathematical proof. Another is the development of techniques for proving that certain statements are in fact not provable. A third is the development of computer tools to carry out automated proof procedures and to assist the human user construct proofs. Still another benefit is that the study of formal logic has educational value for the apprentice mathematician. Generally speaking, those are our goals in trying to develop a model for what we shall call *real-life* logical reasoning.

Of course, one obvious approach to modeling reasoning is to apply formal logic itself, or simple modifications thereof, and this has been attempted on a number of occasions. The most recent significant attempt was in the early work in artificial intelligence in the second half of the twentieth century. By and large, all such attempts have failed. There are various explanations as to why this failure occurred (we outline our own particular take on this in our book [4]), but for the present purpose we need focus only on two issues.

---

[*] Many of the ideas presented in this paper were developed over several years, during which time my research was based at CSLI. Some of the research was supported by an award from the Advanced Research Development Agency, under a subcontract to General Dynamics Advanced Information Systems, as part of ARDA's NIMD Program.

G. Sommaruga (Ed.): Formal Theories of Information, LNCS 5363, pp. 234–252, 2009.

The first issue is that real-life reasoning is rarely about establishing "the truth" about some state of affairs. Rather it is about marshalling evidence to arrive at a conclusion. If the reasoner wants to attach a reliable degree of confidence to their conclusion, she or he must keep track of the sources of all the evidence used, the nature and reliability of those sources, and the reliability of the reasoning steps used in the process. As such, reasoning is better modeled as a process of gathering and processing information.

When you think about it, however, this observation does not amount to a significant departure from the standard model of formal logic, even insofar as logic is viewed as a model of *mathematical* reasoning. Although proofs, both those expressed in formal logic and the kind you find in professional mathematical journals, are often couched in terms of *truths* established, what any mathematical proof really amounts to is an accumulation of evidence — of information that leads to the stated conclusion. Moreover, that conclusion, by virtue of being shown to be *true*, is of interest precisely because it provides us with information! Talk of truth is, then, just a manner of description — one that is often appropriate when discussing proofs of theorems in mathematics (but on few other occasions, courts of law being the most obvious exception where talk of truth is pertinent).

Our second issue (actually a whole list of issues) is considerably more significant, however, and comes not from philosophical reflections on the nature of proof, but on empirical studies of people reasoning in real-life situations. The following set of features are characteristic of much everyday "logical reasoning," yet *formal logic embodies none of them*:

1. Reasoning is often context dependent. A deduction that is justifiable under one set of circumstances may be flat wrong in a different situation.

2. Reasoning is not always linear.

3. Reasoning is often holistic.

4. The information on which the reasoning is based is often not known to be true. The reasoner must, as far as possible, ascertain and remember the source of the evidentiary information used and maintain an estimation of its likelihood of being reliable.

5. Reasoning often involves searching for information to support a particular step. This may involve looking deeper at an existing source or searching for an alternative source.

6. Reasoners often have to make decisions based on incomplete information.

7. Reasoners sometimes encounter and must decide between conflicting information.

8. Reasoning often involves the formulation of a hypothesis followed by a search for information that either confirms or denies that hypothesis.

9. Reasoning often requires backtracking and examining your assumptions.

10. Reasoners often make unconscious use of tacit knowledge, which they may be unable to articulate.

The above list is taken from Richards J. Heuer, Jr.'s classic book *Psychology of Intelligence Analysis* [5][1], popularly known as the "intelligence analyst's bible."

Because of the nature of intelligence analysis, in particular the need to reach concrete conclusions, to document reasoning, and to supply adequate supporting evidence, this activity provides one of the best examples of "real life" logical reasoning outside of mathematics and science. Moreover, in order to improve its intelligence analysis capabilities, the United States intelligence communities have, over the years, carried out several in-depth studies of the way professional analysts work.[2] Heuer was involved in such a study. An intelligence analyst for many years, he returned to university to work on the doctoral dissertation that became his book. It provides an excellent summary of formal reasoning processes outside of mathematics as conducted by a body of professionals trained to do just that. We shall base our model on Heuer's findings.

2   How Does Information Arise?

Since we are approaching reasoning as a specific form of purposeful information gathering and processing, a fundamental question to start with is, how is it possible for something in the world, say a book or a magnetic disk, to store, or represent, information? This question immediately generalizes. For, although we generally think of information as being stored (by way of representations) in things such as books and computer databases, any physical object may store information. In fact, during the course of a normal day, we acquire information from a variety of physical objects, and from the environment. For example, if we see dark clouds in the sky, we may take an umbrella as we leave for work, the state of the sky having provided us with the information that it might rain.

Staying for a moment with that example, how exactly does it come about that dark clouds provide information that it is likely to rain? The answer is that there is a systematic regularity between dark clouds in the sky and rain. Human beings (and other creatures) that are able to recognize that systematic regularity can use it in order to extract information.

In general, then, information can arise by virtue of systematic regularities in the world. People (and certain animals) learn to recognize those regularities, either consciously or subconsciously, possibly as a result of repeated exposure to them. They may then utilize those regularities in order to obtain information from aspects of their environment.

---

[1] Re-published by the United States Central Intelligence Agency in 1999, this book is currently available only in download form from the CIA's website.

[2] Incidentally, it would be unwise to judge the quality of US intelligence analysis by what appear to be some spectacular and costly — in terms of money, human life, and global stability — failures of intelligence decisions by the United States government in the last few years. In all those cases, the problem was not the intelligence analysis, which was in fact highly accurate; rather that, for reasons of political ideology, the government of the day chose to ignore or distort the analysts' recommendations, just as they did with many reports on other issues from the scientific community. That's what can happen when the inmates are put in charge of the most powerful asylum in the world.

What about the acquisition of information from books, newspapers, radio, etc., or from being spoken to by fellow humans? This too depends on systematic regularities. In this case, however, those regularities are not natural in origin like dark clouds and rain. Rather they depend on regularities created by people, the regularities of human language.

In order to acquire information from the words and sentences of English, you have to understand English — you need to know the meanings of the English words and you need a working knowledge of the rules of English grammar. In addition, in the case of written English, you need to know how to read — you need to know the conventions whereby certain sequences of symbols denote certain words. Those conventions of word meaning, grammar, and symbol representation are just that: conventions. Different countries have different conventions: different rules of grammar, different words for the same thing, different alphabets, even different directions of reading — left to right, right to left, top to bottom, or bottom to top.

At an even more local level, there are the conventional information encoding devices that communities establish on an ad hoc basis. For example, a school may designate a bell ring as providing the information that the class should end, or a factory may use a whistle to signal that the shift is over.

The fact is, anything can be used to store information. All it takes to store information by means of some object — or more generally a configuration of objects — is a convention that such a configuration represents that information. In the case of information stored by people, the conventions range from ones adopted by an entire nation (such as languages) to those adopted by a single person (such as a knotted handkerchief). For a non-human example, DNA encodes the information required to create a lifeform (in an appropriate environment).

To make precise these general observations about information, we need to provide a precise, representation-free[3] definition of information, and, second, to examine the regularities, conventions, etc. whereby things in the world represent information. This is what two Stanford University researchers, Jon Barwise and John Perry, set out to do in the late 1970s and early 1980s. The mathematical framework they developed to do this they named Situation Theory, initially described in their book *Situations and Attitudes* [2], with a more developed version of the theory subsequently presented by Devlin in [3]. We shall provide an extremely brief summary of part of situation theory in the following section.

3   Situation Theory

In situation theory, recognition is made of the partiality of information due to the finite, *situated* nature of the agent (human, animal, or machine) with

---

[3] Of course, our theoretical framework will have to have its own representations. The theory we will use adopts the standard application-domain-neutral representation used in science, namely mathematics.

limited cognitive resources. Any agent must employ necessarily limited information extracted from the environment in order to reason and communicate effectively.

The theory takes its name from the mathematical device introduced in order to take account of that partiality. A *situation* can be thought of as a limited part of reality. Such parts may have spatio-temporal extent, or they may be more abstract, such as fictional worlds, contexts of utterance, problem domains, mathematical structures, databases, or Unix directories. The distinction between situations and individuals is that situations have a *structure* that plays a significant role in the theory whereas individuals do not. Examples of situations of particular relevance to the subject matter of this paper will arise as our development proceeds.

The basic ontology of situation theory consists of entities that a finite, cognitive agent individuates and/or discriminates as it makes its way in the world: spatial locations, temporal locations, individuals, finitary relations, situations, types, and a number of other, higher-order entities.

The objects (known as *uniformities*) in this ontology include the following:

- *individuals* — objects such as tables, chairs, tetrahedra, people, hands, fingers, etc. that the agent either individuates or at least discriminates (by its behavior) as single, essentially unitary items; usually denoted in situation theory by $a, b, c, \ldots$
- *relations* — uniformities individuated or discriminated by the agent that hold of, or link together specific numbers of, certain other uniformities; denoted by $P, Q, R, \ldots$
- spatial *locations*, denoted by $l, l', l'', l_0, l_1, l_2$, etc. These are not necessarily like the points of mathematical spaces (though they may be so), but can have spatial extension.
- *temporal locations*, denoted by $t, t', t_0, \ldots$. As with spatial locations, temporal locations may be either points in time or regions of time.
- *situations* — structured parts of the world (concrete or abstract) discriminated by (or perhaps individuated by) the agent; denoted by $s, s', s'', s_0, \ldots$
- *types* — higher order uniformities discriminated (and possibly individuated) by the agent; denoted by $S, T, U, V, \ldots$
- *parameters* — indeterminates that range over objects of the various types; denoted by $\dot{a}, \dot{s}, \dot{t}, \dot{l}$, etc.

The intuition behind this ontology is that in a study of the activity (both physical and cognitive) of a particular agent or species of agent, we notice that there are certain regularities or *uniformities* that the agent either individuates or else discriminates in its behavior.[4]

---

[4] This is true not only of individuals but also of groups, teams, communities. If A and B are engaged in a dialogue or a conversation, or indeed any other form of joint action, they recognize uniformities as individuals in a similar ways. Socially, they negotiate the precise meanings of these, so that they can agree the exact shape of the uniformities that apply in the situation they are in.

For instance, people individuate certain parts of reality as *objects* ('individuals' in our theory), and their behavior can vary in a systematic way according to spatial location, time, and the nature of the immediate environment ('situation types' in our theory).

We note that the ontology of situation theory allows for the fact that different people may discriminate differently. For instance, Russians discriminate as two different colors what Americans classify as merely different shades of blue.

Information is always taken to be information *about* some situation, and is taken to be in the form of discrete items known as *infons*. These are of the form

$$\ll R, a_1, \ldots, a_n, 1 \gg \,, \quad \ll R, a_1, \ldots, a_n, 0 \gg$$

where R is an n-place relation and $a_1, \ldots, a_n$ are objects appropriate for R (often including spatial and/or temporal locations). These may be thought of as the informational item that objects $a_1, \ldots, a_n$ do, respectively, do not, stand in the relation R.

Infons are items of information. They are not things that in themselves are true or false. Rather a particular item of information may be true or false *about a certain part of the world* (a situation).[5]

Given a situation, s, and an infon σ, we write

$$s \models \sigma$$

to indicate that the infon σ is made factual by the situation s, or, to put it another way, that σ is an item of information that is true of s. The official name for this relation is that s *supports* σ.

It should be noted that this approach treats information as a *commodity*. Moreover a commodity that does not have to be true. Indeed, for every positive infon there is a dual negative infon that can be thought of as the opposite informational item, and both of these cannot be true (in the same situation).

A fundamental assumption underlying the situation-theoretic approach to information is that information is not intrinsic to any signal or to any object or configuration of objects in the world; rather information arises from interactions of agents with their environment (including interactions with other agents). The individuals, relations, types, etc. of the situation-theoretic ontology are (third-party) theorist's inventions. For an agent to carry out purposeful, rational activities, however, and even more so for two or more agents to communicate effectively, there must be a substantial agreement first between the way an agent carves up the world from one moment to another, and second between the uniformities of two communicating agents. For instance, if Alice says to Bob, "My car is dirty,"

---

[5] One of the advantages of the framework and notation provided by situation theory is that it allows us to express partial information about complex relations. For example, the relation *eat* presupposes agent, object, instrument, place, time, but much of this information can remain implicit, as in "I'm eating." This makes it possible to choose which aspect of the structure to emphasize in a given instance of interaction. And this choice of emphasis also carries information in its own right, since it is recognized and interpreted as attitude or intent.

and if this communicative act is successful, then the words Alice utters must mean effectively the same to both individuals. In order for a successful information flow to take place, it is not necessary that Alice and Bob share exactly the same concept of "car" or of "dirty," whatever it might mean (if anything) to have or to share an exact concept. Rather, what is required is that their two concepts of "car" and of "dirty" overlap sufficiently. The objects in the ontology of situation theory are intended to be theorist's idealized representatives — prototypes — of the common part of the extensions of individual agent's ontologies. In consequence, the infons are theoretical constructs that enable the theorist to analyze information flow.

Situation theory provides various mechanisms for defining types. The two most basic methods are type-abstraction procedures for the construction of two kinds of types: situation-types and object-types.

Situation-types. Given a SIT-parameter, $\dot{s}$, and a compound infon σ, there is a corresponding *situation-type*

$$[\dot{s} \mid \dot{s} \models \sigma],$$

the *type* of situation in which σ obtains.

This process of obtaining a type from a parameter, $\dot{s}$, and a compound infon, σ, is known as *(situation-) type abstraction*.

For example,

$$[SIT_1 \mid SIT_1 \models \langle\!\langle \text{running}, \dot{p}, LOC_1, TIM_1, 1 \rangle\!\rangle]$$

Object-types. These include the basic types TIM, LOC, IND, REL^n, SIT, INF, TYP, PAR, and POL, as well as the more fine-grained uniformities described below.

Object-types are determined over some initial situation.

Let s be a given situation. If $\dot{x}$ is a parameter and σ is some compound infon (in general involving $\dot{x}$), then there is a type

$$[\dot{x} \mid s \models \sigma],$$

the *type* of all those objects x to which $\dot{x}$ may be anchored in the situation s, for which the conditions imposed by σ obtain.

This process of obtaining a type $[\dot{x} \mid s \models \sigma]$ from a parameter, $\dot{x}$, a situation, s, and a compound infon, σ, is called *(object-) type abstraction*.

The situation s is known as the *grounding* situation for the type. In many instances, the grounding situation, s, is the world or the environment we live in (generally denoted by w).

For example, the *type* of all people could be denoted by

$$[IND_1 \mid w \models \langle\!\langle \text{person}, IND_1, \dot{l}_w, \dot{t}_{now}, 1 \rangle\!\rangle]$$

Again, if s denotes Jon's environment (over a suitable time span), then

$$[\dot{e} \mid s \models \langle\!\langle \text{sees}, \text{Jon}, \dot{e}, LOC_1, TIM_1, 1 \rangle\!\rangle]$$

denotes the type of all those situations Jon sees (within s).

This is a case of an object-type that is a type of situation.

This example is not the same as a *situation-type*. Situation-types classify situations according to their internal structure, whereas in the type

$$[\dot{e} \mid s \models \langle\langle\text{sees, Jon, } \dot{e}, LOC_1, TIM_1, 1\rangle\rangle]$$

the situation is typed from the outside.

Types and the type abstraction procedures provide a mechanism for capturing the fundamental process whereby a cognitive agent classifies the world. Applying the distinction between situation types and object types to interaction phenomena, we may say that we all recognize that the relationship between situation-type *fire* and the situation-type *smoke* obtains only if both are in the same place at the same time. This is then a part of the shared knowledge among members of the same group or community that is often assumed and therefore rarely articulated. Situation theory offers a mechanism for articulating these assumptions by means of defined constraints. *Constraints* provide the situation theoretic mechanism that captures the way that agents make inferences and act in a rational fashion. Constraints are linkages between situation types. They may be natural laws, conventions, logical (i.e., analytic) rules, linguistic rules, empirical, law-like correspondences, etc.

For example, humans and other agents are familiar with the constraint:

Smoke means fire.

If S is the type of situations where there is smoke present, and S' is the type of situations where there is a fire, then an agent (e.g. a person) can pick up the information that there is a fire by observing that there is smoke (a type S situation) and being aware of, or *attuned to*, the constraint that links the two types of situation.

This constraint is denoted by

$$S \Rightarrow S'$$

(This is read as "S *involves* S'.")

Another example is provided by the constraint

FIRE *means fire.*

This constraint is written

$$S'' \Rightarrow S'$$

It links situations (of type S'') where someone yells the word FIRE to situations (of type S') where there is a fire.

Awareness of the constraint

FIRE *means fire*

involves knowing the meaning of the word FIRE and being familiar with the rules that govern the use of language.

The three types that occur in the above examples may be defined as follows:

$$S = [\dot{s} \mid \dot{s} \models \langle\!\langle \text{smokey}, \dot{t}, 1 \rangle\!\rangle]$$

$$S' = [\dot{s} \mid \dot{s} \models \langle\!\langle \text{firey}, \dot{t}, 1 \rangle\!\rangle]$$

$$S'' = [\dot{u} \mid \dot{u} \models \langle\!\langle \text{speaking}, \dot{a}, \dot{t}, 1 \rangle\!\rangle \wedge \langle\!\langle \text{utters}, \dot{a}, \text{fire}, \dot{t}, 1 \rangle\!\rangle]$$

Notice that constraints link types, not situations. However, any particular instance where a constraint is utilized to make an inference or to govern/influence behavior will involve specific situations (of the relevant types). Constraints function by capturing various regularities across actual situations.

A constraint

$$C = [S \Rightarrow S']$$

allows an agent to make a logical inference, and hence facilitates information flow, as follows. First the agent must be able to discriminate the two types S and S'. Second, the agent must be aware of, or behaviorally attuned to, the constraint. Then, when the agent finds itself in a situation s of type S, it knows that there must be a situation s' of type S'. We may depict this diagrammatically as follows:

$$S \overset{C}{\Longrightarrow} S'$$
$$s : S \uparrow \qquad \uparrow s' : S'$$
$$s \overset{\exists}{\longrightarrow} s'$$

For example, suppose $S \Rightarrow S'$ represents the constraint *smoke means fire*. Agent $\mathcal{A}$ sees a situation s of type S. The constraint then enables $\mathcal{A}$ to conclude correctly that there must in fact be a fire, that is, there must be a situation s' of type S'. (For this example, the constraint $S \Rightarrow S'$ is most likely reflexive, in that the situation s' will be the same as the encountered situation s.)

A particularly important feature of this analysis is that it separates clearly the two very different kinds of entity that are crucial to the creation and transmission of information: one the one hand the abstract types and the constraints that link them, and on the other hand the actual situations in the world that the agent either encounters or whose existence it infers.

For further details of situation theory, the reader should consult [3], upon which the above account was based.

4   A Situation-Theoretic Model of Human Reasoning

Our framework views reasoning as a temporal cognitive process that acts not on statements σ (as in the model of a mathematical proof) but on entities of the form

$$s \models_{\tau_1, \tau_2, \ldots} \sigma$$

where:

1. σ is a statement (or fact);
2. s is a *situation* which provides support or context of origin for σ; and
3. $\tau_1, \tau_2, \ldots$ are the *indicators*[6] of σ, i.e., the specific items of information in s that the reasoner takes as justification of σ.

We call an entity of the form $s \models_{\tau_1, \tau_2, \ldots} \sigma$ a *basic reasoning element*.

Within our framework, a process of reasoning to decide an issue $\mathcal{I}$ can be represented like this:

$$
\begin{array}{ccc}
& \mathcal{I} & \\
\hline
s_1 & \models_{\tau_1, \ldots} & \sigma_1 \\
s_2 & \models_{\tau_2, \ldots} & \sigma_2 \\
s_3 & \models_{\tau_3, \ldots} & \sigma_3 \\
& \vdots & \cdot \\
\hline
s & \models_{\tau_1, \ldots, \tau_2, \ldots, \tau_3, \ldots} \sigma
\end{array}
$$

where each basic reasoning element either supplies evidence for the reasoning or else follows from one or more previous elements by a logical deduction rule.

Analogous to the concept of a mathematical proof (sequence), we define (subject to some technical modifications) an *evidential reasoning process* as a finite sequence $\rho_1, \rho_2, \ldots, \rho_n$ of entities of the above form such that each ρ_i is either *evidential* (i.e., an input to the reasoning process) or else the result of applying some logical rule of reasoning to one or more of $\rho_1, \ldots, \rho_{i-1}$. Here is the formal development of this notion.

By an *evidential reasoning element* we mean a 1×3 matrix of the form

$$\boxed{\text{FACT} \mid \text{SUPPORT} \mid \text{INDIC}(1), \ \text{INDIC}(2), \ldots}$$

such that

$$\text{SUPPORT} \models_{\text{INDIC}(1), \text{INDIC}(2), \ldots} \text{FACT}$$

By an *evidential reasoning step* we shall mean a finitary array of the form

OPERATOR	FACT_1	SUPPORT_1	$\text{INDIC}_1(1), \text{INDIC}_1(2), \ldots$
	FACT_2	SUPPORT_2	$\text{INDIC}_2(1), \text{INDIC}_2(2), \ldots$
		$\cdots$	
	FACT_k	SUPPORT_k	$\text{INDIC}_k(1), \text{INDIC}_k(2), \ldots$
OUTPUT	FACT_{k+1}	SUPPORT_{k+1}	$\text{INDIC}_{k+1}(1), \text{INDIC}_{k+1}(2), \ldots$

where each row

$$\boxed{\text{FACT}_i \mid \text{SUPPORT}_i \mid \text{INDIC}_i(1), \text{INDIC}_i(2), \ldots}$$

is an evidential reasoning element. The index k depends on the operator OPERATOR, and is called the *arity* of the operator.

---

[6] Our use of the term "indicators" with this meaning comes from social science.

The idea is that a basic evidential reasoning step consists of the application of the logical operator to one or more constituents of the evidential reasoning elements in its *scope* (the first k elements listed) to produce the *output element* in the final row.

An *evidential reasoning process* is a finite sequence $\rho_1, \ldots \rho_n$ of basic reasoning steps such that each element is either *evidential* (i.e., an input to the reasoning process) or else the output of some previous (in the sequence) evidential reasoning step, or else is the special element STOP, which is the final element in the process. (STOP is a failure condition; we describe it later.)

The sequence of elements in an evidential reasoning process are not intended to provide a temporal model of the actual steps carried out by a reasoner. Rather, an evidential reasoning process models the logical flow of the reasoning as it leads to the conclusion. As we mentioned earlier, much real-life reasoning is not linear. However, our model is such that any linear progression of steps in the actual reasoning a human carries out will be mapped to a linear ordering of the corresponding basic reasoning elements in the model.

The actual operators that arise in any particular instance of reasoning will depend on the specific circumstances that pertain in that application. In this document we simply indicate the general form of some of the more generic operations that are likely to be used in any instance.

For example, among the operators are some that correspond to classical logic. Since classical logic ignores context, we have to exercise care in porting classical logic operators into our calculus. This means that our rules all have restrictions on when they may be applied. We start with the following two rules, each of which involves a binary reasoning operator:

Evidential Conjunction Rule

CONJOIN	σ	s	$\tau_1, \tau_2, \ldots$
	θ	t	$\gamma_1, \gamma_2, \ldots$
OUTPUT	$\sigma \wedge \theta$	$s \cup t \cup \{\delta\}$	$\delta, \tau_1, \tau_2, \ldots, \gamma_1, \gamma_2, \ldots$

where $\delta = \mathrm{Con}\{\tau_1, \tau_2, \ldots, \gamma_1, \gamma_2, \ldots\}$, the assertion that the set $\{\tau_1, \tau_2, \ldots, \gamma_1, \gamma_2, \ldots\}$ is logically consistent (i.e., has no internal contradictions), and where the rule may be applied only if δ is valid. The restriction that δ is called the *indicator consistency condition* for the rule. If this condition is not satisfied, the rule produces the output STOP. (We consider later what happens when the STOP element is generated.)

Evidential Modus Ponens Rule

MP	σ	s	$\tau_1, \tau_2, \ldots$
	$\sigma \to \theta$	t	$\gamma_1, \gamma_2, \ldots$
OUTPUT	θ	$s \cup t \cup \{\delta\}$	$\delta, \tau_1, \tau_2, \ldots, \gamma_1, \gamma_2, \ldots$

where $\delta = \mathrm{Con}\{\tau_1, \tau_2, \ldots, \gamma_1, \gamma_2, \ldots\}$, and where the rule may be applied only if δ. If this condition is not satisfied, the rule produces the output STOP.

We need to exercise care in using these two rules. If the supports s and t are identical, there is in general no problem, nor if one support is contained within the other. In either of these cases, the indicator consistency condition can generally be assumed to be automatically satisfied, since reasoning generally proceeds under the tacit assumption that each individual source is internally consistent. (If, however, the reasoner suspects — or comes to suspect — that one of the supports used in the reasoning is internally inconsistent, then resolving that inconsistency becomes part of the reasoning process. This is a particular case of the following general observation concerning reasoning.)

The idea behind our approach is this. Coupling a fact σ with its support s in our framework does two things: (i) it acknowledges that σ does come from a particular source, and (ii) it provides a record of that source. Explicitly listing the indicators $\tau_1, \tau_2, \ldots$ with σ and s puts on record the particular items of information in s that the reasoner believes are salient in supporting σ, and uses to justify making use of σ in the reasoning. When an unexpected or troublesome conclusion is reached, or when the reasoning fails to yield a conclusion, it may be necessary to re-examine the veracity of some of the facts used in the reasoning, and that may involve reconsideration of the indicator already identified, or a search for indicators hitherto ignored. In an extreme case, the reasoner may have to question an entire source, perhaps rejecting it and looking for evidence elsewhere.

There are two unary reasoning operators associated with the indicators in a reasoning element: EVAL-INDIC, which checks the indicators already identified for veracity, and FACTORIZE, which identifies new items of information in the support that are salient to the use of the fact in the reasoning process. The rules associated with these operators are:

Indicator Evaluation Rule

EVAL-INDIC	σ	s	$\tau_1, \tau_2, \ldots$
OUTPUT::	σ	s	$\tau_1, \tau_2, \ldots$
		STOP	

where the notation here (note the double-colon after OUTPUT) indicates that the output of the rule is exactly one of the two elements

$$\boxed{\sigma \mid s \mid \tau_1, \tau_2, \ldots}$$

and

STOP

The former output is obtained if the evaluation of $\tau_1, \tau_2, \ldots$ affirms their veracity; the output is STOP if the evaluation determines that one of these indicators is in fact not valid, or at least is in doubt.

Thus, the evidential reasoning step generated by an application of the Indicator Evaluation Rule is of one of the two forms:

EVAL-INDIC	σ	s	$\tau_1, \tau_2, \ldots$
OUTPUT	σ	s	$\tau_1, \tau_2, \ldots$

EVAL-INDIC	σ	s	$\tau_1, \tau_2, \ldots$
OUTPUT			STOP

Indicators Extension Rule

EXTEND-INDICS	σ	s	$\tau_1, \tau_2, \ldots$
OUTPUT	σ	s	$\tau_1, \tau_2, \ldots, \gamma_1, \gamma_2, \ldots$

where $\gamma_1, \gamma_2, \ldots \in s$.

This rule implies that

$$s \models_{\tau_1, \tau_2, \ldots, \gamma_1, \gamma_2, \ldots} \sigma$$

The intuition is that the reasoner identifies additional information (additional indicators) that she or he judges to contribute to the acceptance of the fact σ under consideration.

Use of the following rule, which involves the unary operator EVAL-SUPPORT, indicates a suspicion that the reasoning process has a serious flaw.

Support Evaluation Rule

EVAL-SUPPORT	σ	s	$\tau_1, \tau_2, \ldots$
OUTPUT::	σ	s	$\tau_1, \tau_2, \ldots$
		STOP	

The former output is obtained if the evaluation of s affirms its internal consistency and reliability; the output is STOP if the evaluation determines that s is inconsistent or unreliable, or at least that the consistency or reliability of s is in serious doubt.

Thus, the evidential reasoning step generated by an application of the Support Evaluation Rule is of one of the two forms:

EVAL-SUPPORT	σ	s	$\tau_1, \tau_2, \ldots$
OUTPUT	σ	s	$\tau_1, \tau_2, \ldots$

EVAL-SUPPORT	σ	s	$\tau_1, \tau_2, \ldots$
OUTPUT			STOP

When a reasoning step produces the output STOP, the reasoner has to backtrack and examine the process so far. If it is not possible to make any changes to any previous steps, then the reasoning process breaks down. In such a case, the available information is either contradictory or else simply not adequate to resolve the target issue.

A common step in reasoning is to decide between two or more different possibilities, which may or may not be mutually exclusive. The exact mechanism by which the comparison is made will vary from case to case, but functionally such an operation produces the following basic reasoning step:

Selection Rule

SELECT	σ_1	s_1	$\tau_1(1), \tau_1(2), \ldots$
	σ_2	s_2	$\tau_2(1), \tau_2(2), \ldots$
			$\ldots$
	σ_n	s_n	$\tau_n(1), \tau_n(2), \ldots$
OUTPUT	σ_i	$s_i \cup s$	$\gamma, \delta, \tau_i(1), \tau_i(2), \ldots$

for some i, $1 \leq i \leq n$, where s is the very reasoning process the agent is carrying out (and which we are capturing with our calculus), $\gamma \in s$ is the fact that this particular selection has been made, and $\delta \in s$ is the criterion for making the selection.

Note that the output of a selection step carries a record of the selection having been made and of how it was made.

In practice, making a selection may involve examination of the supports and the indicators associated with the facts being compared, possibly leading to additional factorization for some facts or other operations. Such factorizations, or other steps, will be captured in our model by being represented explicitly as earlier steps in the process sequence.

Sometimes during the course of reasoning, the reasoner believes it is necessary to expand the scope of the domain from which particular facts were obtained, perhaps with a view to finding additional indicators to strengthen confidence in the fact or to replace the fact with a stronger version. This is captured by the following rules, often used in successively in conjunction, together with the indicators extension rule.

Support Expansion Rule

EXPAND-SUPPORT	σ	s	$\tau_1, \tau_2, \ldots$
OUTPUT	σ	s'	$\tau_1, \tau_2, \ldots$

where $s \subseteq s'$.

Strengthen Fact Rule

STRENGTHEN-FACT	σ	s	$\tau_1, \tau_2, \ldots$
OUTPUT	σ'	s	$\tau_1, \tau_2, \ldots$

where $s \models_{\tau_1, \tau_2, \ldots} \sigma' \to \sigma$.

Multiple Views Uniformization Rule

Reasoners sometimes view more than one data source in order to use their experience and tacit knowledge to synthesize a conclusion that may not follow directly from the different sources by logical reasoning. To capture such actions, we could add an operator that provides some form of merge or unification for simultaneous views of information from different sources. However, the evidential conjunction rule that we already have will handle many cases of multiple views of data.

In circumstances where two views of a data item σ can be regarded as providing two indicator sets for the same fact within the same context:

$$s \models_{\tau_1, \tau_2, \ldots} \sigma \quad \text{and} \quad s \models_{\gamma_1, \gamma_2, \ldots} \sigma$$

we can apply the following operator:

MV UNIF	σ	s	$\tau_1, \tau_2, \ldots$
	σ	s	$\gamma_1, \gamma_2, \ldots$
OUTPUT	σ	s	$\delta, \tau_1, \tau_2, \ldots, \gamma_1, \gamma_2, \ldots$

where δ is the fact that this unification has taken place.

Subtasking

Reasoners often need to break a particular task into subtasks. Typically, this entails defining a set of subtasks that together will complete the given task, and then working on each subtask in turn. Alternatively, the reasoner may decide to abandon (perhaps just for the time being) the current goal and concentrate solely on some subtask, which then becomes the new goal.

The framework as described so far can handle the individual steps in each subtask analysis, and can track choices of subtasks as localized reasoning contexts. But we have not introduced an operator for subtask selection or for breaking a task into a sufficient group of subtasks. Instead, we have left this as a meta-level operation. We did so in order to avoid making our technical machinery more complicated than it already is. Since our primary aim is to provide a framework to aid human reasoners, not a blueprint for an automated reasoning system, we feel this is a reasonable choice. But before moving on let's take a brief look at what would be required to modify our framework to incorporate subtasking.

Within our current framework, a process reasoning to decide an issue $\mathcal{I}$ is represented like this:

$$
\begin{array}{ccc}
& \mathcal{I} & \\
\hline
s_1 & \models_{\tau_1, \ldots} & \sigma_1 \\
s_2 & \models_{\tau_2, \ldots} & \sigma_2 \\
s_3 & \models_{\tau_3, \ldots} & \sigma_3 \\
& \vdots & \\
\hline
s & \models_{\tau_1, \ldots, \tau_2, \ldots, \tau_3, \ldots} \sigma
\end{array}
$$

The issue $\mathcal{I}$ is kept constant throughout our development. In order to incorporate subtask selection, we could introduce a mechanism to represent the selection of a subtask $\mathcal{J}$ of $\mathcal{I}$ or else the division of $\mathcal{I}$ into a collection of subtasks $\mathcal{J}_1, \mathcal{J}_2, \ldots, \mathcal{J}_n$. The framework would need to keep track of the supports and the indicators, both when the subtask(s) is (are) selected and when the completion of all the tasks in a subdivision results in the completion of the original task. This is all very straightforward.

5   Some Special Cases

To get a sense of how our framework operates, we show how it applies to some familiar special cases or models for reasoning

Mathematical reasoning. First of all, let's take the case of mathematics, where $\sigma_1, \ldots, \sigma_n$ are statements about some mathematical structure $\mathcal{M}$, say a group or a field. We may assume that $\sigma_1, \ldots, \sigma_n$ are written in the first-order language for $\mathcal{M}$. In that case, each of the expressions $s_i \models \sigma_i$ denotes a standard proposition of classical Tarski-based model theory. In this case, by the Completeness Theorem of first-order predicate logic, $s = \mathcal{M}$ and the deduction takes the form

$$\frac{\begin{array}{c} \mathcal{I} \\ \hline \mathcal{M} \models \sigma_1 \\ \mathcal{M} \models \sigma_2 \\ \vdots \\ \mathcal{M} \models \sigma_n \end{array}}{\mathcal{M} \models \sigma}$$

If this reasoning is valid, then we must have

$$\mathcal{I} = ![\mathcal{M} \models \sigma]?$$

where an expression of the form $!P?$ for some proposition P denotes the goal "Determine whether P true or false." That is, the goal is to determine whether or not σ is true of $\mathcal{M}$.

The completeness theorem also tells us that (if the deduction is valid), σ follows from $\sigma_1, \ldots, \sigma_n$ by the rules of logic alone.

Reasoning from a common source. Another special case is where all of the information $\sigma_1, \ldots, \sigma_n$ comes from the same source, $\mathcal{S}$. In this case the conclusion support s is also $\mathcal{S}$, and the deduction takes the form:

$$\frac{\begin{array}{c} \mathcal{I} \\ \hline \mathcal{S} \models \sigma_1 \\ \mathcal{S} \models \sigma_2 \\ \vdots \\ \mathcal{S} \models \sigma_n \end{array}}{\mathcal{S} \models \sigma}$$

For a valid process, we must have

$$\mathcal{I} = !\,\sigma?$$

(Determine whether to do σ, or else determine whether σ is true.)

Bayesian inference. In some cases, knowledge of the source of each data item σ_i may be converted into a numerical probability of the reliability of σ_i, i.e. the probability that σ_i is true. In such a situation, we may be able to apply Bayes' Theorem repeatedly in order to obtain a conclusion σ and assign a probability to σ. In this case, the function F is a numerical function based upon Bayes' Theorem and the function H is an instance of Bayesian inference. This kind of reasoning is quite common, particularly in intelligence gathering.

We may represent a Bayesian reasoning process using the original notation

$$
\frac{
\begin{array}{c}
\mathcal{I} \\
\hline
s_1 \models \sigma_1 \\
s_2 \models \sigma_2 \\
\vdots \\
s_n \models \sigma_n
\end{array}
}{s \models \sigma}
$$

with the understanding that each of $s_1,, \ldots, s_n, s$ is a number between 0 and 1 inclusive, and each expression $s_i \models \sigma_i$ should be interpreted as a probability statement $p(\sigma_i) = s_i$, and similarly for $s \models \sigma$.

6   Summary and Discussion

The basis for our method is to view reasoning as a temporal cognitive process that acts on entities of the form

$$
s \models_{\tau_1, \tau_2, \ldots} \sigma
$$

where σ is a statement (or fact), s is its support or context of origin, and τ_1, $\tau_2, \ldots$ are its indicators, the specific items of information in s that the reasoner takes as justification of σ.

We analyze reasoning so described in terms of a number of basic reasoning steps, an illustrative example being the Evidential Modus Ponens Rule:

MP	σ	s	$\tau_1, \tau_2, \ldots$
	$\sigma \to \theta$	t	$\gamma_1, \gamma_2, \ldots$
OUTPUT	θ	$s \cup t \cup \{\delta\}$	$\delta, \tau_1, \tau_2, \ldots, \gamma_1, \gamma_2, \ldots$

where $\delta = \mathrm{Con}\{\tau_1, \tau_2, \ldots, \gamma_1, \gamma_2, \ldots\}$, and where the rule may be applied only if δ.

We list a number of such rules, but acknowledge that many applications will involve rules not listed here. Our framework is designed to allow for such additional rules to be incorporated.

Readers familiar with situation theory will have observed that our present framework amounts to making explicit in the model the features of the context situation — our *indicators* — that provide direct support for the items of information considered in the reasoning — what we call the *facts*. Moreover, we model (aspects of) the process of reasoning, not just the sequence of facts and their situational supports. By making this additional salient information explicit in the model, we can obtain a finer grained analysis than is possible in situation theory, that requires much less ad hocing when we carry out an analysis of a specific reasoning process. In our framework, the Evidential Modus Ponens rule performs the task that was handled by constraints in situation theory. Our decision to ignore much of the machinery for handling situation-theoretic constraints was based on pragmatic grounds, with a view to the kinds of reasoning we are attempting to model.

Although our primary goal is to develop a framework that aids understanding, we are aware that any enterprise such as ours has the potential of forming the basis for the specification of reasoning protocols or the design of reasoning support tools. The model we have developed would result in protocols or support tools that:

1. Force explicit identification and tracking of sources.
2. Force explicit identification and tracking of supporting information (the indicators).
3. Force regular reconsideration of the reasoning process itself.
4. Allow for backtracking when a problem is encountered, without the necessity of starting over afresh.

Above all, our framework makes it clear that reasoning involves three components: facts, sources, and indicators. Real-life reasoning typically involves all three. Any protocol or tool developed in line with our model should provide the user with regular prompts to check all three components. Many examples of failures in human reasoning and analysis have resulted from a neglect of one or more of the three basic components.

Jon Barwise

I think it is appropriate to end with a quotation from my former friend and colleague Jon Barwise, whose untimely death in 2000 deprived the world of one of the most innovative logicians of the twentieth century. In his collected work *The Situation in Logic* [1], Barwise wrote [pp.xv–xvi]:

> Back in the days before I became interested in the situated aspects of logic, I sometimes used to wonder how logicians felt in the first quarter of this century. Did they *feel* confused. Reading the literature of that period, one senses the extent to which they were groping toward the view of logic that eventually emerged, but also the extent to which they were still in the dark about what was central and what was peripheral. One also realizes that they were just missing certain key distinctions. In other words, they were confused. It was only with the pioneering work of Gödel, Church, Turing, Tarski, and Kleene in the 1930's that the modern conception of logic really took hold.
>
> I now feel I have some idea of how logicians must have felt in that period before the really seminal work, since I feel we are in an analogous

stage now ... As we try to let go of some of the simplifying idealizations made in standard logic, we too are groping for the key notions, and probably missing some key distinctions. In giving up these simplifying assumptions, there are many things to be rethought, many choices to be made, and many things to be tried. It is an exciting time, if you have the patience for that sort of thing, and a taste for the basic task of conceptual clarification. But it is also frustrating ...

... There is only one point about which I am really certain. That is that the view of language and logic as situated activities is an important one, and that situating logic is a task that must be carried out if we are to come to grips with some of the problems that currently vex the field.

I say Amen to that.

References

1. Barwise, J.: The Situation in Logic. CSLI Lecture Notes 17 (1989)
2. Barwise, J., Perry, J.: Situations and Attitudes. Bradford Books, MIT Press (1983)
3. Devlin, K.: Logic and Information. Cambridge University Press, Cambridge (1991)
4. Devlin, K.: Goodbye Descartes: The End of Logic and the Search for a New Cosmology of the Mind. John Wiley, Chichester (1997)
5. Heuer Jr., R.J.: Psychology of Intelligence Analysis. Central Intelligence Agency (1999), http://www.odci.gov/csi/books/19104/

One or Many Concepts of Information?*

Giovanni Sommaruga

University of Fribourg
Department of Informatics DIUF
Bd de Pérolles 90
1700 Fribourg, Switzerland
giovanni.sommaruga@unifr.ch
http://diuf.unifr.ch/tcs/

1   Conceptual Considerations

One of the purposes of this paper is to inquire whether the theoretical (mathematical - logical) study of information can be tracked down to a single concept of information, or whether this study infallibly leads to several distinct concepts (of information). The philosophical interest of this inquiry is, to use an old and slightly infamous concept, self-evident.

1.1   A Few Distinctions to Start with

Let me start by distinguishing 3 kinds of concepts:

a) ordinary language concepts
b) informal theoretical concepts
c) formal theoretical concepts

How does one get from a) to b) or c)? By means of an explication, that is, by a certain way of making the meaning of an ordinary language expression precise (according to Carnap, this way has to satisfy certain well-known criteria).

What is the relationship between b) and c)? There are at least 2 ways to conceive of it:

1. b) is an informal description of a concept embodied in a formal (mathematical) theory. In this case, the relationship between b) and c) is totally unproblematic.
2. c) is to be the formalisation of b). In this case, the relationship is more problematic. (More about it later.)

1.2   Reflections on Informal Theoretical Concepts of Information

How many Concepts of Information? The not entirely unserious question to be answered here is:

(Q) How many informal theoretical concepts of information are there?

---

* I dedicate this paper to my good friend, the economist, computer scientist and philosopher Ambros Lüthi.

G. Sommaruga (Ed.): Formal Theories of Information, LNCS 5363, pp. 253–267, 2009.

Floridi distinguishes 3 types of answer ('approaches') to this question:[1]

$\longrightarrow$ the reductionist answer : 1
$\longrightarrow$ the antireductionist answer : many very different ones
$\longrightarrow$ the nonreductionist answer : here 3 subcases are to be distinguished
$\longrightarrow\longrightarrow$ the centralized answer : essentially 1
$\longrightarrow\longrightarrow$ the multicentered answer : few
$\longrightarrow\longrightarrow$ the completely decentralized answer : quite a few

There is a question analogous to **(Q)** regarding the ordinary language concepts of information, namely:

(Q+) How many ordinary language concepts of information are there?

What do the 3 types of answer to question **(Q)** imply concerning this latter question **(Q+)**? Are there at least as many ordinary language concepts of information as there are informal theoretical explicata? Or is it rather the other way round (are there at most as many)? These questions won't be followed up in the sequel.

Floridi's point of view. Floridi's own point of view is a nonreductionist epistemically centralized type of answer (Floridi 2003b:42). What this point of view amounts to, can well be illustrated by something he calls 'an informational map':[2]

Within the network of concepts called 'information', the central concept is the one of factual or epistemically oriented (semantic) information. It is defined by the following
Special Def. of (factual) Information **(SDI)**:
σ is an instance of truthful (factual) information iff σ is mwfd + t.
This definition presupposes in turn the
General Def. of Information **(GDI)**:
σ is an instance of information (in the sense of objective semantic content) iff σ is wfd + m.

As a matter of fact, there seems to be a certain ambiguity in Floridi's point of view:[3] it appears that sometimes semantic information (content) is the central concept, sometimes factual semantic information, and at times it is truthful (factual) semantic information. As will be seen in sect. 3.2, some kind of propensity towards a certain ambiguity is fairly common in this context.

According to the informational map above, factual information is alethically neutral, i.e. neutral w.r.t. truth-values. Floridi points out that this is a point of controversy among information theorists: some take information to be truthful by definition, others take it to be true or false and, at any rate, the truth-value not to be part of the definition of information (Floridi 2003b:45f).

---

[1] (Floridi 2003b:40f) and (Floridi 2005a:).
[2] Cf. Floridi's article in this volume, sect. 2.
[3] At least in 2 of his articles, but not in his article (Floridi 2005b).

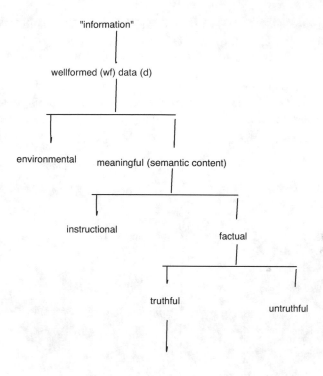

An informational map (Floridi)

Floridi's thesis If I have understood Floridi's point of view correctly, he upholds something like the following thesis:

(FT) The concept of semantic (factual) truthful information is the most appropriate informal theoretical concept of information.

He appears to justify **(FT)** as follows:

i) This informal theoretical concept explicates or corresponds the best to the most frequent common sense understanding of the word 'information'.

ii) It plays a very important epistemological role, since it provides a necessary condition for knowledge (it is actually sometimes downright confused with knowledge).

As a consequence of **(FT)**, the concept of semantic (factual) truthful information serves as a standard for the 'measurement' or rather assessment of the relative appropriateness of other informal theoretical concepts of information such as:
    Examples

– the algorithmic concept of information (as the size of a computer program necessary to generate certain wellformed data)

- the situation-theoretical concept of information (e.g. as the information content carried by a fact and relative to a constraint)
- the algebraic concept of information (as the information content relative to a question)
- the theoretical concepts of information in the modal or inferential approach (the information content of a sentence as the class of possible states of the universe (possible worlds) which are excluded by the resp. sentence)

These observations suggest another diagram of concepts:

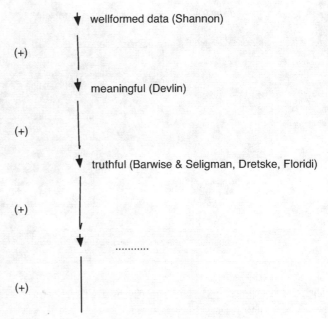

an additive conceptual diagram

As this diagram shows the intension of the informal theoretical concept gets increasingly larger or richer. And **(FT)** allows to assess the degree of deviation of one of these concepts from the most appropriate one.

1.3   Reflections on Formal Theoretical Concepts of Information

The formal theoretical concept of information of statistical IT (information theory) as well as the one of algorithmic IT both simply refer to well-formed data (which needn't be meaningful).[4] What does that mean? The formal theoretical

---

[4] The term 'well-formed' might actually be misleading and ought then to be replaced by the term 'formed' or rather 'structured' if 'well-formed' should imply 'inductively generated'.

concept of information of strongly semantical IT as well as the one of algebraic IT both refer to truthful meaningful well-formed data. What does that mean?

This indicates - what comes as no surprise - that the just mentioned informal theoretical concepts of information (e.g. as well-formed data or as truthful meaningful well-formed data) are underdetermined: they allow for different formalisations. In other words, different formal theoretical concepts can correspond to one and the same informal theoretical concept of information. And what does that mean?

In order to simplify things, let's make the following assumption:

(A) To every formal theory of information corresponds or in every formal theory of information is embodied a unique formal theoretical concept of information. (This presumably means that the resp. formal theories are what logicians call categorical (or monomorphical).)

Possible relations between two formal theories of information. Let us ask a **(Q)**-type question with regards to formal theoretical concepts of information:

(Q*) How many formal theoretical concepts of information are there?

To answer **(Q*)** let's distinguish the following ways 2 formal theories of information can relate to each other:

Let 2 formal theories be called totally coextensional if they have the same extension, i.e. they have the same (set of) theorems; let them be called partially coextensional, if they share some of their extension, but not the whole one; let them be non-coextensional if their extensions are disjoint. Let 2 formal theories be compatible if the union of their extensions is consistent; o.w. they are called incompatible. The following 6 relations are possible, where case 0 is completely uninteresting (and therefore 'zero' – it is plain inconsistency):

	totally coextensional	partially coextensional	non-coextensional
incompatible	0	2	4
compatible	1	3	5

2 Application of This Conceptual Apparatus to Theories of Information

Applying this conceptual apparatus to the 2 mathematical theories of statistical and algorithmic IT, one can ask the question: Which category do these 2 theories belong to (or in which type of relation do they stand)?

Likewise one can ask the questions: Which category do the 2 formal theories of algebraic IT[5] and (strongly) semantical IT[6] belong to? or Which category do

---

[5] Cf. Kohlas' and Schneuwly's contribution to this volume.

[6] E.g. in the form of van Rooij's logic of questions and answers, cf. his article in this volume.

the 3 formal theories of (strongly) semantical IT, situation-theoretical IT and Barwise-Seligman IT belong to?[7]

2.1 A Digression on Facts and 'Wishful Thinking'

So far there has been talk about questions and facts. Lets turn for a moment to 'wishful thinking':

Suppose we formulate the following thesis:

(T) The 'essential' meaning of 'information' as truthful meaningful wellformed data is captured by the algebraic IT which is totally coextensional with the (strongly) semantical, (veridical) situation-theoretical and Barwise-Seligman ITs (and possibly other ITs).[8]

(T) is some kind of an information-theoretical analogue to Church's Thesis **(CT)**. Its handicap or one of its handicaps is that the informal theoretical concept of information as truthful meaningful wellformed data is still much vaguer than the analogous one of effectively computable function in **(CT)**. A consequence of this is that **(T)** might not or won't be suitable for use in informal arguments and proofs in information theory the way **(CT)** is in recursion theory.

What is the appeal, the wishfulness (or rather desirability) of such a thesis **(T)**?

The appeal is first and foremost a philosophical one: **(T)** would most likely show that the various equivalent (or totally coextensional) formalisations are appropriate formalisations of the informal theoretical concept of information as truthful meaningful wellformed data. It would mean that there is one formal theoretical concept only corresponding to the just mentioned informal theoretical concept and that the former is clearly a sharpening of the latter. To thesis **(T)** applies what Chaitin wrote about a useful theory: it is a compression of data. And, as he added: comprehension is compression (and at least often, I'd say, vice versa). **(T)** would equally point in the direction of a unification of 'information theory' by connecting similar or related ideas and approaches which have to a great extent been discovered or developed independently from each other and which undoubtedly all have to do with information as truthful meaningful wellformed data.

Needless to add, 'totally coextensional' doesn't mean 'cointensional', and that is why the different totally coextensional theories of information would still reflect or highlight different aspects or features of the informal theoretical concept (as is the case with **(CT)**).

---

[7] This is a bit of a loose way of writing: The term 'situation-theoretical IT' can refer to Barwise and Perry's original veridical version or to Devlin's and Barwise and Perry's later alethically neutral version of this IT.

[8] J. Seligman drew my attention to the point that there is a descriptive and a normative understanding of **(T)**. Its primary understanding here is descriptive. Should there be or show up a formal IT which doesn't fit in with thesis **(T)**, its understanding might become normative (in the sense of leading to the question: What is wrong with this new formal IT that it doesn't fit thesis **(T)**?).

2.2   Digression Continued

Suppose thesis **(T)** is wishful but not true; i.e. it can be falsified by proving that 2 of the relevant formal theories of information belong to one of the categories 2-5 rather than to 1.

This could mean lots of things. It could e.g. mean that there is a family of meanings of 'information' and one could invoke Wittgenstein's term of family-resemblance (rather than e.g. claim downright ambiguity of the word 'information'). Family-resemblance of a concept means that the concept consists 'of a complex network of overlapping and criss-crossing similarities, just as the different members of a family resemble each other in different respects' (but there isn't one common single feature) (Glock 1996:121). Invoking family-resemblance could be done for good reasons: According to Wittgenstein some branches of a family-resemblance concept allow for an analytical definition, or alternatively, for a formalisation. Another reason is, that even the informal theoretical term of information (as meaningful well-formed data) might refer, just like Wittgenstein's own examples of family-resemblance concepts 'language' and 'proposition', to a variety of different but related phenomena.(cf. Glock 1996:120-124)

2.3   Another Digression on Facts and 'Wishful Thinking'

Even the situation of formal concepts of information as wellformed data (wd) is of considerable interest: The original concept is no doubt the one of Shannon entropy $H(X)$ of the random variable X, i.e. the amount of uncertainty and information of the scheme of choice presented by a random variable X (where a random variable presents a probabilistic scheme of choice where one of its possible values $x_i \in S = \{x_1, \ldots, x_m\}$ is chosen with probability $P_X(x_i)$ ($\sum_{i=1}^{m} P_X(x_i) = 1$ and $0 \le P_X(x_i) \le 1$, for $i = 1, \ldots, m$)):

$$H(X) = -\sum_{i=1}^{m} P_X(x_i) \log_2 P_X(x_i) = E\left[-\log_2 P_X(X)\right]$$

Cover & Thomas note that in a certain sense, $\log_2 \frac{1}{P_X(x_i)}$ is the descriptive complexity of the event $X = x_i$, since $\lceil \log_2 \frac{1}{P_X(x_i)} \rceil$ is the number of bits required to describe x_i by a Shannon code. They observe that the descriptive complexity of such an object x_i depends on the probability distribution P_X, it is thus a relative concept (Cover and Thomas 1991:144).

Another famous concept is of course the one of Kolmogorov complexity (plain or prefix) of an object x ($x \in \{0, 1\}^*$), also called the algorithmic complexity of x by Cover & Thomas, which is the shortest binary computer program p which describes or rather produces x by means of a universal computer $\mathcal{U}$:

$$K_{\mathcal{U}}(x) = \min\{l(p) | \mathcal{U}(p) = x\}$$

And the conditional Kolmogorov complexity $K_{\mathcal{U}}(x|y)$ is defined by

$$K_{\mathcal{U}}(x|y) = \min\{l(p) | \mathcal{U}(p, y) = x\}$$

where y is taken to be the length $l(x)$ of x (Cover and Thomas 1991:147ff).[9]

Cover & Thomas observe that in the definition of Kolmogorov complexity of an object x, a probability distribution P plays no role whatsoever; it is thus an absolute concept (unlike the one of descriptive complexity). Moreover, the definition of Kolmogorov complexity is also computer independent - this provides for the universality of Kolmogorov complexity (Cover and Thomas 1991:144).

Now, there is the following remarkable relationship between the entropy $H(X)$ of the random variable X and the expectation of the (conditional) Kolmogorov complexity $K_{\mathcal{U}}(x^n|n)$ of an object x^n: the expectation of $K_{\mathcal{U}}(x^n|n)$ differs from $H(X)$ merely by a constant c_P depending on the probability distribution P.

Let x^n be a string of length n of elements $x_i \in S$; and let $P_X(x^n) = \prod_{i=1}^{n} P_X(x_i)$. Then there exists a constant c_P s.t.[10]

$$H(X) = \frac{1}{n} \sum_{i=1}^{n} P_X(x^n) K_{\mathcal{U}}(x^n|n) - c_P$$

In other terms, the expected value of Kolmogorov complexity of a random sequence tends to its Shannon entropy (Cover and Thomas 1991:154f).

As a matter of fact, there exists another universal complexity measure, i.e. another formal concept of information as wellformed data, which serves as sort of an intermediate between Kolmogorov complexity and Shannon entropy, namely the universal probability of an object (a string) x: on the one hand universal probability is essentially equivalent to Kolmogorov complexity, on the other hand it exhibits the same basic form as Shannon entropy.

Cover & Thomas define the universal probability Pr of an object (string) x as the probability that a computer program p randomly drawn as a sequence of fair coin flips $p_1, p_2, \ldots$ produces x by means of a universal computer $\mathcal{U}$:

$$Pr(\mathcal{U}(p) = x) = \sum p : \mathcal{U}(p) = x 2^{-l(p)} = P_{\mathcal{U}}(x)$$

They observe that the concept of universal probability is essentially determined by the one of Kolmogorov complexity: There exists a constant c, independent of x, s.t. for all strings x:

$$P_{\mathcal{U}}(x) = c \cdot 2^{-K_{\mathcal{U}}(x)} \qquad \text{or} \qquad K_{\mathcal{U}}(x) = \log_2 \frac{1}{P_{\mathcal{U}}(x)} + c'$$

Cover & Thomas show that Pr is – just as Kolmogorov complexity – a universal probability distribution and it is thus likewise an absolute concept (i.e. independent of any particular probability distribution P) (Thomas and Cover 1991:160ff, 169f).

---

[9] It is somewhat confusing that Cover & Thomas use K to denote plain Kolmogorov complexity (which Li & Vitanyi denote by C), while Li & Vitanyi use the same letter to denote prefix Kolmogorov complexity. Cf. (Li and Vitanyi 1997:194)

[10] There actually exists a constant c'_P s.t. $c_P = \frac{|S| \log_2 n + c'_P}{n}$.

From all of this, the following conclusions can be drawn:

First, even though the concepts of Shannon entropy and Kolmogorov complexity have been introduced and developed in different contexts (the theory of communication and algorithmic complexity theory resp.) and for different purposes, it is a fortunate coincidence that they both are equivalent up to an additive constant (reflecting the choice of the reference machine).[11]

The third formal concept of information as wellformed data, namely the universal probability of an object, is a concept derived from the one of Kolmogorov complexity. The fact that there are three formal concepts of information as wellformed data, all essentially equivalent with each other, provides a certain evidence for the view that there might be a thesis (**T'**) about the concept of information as wellformed data analogous to thesis (**T**).[12]

The second conclusion is due to the following observations by Cover & Thomas: They draw attention to a striking similarity between $H(X)$ and $\log \frac{1}{P_X(x)}$ in information theory[13], and $K_\mathcal{U}(x)$ and $\log_2 \frac{1}{P_\mathcal{U}(x)}$ in algorithmic complexity theory. The Shannon code length assignment $l(x_i) = \lceil \log_2 \frac{1}{P_X(x_i)} \rceil$ achieves an average description length $H(X)$; while in Kolmogorov complexity theory $\log_2 \frac{1}{P_\mathcal{U}(x)}$ is almost equal to the algorithmic complexity $K_\mathcal{U}(x)$ of x. From this they conclude that $\log_2 \frac{1}{P(x)}$ looks like the fundamental or natural form of the descriptive complexity of a string x in algorithmic as well as probabilistic settings (Cover and Thomas 1991:170). That is, $\log_2 \frac{1}{P(x)}$ might be the fundamental form of the coextensional formal concepts of information as well-formed data.

A third and last conclusion is also due to Cover & Thomas: Chronologically, the concept of Shannon entropy is prior to the one of Kolmogorov complexity.[14] However, as was mentioned earlier on, the concept of Kolmogorov complexity is an absolute concept (unlike the one of Shannon entropy), independent of any, or equivalently, universally good for all probability distributions P. Shannon's relative concept of descriptive complexity of an object can therefore be considered as a relativization of the concept of Kolmogorov complexity to particular probability distributions. In this sense, the formal concept of Kolmogorov complexity is conceptually prior to the one of Shannon entropy (Cover and Thomas 1991:144).

2.4   A Third Digression on Facts and 'Wishful Thinking'

Since thesis (**T**) is about meaningful well-formed data mwd, or what is often called semantical information, the proof of some sort of an equivalence of a

---

[11] Li & Vitanyi just notice that it would have been troublesome had this not been so, as both are intended to express the content of information (Li and Vitanyi 1997:190). And Cover & Thomas call this approximate equality an amazing fact (Cover and Thomas 1991:144) Indeed, if one restricts the theory of Kolmogorov complexity to prefix-free programs (cf. the concept of prefix Kolmogorov complexity), the entire theory is formally equivalent to or coextensional with Shannon's information theory.

[12] For a hint in this direction, cf. (Li and Vitanyi 1997:525).

[13] Remember $H(X) = E[\log_2 \frac{1}{P_X(X)}]$.

[14] (Shannon 1948), (Kolmogorov 1965, 1968).

(strongly) semantical IT with the algebraic IT is of particular interest. It might provide evidence that our wishful thinking is not without at least a grain of truth.

One variant of a semantical IT is van Rooij's logic of questions and answers or at least the starting point of his logic.[15] The demonstration of this sort of equivalence will be carried out by demonstrating that it as well as the first-order predicate logic as information algebra (an algebraic IT) both are examples of a more general structure, exhibited by Jeremy Seligman, namely the Gentzen-theory-based information algebra, or GTBIA for short. The so-called Gentzen theories and the links determined by Gentzen theories are a fundamental component in Seligman's signalling/indicating model of information flow.[16] Thus, there might be something like an embedding of van Rooij's logic of questions and answers and Kohlas' information algebra in the Barwise-Seligman theory of information and its signalling/indicating model in particular.[17]

As a first step, the structure of a Gentzen-theory-based information algebra GTBIA will be introduced.[18]

Let $\langle \Sigma, \vdash \rangle$ be a Gentzen theory (i.e., closed under Identity, Weakening and Cut). And suppose that it is extensional: if $\sigma \vdash \tau$ and $\tau \vdash \sigma$ then $\sigma = \tau$. For $\sigma, \tau \in \Sigma$, say $\sigma \& \tau \in \Sigma$ is the conjunction of σ and τ iff for all $\Gamma, \Delta \subseteq \Sigma$,

(&L) $\sigma \& \tau, \Gamma \vdash \Delta$ iff $\sigma, \tau, \Gamma \vdash \Delta$
(&R) $\Gamma \vdash \Delta, \sigma$ and $\Gamma \vdash \Delta, \tau$ iff $\Gamma \vdash \Delta, \sigma \& \tau$

(Note that conjunctions, if they exist, are unique, by extensionality.)

A question of the theory is a subset $q \subset \Sigma$ such that $\vdash q$ and $\sigma, \tau \vdash$ for each $\sigma \neq \tau \in q$. Say that q_1 is a strong refinement of q_2, written $q_1 \sqsubseteq q_2$, iff for each $\sigma \in q_1$ there is a $\tau \in q_2$ such that $\sigma \vdash \tau$. (This is a partial order.) The restriction of q to σ, written $q|_\sigma$ is the set $\{\tau \in q \mid \sigma, \tau \nvdash\}$.

Now say that $[q]\sigma \in \Sigma$ is the answer that σ gives to q iff for all $\Gamma, \Delta \subseteq \Sigma$,

([q]L) $[q]\sigma, \Gamma \vdash \Delta$ iff $\tau, \Gamma \vdash \Delta$ for each $\tau \in q|_\sigma$
([q]R) $\Gamma \vdash \Delta, [q]\sigma$ iff $\Gamma \vdash \Delta, q|_\sigma$

(Again, answers, if they exist, are unique, by extensionality.)

Now suppose that Q is a set of questions closed under joins (least upper bounds in the $\sqsubseteq$ order) and A is a subset of Σ closed under conjunctions and answers to questions in Q, i.e., if $\sigma, \tau \in A$ and $q \in Q$ then both $\sigma \& \tau$ and $[q]\sigma$ are in A.

---

[15] His logic actually starts off as a semantical IT and soon turns into some sort of a pragmatical IT large parts of which at least can however still be accomodated within a semantical framework. i.e. in terms of a semantical IT.

[16] Cf. Seligman's contribution to this volume, sect. 3.

[17] This latter speculative remark won't be elaborated in the sequel.

[18] I owe the following comments to Jeremy Seligman. Since this point (namely the relationship of the Barwise-Seligman theory to Kohlas' information algebra and to van Rooij's logic of questions and answers) is of considerable importance for my argument, I'm very grateful to him for these comments and his very generous support.

Proposition. By defining $\sigma \otimes \tau$ to be the conjunction $\sigma \& \tau$ and $\sigma^{\Rightarrow q}$ to be the answer $[q]\sigma$, we get an information algebra $\langle A, \otimes, Q, \sqsubseteq, \Rightarrow \rangle$.

This proposition is demonstrated by a fairly straightforward verification that $\langle A, \otimes, Q, \sqsubseteq, \Rightarrow \rangle$ satisfies the axioms of an information algebra.[19]

Call any information algebra of this form a 'Gentzen-theory-based information algebra', or GTBIA for short. This algebra interprets the conjunction of the Gentzen theory as combination of information, as expected, and uses questions as a filtering mechanism: $\sigma^{\Rightarrow q} = \tau^{\Rightarrow q}$ just in case any difference between σ and τ is irrelevance to the answering of q.

Two examples of GTBIAs:

1. For a language L of first-order predicate logic with a set V of individual variables, and a model M with domain D, let $\Sigma = \wp(D^V)$, the powerset of the set of assignment functions, and let $\Gamma \vdash \Delta$ iff $\bigcap \Gamma \subseteq \bigcup \Delta$, which is an extensional Gentzen theory. For each formula φ of L, let $[\![\varphi]\!]$ be the set of assignments that satisfy φ in M. For each set $X \subseteq V$ of individual variables and $\alpha, \beta \in D^V$, define the equivalence relation $\alpha \sim_X \beta$ iff for each $x \in X$, $\alpha(x) = \beta(x)$, and let q_X be the set of $\sim_X$-equivalence classes, which is a question in $\langle \Sigma, \vdash \rangle$.

   Now, let A be the set of $[\![\varphi]\!]$ for each formula φ in L and let Q be the set of q_X for each finite set of variables X. Then Q is closed under joins because $q_X \sqcup q_Y = q_{X \cap Y}$. Also A is closed under conjunction because $[\![\varphi \wedge \psi]\!] = [\![\varphi]\!] \& [\![\psi]\!]$, and under answers because $[q_X][\![\varphi]\!] = [\![\exists \overline{X} \varphi]\!]$ where $\overline{X}$ is the set of free variables of φ not in X. Then $\langle \{[\![\varphi]\!] \mid \varphi \in L\}, \wedge, \{q_X \mid \text{finite } X \subset V\}, \sqsubseteq, \exists \rangle$ is a GTBIA and shows first-order predicate logic to be a GTBIA.[20]

2. The language LQ of van Rooij's logic of questions and answers extends L by recursively adding a modal operator $?\varphi$ for each formula φ.[21] Evaluate $\psi \in LQ$ in a modal model with possible worlds W and constant domain D, and with $?\varphi$ interpreted by the following accessibility relation
   $u \sim_\varphi v$ iff for each assignment function $\alpha \in D^V$, α satisfies φ in world u iff α satisfies φ in world v[22]
   This is clearly an equivalence relation, so $?\varphi$ is an S5 diamond. Now let $\Sigma = \wp(W)$ and let $\Gamma \vdash \Delta$ iff $\bigcap \Gamma \subseteq \bigcup \Delta$, which is an extensional Gentzen

---

[19] Cf. Kohlas and Schneuwly's article in this volume, sect. 2.

[20] In (J. Langel and J. Kohlas 2007), the language L of first-order logic is taken to be the language of a many-sorted first-order predicate logic. Technically, this is of no significance whatsoever.

[21] This is a sloppy or even a not entirely correct claim: LQ is closely related to the language of van Rooij's logic, but it isn't its language (its language doesn't involve modalities; LQ rather is a natural extension of van Rooij's language). It looks like everything van Rooij says about questions and answers can be said in LQ. But LQ allows iteration of the question-forming modality which is something, van Rooij doesn't consider.

[22] In other words, $u \models ?\varphi \psi [\alpha]$ iff there is a $v \in W$ such that $u \sim_\varphi v$ and $v \models \psi [\alpha]$. If van Rooij presented the Kripke semantics of his logic as explicitly as it is presented here, he (presumably) would have an equivalent definition.

theory. For each sentence φ of L, let $[\![\varphi]\!]$ be the set of worlds at which φ is satisfiable. For each φ, let q_φ be the set of $\sim_\varphi$-equivalence classes, which is a question in $\langle \Sigma, \vdash \rangle$. Then, as before, let A be the set of $[\![\varphi]\!]$ for each formula φ in LQ and let Q be the set of q_φ for each formula φ in LQ. Then Q is closed under joins because $q_\varphi \sqcup q_\psi = q_{\varphi \cap \psi}$. Also A is closed under conjunction because $[\![\varphi \wedge \psi]\!] = [\![\varphi]\!] \& [\![\psi]\!]$, as before, and under answers because $[q_\varphi][\![\psi]\!]$ is the union of those $\sim_\varphi$ equivalence classes that intersect with $[\![\psi]\!]$, which is $[\![?\varphi\psi]\!]$ and so in A. Then the construction above yields a GTBIA, that is, $\langle \{[\![\varphi]\!] \mid \varphi \in LQ\}, \wedge, \{q_\varphi \mid \varphi \in LQ\}, \sqsubseteq, \cup \rangle$ is a GTBIA and shows van Rooij's logic of questions and answers (or something close to it) to be a GTBIA.

3   Some Conclusions from These Conceptual Analyses and Considerations

3.1   The Upshot

The upshot of these conceptual analyses is the following. An examination of the possibility to make out a reductionist approach in the theoretical study of information yields a negative result: There is no such reductionist way. Thus, the answer to the initial question **(Q)** [How many informal theoretical concepts of information are there?] is: several. The answer to question **(Q*)** [How many formal theoretical concepts of information are there?] is: at least several.

Now, the nonreductionist approach appears to be the right one, as the antireductionist approach is just a lazy answer to the question: how many; since it abstains from any consideration of connections, similarities, networks between different concepts of information. And the next question is, following Floridi's scheme of options, whether it is a centralized or multi-centered approach.[23]

3.2   Further Evidence for These Conclusions

The first prima facie conclusion seems to be what might be called a multi-centralized approach. Evidence for this prima facie conclusion is provided by the fact – already alluded to earlier on – that there is hardly one formal theory of information based on a single informal theoretical concept of information. However, a second look at this conclusion and a closer inspection and analysis of the resp. concepts should or will show that this multi-centralized approach can actually be so to speak 'reduced' to a centralized one.

3 examples:

i) The situation-theoretical IT: In (Israel and Perry 1990) the term 'information' is used ambiguously meaning

---

[23] The completely decentralized one drops out for the same reason as that speaking against the antireductionist approach. As a matter of fact, the difference between the antireductionist approach and the completely decentralized nonreductionist one is not entirely clear.

a) information content carried by a fact and relative to a constraint.
b) (just) information content.

(Devlin 1991) distinguishes 3 meanings of the term 'information', namely

a) true infonic proposition
b) infonic proposition (i.e. information content)
c) parameterfree infon

ii) the algebraic IT: (Kohlas 2003) distinguishes 2 informal theoretical concepts
   of information, namely
a) a relative one: an information item relative to a domain (question)
b) an absolute one: a domainfree information item (which is an equivalence
   class of information items relative to a domain)

This example is insofar less telling as the two concepts are shown to be formally
equivalent; and moreover, Kohlas makes it clear that the relative concept is the
primary one, whereas the absolute concept is derivative and secondary.

   There is, however, a more telling ambiguity, namely whether the term 'information' refers to

a) a 'body of information' (i.e. a set of information items)
   or
b) an individual information item

And if it is taken to refer to an individual information item, there appears to be
another ambiguity: since an information item is conceived of as an answer to a
question, the term 'information item' can either be taken to refer to

b1) something like an ordered pair of question and answer (remember: informa-
    tion is a relative concept)
   or
b2) to one member of this pair, namely to the answer

iii) The statistical IT: In (Shannon and Weaver 1949) one can find at least two
    meanings of 'information' which obviously cannot be said to be the same:
a) information (from the point of view of the communication engineer) is conceived of as a choice of one message from a set of possible messages
b) information is the decrease (or change) of (in) uncertainty induced by the
   production of a particular event

3.3   Final Conclusion

Lets come back to question (**Q**) or rather ask a question following from it:

(**Q**^{f}) Is the multi-centered or the centralized answer to question (**Q**) the more
    appropriate one?

For the multi-centered answer one might argue as follows: There appears to be something like a center at the syntactical level (cf. thesis **(T')**). And there might be something like a center on the semantical level (cf. thesis **(T)**). One could even be tempted to speculate about the existence of a further center on a 'higher', pragmatical level, a level including epistemic, social, political or other considerations.[24]

It is possible to argue for the centralized answer, and more seems to speak in favor of this latter answer: The center on the syntactical level (e.g. Shannon's statistical IT) is an integrated part of probably every formal IT on the semantical level (i.e. it doesn't just exist side by side and unrelated with a center conjectured on the semantical level).[25] The same line of thought as before could be carried on in the following way: it is possible that, due to the development of information science and information philosophy, a new center of formal ITs will be formed on some sort of pragmatical level and that the center of formal ITs on the semantical level will be an integrated part of this new center. Furthermore, one could add the observation – already contained in a philosophical statement by Seligman – that a center concept on the semantical level is the reduction of uncertainty by gaining information. If uncertainty is conceptually related or tied to (asking a) question, another way of phrasing this center concept is in terms of questions and answers. Again, the model of question and answer plays a crucial role in Kohlas', van Rooij's and Seligman's contribution to this volume.

All this of course is 'information science fiction'. But it is a philosophical-speculative attempt to not only understand where the research on information in the form of formal ITs comes from, but also where it may be heading to.

References

[Cover/Thomas (1991)]	Cover, T.M., Thomas, J.A.: Elements of Information Theory. Wiley, New York (1991)
[Devlin (1991)]	Devlin, K.: Logic and Information. Cambridge University Press, Cambridge (1991)
[Floridi (2003a)]	Floridi, L.: The Blackwell Guide to the Philosophy of Computing and Information. Blackwell, Oxford (2003a)
[Floridi (2003b)]	Floridi, L.: Information' in Floridi (2003a), pp. 40–61 (2003b)
[Floridi (2005a)]	Floridi, L.: Information, Semantic Conceptions Of. In: Zalta, E.N. (ed.) Stanford Encyclopedia of Philosophy (2005a), http://plato.stanford.edu/entries/information-semantic/

---

[24] A hint at such pragmatic theorizing about information can already be found in (Floridi 2005a:41 and also 57); cf. his reference to philosophers such as Baudrillard, Foucault, Lyotard, McLuhan, Rorty and Derrida. Needless to say, the latters' kind of theorizing is not (yet) of the form of a theory intended here.

[25] A hint at this type of relationship is given in (Floridi 2005a:53); cf. MTC as a rigorous constraint.

Moreover, the statistical IT is extensively treated in Kohlas', van Rooij's and Seligman's contribution to this volume.

[Floridi (2005b)]   Floridi, L.: Is Information Meaningful Data? Philosophy
                            and Phenomenological Research 70(2), 351–370 (2005b)
[Glock (1996)]   Glock, H.-J.: A Wittgenstein Dictionary. Blackwell, Oxford
                            (1996)
[Hanson (1990)]   Hanson, P.: Information, Language and Cognition. Univer-
                            sity of British Columbia Press, Vancouver (1990)
[Israel/Perry (1990)]   Israel, D., Perry, J.: What is information? in Hanson, pp.
                            1–19 (1990)
[Kohlas (2003)]   Kohlas, J.: Information Algebras. Generic Structures for
                            Inference. Springer, London (2003)
[Kolmogorov (1965)]   Kolmogorov, A.N.: Three approaches to the quantita-
                            tive definition of information. Problems Inform. Transmis-
                            sion 1(1), 1–7 (1965)
[Kolmogorov (1968)]   Kolmogorov, A.N.: Logical basis for information theory and
                            probability theory. IEEE Trans. Inform. Theory IT-14(5),
                            662–664 (1968)
[Langel/Kohlas (2007)]   Langel, J., Kohlas, J.: Algebraic Structure of Semantic In-
                            formation and Questions. Predicate Logic: an Information
                            Algebra (Technical Report, Fribourg University) (2007)
[Li/Vitányi (1997)]   Li, M., Vitányi, P.: An Introduction to Kolmogorov Com-
                            plexity and its Applications. Springer, New York (1997)
[Shannon (1948)]   Shannon, C.: The Mathematical Theory of Communica-
                            tion. Bell System Technical Journal 27, 379–423, 623-656
                            (1948)
[Shannon/Weaver (1949)]   Shannon, C., Weaver, W.: The Mathematical Theory of
                            Communication. University of Illinois Press, Urbana (1949)

Author Index

Printed in the United States
By Bookmasters